AF707627

IET ENERGY ENGINEERING SERIES 196

Cyber Security for Microgrids

Other volumes in this series:

Volume 1 **Power Circuit Breaker Theory and Design** C.H. Flurscheim (Editor)
Volume 4 **Industrial Microwave Heating** A.C. Metaxas and R.J. Meredith
Volume 7 **Insulators for High Voltages** J.S.T. Looms
Volume 8 **Variable Frequency AC Motor Drive Systems** D. Finney
Volume 10 **SF_6 Switchgear** H.M. Ryan and G.R. Jones
Volume 11 **Conduction and Induction Heating** E.J. Davies
Volume 13 **Statistical Techniques for High Voltage Engineering** W. Hauschild and W. Mosch
Volume 14 **Uninterruptible Power Supplies** J. Platts and J.D. St Aubyn (Editors)
Volume 15 **Digital Protection for Power Systems** A.T. Johns and S.K. Salman
Volume 16 **Electricity Economics and Planning** T.W. Berrie
Volume 18 **Vacuum Switchgear** A. Greenwood
Volume 19 **Electrical Safety: a guide to causes and prevention of hazards** J. Maxwell Adams
Volume 21 **Electricity Distribution Network Design, 2nd Edition** E. Lakervi and E.J. Holmes
Volume 22 **Artificial Intelligence Techniques in Power Systems** K. Warwick, A.O. Ekwue and R. Aggarwal (Editors)
Volume 24 **Power System Commissioning and Maintenance Practice** K. Harker
Volume 25 **Engineers' Handbook of Industrial Microwave Heating** R.J. Meredith
Volume 26 **Small Electric Motors** H. Moczala *et al.*
Volume 27 **AC-DC Power System Analysis** J. Arrillaga and B.C. Smith
Volume 29 **High Voltage Direct Current Transmission, 2nd Edition** J. Arrillaga
Volume 30 **Flexible AC Transmission Systems (FACTS)** Y-H. Song (Editor)
Volume 31 **Embedded generation** N. Jenkins *et al.*
Volume 32 **High Voltage Engineering and Testing, 2nd Edition** H.M. Ryan (Editor)
Volume 33 **Overvoltage Protection of Low-Voltage Systems, Revised Edition** P. Hasse
Volume 36 **Voltage Quality in Electrical Power Systems** J. Schlabbach *et al.*
Volume 37 **Electrical Steels for Rotating Machines** P. Beckley
Volume 38 **The Electric Car: Development and future of battery, hybrid and fuel-cell cars** M. Westbrook
Volume 39 **Power Systems Electromagnetic Transients Simulation** J. Arrillaga and N. Watson
Volume 40 **Advances in High Voltage Engineering** M. Haddad and D. Warne
Volume 41 **Electrical Operation of Electrostatic Precipitators** K. Parker
Volume 43 **Thermal Power Plant Simulation and Control** D. Flynn
Volume 44 **Economic Evaluation of Projects in the Electricity Supply Industry** H. Khatib
Volume 45 **Propulsion Systems for Hybrid Vehicles** J. Miller
Volume 46 **Distribution Switchgear** S. Stewart
Volume 47 **Protection of Electricity Distribution Networks, 2nd Edition** J. Gers and E. Holmes
Volume 48 **Wood Pole Overhead Lines** B. Wareing
Volume 49 **Electric Fuses, 3rd Edition** A. Wright and G. Newbery
Volume 50 **Wind Power Integration: Connection and system operational aspects** B. Fox *et al.*
Volume 51 **Short Circuit Currents** J. Schlabbach
Volume 52 **Nuclear Power** J. Wood
Volume 53 **Condition Assessment of High Voltage Insulation in Power System Equipment** R.E. James and Q. Su
Volume 55 **Local Energy: Distributed generation of heat and power** J. Wood
Volume 56 **Condition Monitoring of Rotating Electrical Machines** P. Tavner, L. Ran, J. Penman and H. Sedding
Volume 57 **The Control Techniques Drives and Controls Handbook, 2nd Edition** B. Drury
Volume 58 **Lightning Protection** V. Cooray (Editor)
Volume 59 **Ultracapacitor Applications** J.M. Miller
Volume 62 **Lightning Electromagnetics** V. Cooray
Volume 63 **Energy Storage for Power Systems, 2nd Edition** A. Ter-Gazarian
Volume 65 **Protection of Electricity Distribution Networks, 3rd Edition** J. Gers
Volume 66 **High Voltage Engineering Testing, 3rd Edition** H. Ryan (Editor)
Volume 67 **Multicore Simulation of Power System Transients** F.M. Uriate
Volume 68 **Distribution System Analysis and Automation** J. Gers
Volume 69 **The Lightening Flash, 2nd Edition** V. Cooray (Editor)
Volume 70 **Economic Evaluation of Projects in the Electricity Supply Industry, 3rd Edition** H. Khatib
Volume 72 **Control Circuits in Power Electronics: Practical issues in design and implementation** M. Castilla (Editor)
Volume 73 **Wide Area Monitoring, Protection and Control Systems: The enabler for Smarter Grids** A. Vaccaro and A. Zobaa (Editors)
Volume 74 **Power Electronic Converters and Systems: Frontiers and applications** A. M. Trzynadlowski (Editor)
Volume 75 **Power Distribution Automation** B. Das (Editor)
Volume 76 **Power System Stability: Modelling, analysis and control** A.A. Sallam and B. Om P. Malik
Volume 78 **Numerical Analysis of Power System Transients and Dynamics** A. Ametani (Editor)
Volume 79 **Vehicle-to-Grid: Linking electric vehicles to the smart grid** J. Lu and J. Hossain (Editors)
Volume 81 **Cyber-Physical-Social Systems and Constructs in Electric Power Engineering** S. Suryanarayanan, R. Roche and T.M. Hansen (Editors)
Volume 82 **Periodic Control of Power Electronic Converters** F. Blaabjerg, K.Zhou, D. Wang and Y. Yang

Volume 86 **Advances in Power System Modelling, Control and Stability Analysis** F. Milano (Editor)
Volume 87 **Cogeneration: Technologies, Optimisation and Implentation** C. A. Frangopoulos (Editor)
Volume 88 **Smarter Energy: from Smart Metering to the Smart Grid** H. Sun, N. Hatziargyriou, H. V. Poor, L. Carpanini and M. A. Sánchez Fornié (Editors)
Volume 89 **Hydrogen Production, Separation and Purification for Energy** A. Basile, F. Dalena, J. Tong and T. N.Veziroğlu (Editors)
Volume 90 **Clean Energy Microgrids** S. Obara and J. Morel (Editors)
Volume 91 **Fuzzy Logic Control in Energy Systems with Design Applications in Matlab/Simulink®** İ. H. Altaş
Volume 92 **Power Quality in Future Electrical Power Systems** A. F. Zobaa and S. H. E. A. Aleem (Editors)
Volume 93 **Cogeneration and District Energy Systems: Modelling, Analysis and Optimization** M. A. Rosen and S. Koohi-Fayegh
Volume 94 **Introduction to the Smart Grid: Concepts, technologies and evolution** S.K. Salman
Volume 95 **Communication, Control and Security Challenges for the Smart Grid** S. M. Muyeen and S. Rahman (Editors)
Volume 96 **Industrial Power Systems with Distributed and Embedded Generation** R Belu
Volume 97 **Synchronized Phasor Measurements for Smart Grids** M.J.B. Reddy and D.K. Mohanta (Editors)
Volume 98 **Large Scale Grid Integration of Renewable Energy Sources** A. Moreno-Munoz (Editor)
Volume 100 **Modeling and Dynamic Behaviour of Hydropower Plants** N. Kishor and J. Fraile-Ardanuy (Editors)
Volume 101 **Methane and Hydrogen for Energy Storage** R. Carriveau and D. S-K. Ting
Volume 104 **Power Transformer Condition Monitoring and Diagnosis** A. Abu-Siada (Editor)
Volume 106 **Surface Passivation of Industrial Crystalline Silicon Solar Cells** J. John (Editor)
Volume 107 **Bifacial Photovoltaics: Technology, applications and economics** J. Libal and R. Kopecek (Editors)
Volume 108 **Fault Diagnosis of Induction Motors** J. Faiz, V. Ghorbanian and G. Joksimović
Volume 109 **Cooling of Rotating Electrical Machines:** Fundamentals, modelling, testing and design D. Staton, E. Chong, S. Pickering and A. Boglietti
Volume 110 **High Voltage Power Network Construction** K. Harker
Volume 111 **Energy Storage at Different Voltage Levels: Technology, integration, and market aspects** A. F. Zobaa, P. F. Ribeiro, S. H. A. Aleem and S. N. Afifi (Editors)
Volume 112 **Wireless Power Transfer: Theory, Technology and Application** N.Shinohara
Volume 114 **Lightning-Induced Effects in Electrical and Telecommunication Systems** Y. Baba and V. A. Rakov
Volume 115 **DC Distribution Systems and Microgrids** T. Dragičević, F.Blaabjerg and P. Wheeler
Volume 116 **Modelling and Simulation of HVDC Transmission** M. Han (Editor)
Volume 117 **Structural Control and Fault Detection of Wind Turbine Systems** H.R. Karimi
Volume 119 **Thermal Power Plant Control and Instrumentation: The control of boilers and HRSGs, 2nd Edition** D. Lindsley, J. Grist and D. Parker
Volume 120 **Fault Diagnosis for Robust Inverter Power Drives** A. Ginart (Editor)
Volume 121 **Monitoring and Control using Synchrophasors in Power Systems with Renewables** I. Kamwa and C. Lu (Editors)
Volume 123 **Power Systems Electromagnetic Transients Simulation, 2nd Edition** N. Watson and J. Arrillaga
Volume 124 **Power Market Transformation** B. Murray
Volume 125 **Wind Energy Modeling and Simulation Volume 1: Atmosphere and plant** P.Veers (Editor)
Volume 126 **Diagnosis and Fault Tolerance of Electrical Machines, Power Electronics and Drives** A.J. M. Cardoso
Volume 128 **Characterization of Wide Bandgap Power Semiconductor Devices** F. Wang, Z. Zhang and E.A. Jones
Volume 129 **Renewable Energy from the Oceans: From wave, tidal and gradient systems to offshore wind and solar** D. Coiro and T. Sant (Editors)
Volume 130 **Wind and Solar Based Energy Systems for Communities** R. Carriveau and D. S-K. Ting (Editors)
Volume 131 **Metaheuristic Optimization in Power Engineering** J. Radosavljević
Volume 132 **Power Line Communication Systems for Smart Grids** I.R.S Casella and A. Anpalagan
Volume 134 **Hydrogen Passivation and Laser Doping for Silicon Solar Cells** B. Hallam and C. Chan (Editors)
Volume 139 **Variability, Scalability and Stability of Microgrids** S. M. Muyeen, S. M. Islam and F. Blaabjerg (Editors)
Volume 143 **Medium Voltage DC System Architectures** B. Grainger and R. D. Doncker (Editors)
Volume 145 **Condition Monitoring of Rotating Electrical Machines** P. Tavner, L. Ran, C. Crabtree
Volume 146 **Energy Storage for Power Systems, 3rd Edition** A.G. Ter-Gazarian
Volume 147 **Distribution Systems Analysis and Automation 2nd Edition** J. Gers
Volume 151 **SiC Power Module Design: Performance, robustness and reliability** A. Castellazzi and A. Irace (Editors)
Volume 152 **Power Electronic Devices: Applications, failure mechanisms and reliability** F Iannuzzo (Editor)
Volume 153 **Signal Processing for Fault Detection and Diagnosis in Electric Machines and Systems** M. Benbouzid (Editor)
Volume 155 **Energy Generation and Efficiency Technologies for Green Residential Buildings** D. Ting and R. Carriveau (Editors)
Volume 156 **Lithium-ion Batteries Enabled by Silicon Anodes** C. Ban and K. Xu (Editors)
Volume 157 **Electrical Steels, 2 Volumes** A. Moses, K. Jenkins, Philip Anderson and H. Stanbury
Volume 158 **Advanced Dielectric Materials for Electrostatic Capacitors** Q Li (Editor)
Volume 159 **Transforming the Grid Towards Fully Renewable Energy** O. Probst, S. Castellanos and R. Palacios (Editors)
Volume 160 **Microgrids for Rural Areas: Research and case studies** R.K. Chauhan, K. Chauhan and S.N. Singh (Editors)
Volume 161 **Artificial Intelligence for Smarter Power Systems: Fuzzy Logic and Neural Networks** M. G. Simoes
Volume 166 **Advanced Characterization of Thin Film Solar Cells** N. Haegel and M Al-Jassim (Editors)

Cyber Security for Microgrids

Edited by
Subham Sahoo, Frede Blaabjerg and
Tomislav Dragicevic

The Institution of Engineering and Technology

Published by The Institution of Engineering and Technology, London, United Kingdom

The Institution of Engineering and Technology is registered as a Charity in England & Wales (no. 211014) and Scotland (no. SC038698).

© The Institution of Engineering and Technology 2022

First published 2022

This publication is copyright under the Berne Convention and the Universal Copyright Convention. All rights reserved. Apart from any fair dealing for the purposes of research or private study, or criticism or review, as permitted under the Copyright, Designs and Patents Act 1988, this publication may be reproduced, stored or transmitted, in any form or by any means, only with the prior permission in writing of the publishers, or in the case of reprographic reproduction in accordance with the terms of licences issued by the Copyright Licensing Agency. Enquiries concerning reproduction outside those terms should be sent to the publisher at the undermentioned address:

The Institution of Engineering and Technology
Futures Place
Kings Way, Stevenage
Herts, SG1 2UA, United Kingdom

www.theiet.org

While the authors and publisher believe that the information and guidance given in this work are correct, all parties must rely upon their own skill and judgement when making use of them. Neither the author nor publisher assumes any liability to anyone for any loss or damage caused by any error or omission in the work, whether such an error or omission is the result of negligence or any other cause. Any and all such liability is disclaimed.

The moral rights of the author to be identified as author of this work have been asserted by him in accordance with the Copyright, Designs and Patents Act 1988.

British Library Cataloguing in Publication Data
A catalogue record for this product is available from the British Library

ISBN 978-1-83953-331-0 (hardback)
ISBN 978-1-83953-332-7 (PDF)

Typeset in India by Exeter Premedia Services Private Limited

Cover image: Andy A. Widmer / EyeEm via Getty Images

Contents

About the Editors

Subham Sahoo is an assistant professor in the Department of Energy at Aalborg University, Denmark. With his PhD from the Indian Institute of Technology Delhi, he worked as a visiting student with the Department of Electrical and Electronics Engineering, Cardiff University in 2017, and as a postdoctoral researcher at the National University of Singapore. His research focuses on cyber-security and applications of AI in renewable energy generation and microgrids. He has (co-)authored 50 publications.

Frede Blaabjerg is a full professor in the Department of Energy at Aalborg University, Denmark. Prior experience includes work with ABB. He is honoris causa at University Politehnica Timisoara (UPT), Romania and Tallinn Technical University (TTU), Estonia. His research interests include power electronics and its applications, including wind turbines, PV systems, and reliability. He has published more than 600 journal papers and edited or co-authored several monographs.

Tomislav Dragicevic is a professor at the Department of Electrical Engineering at the Technical University of Denmark. His research is in application of advanced control, optimization and AI to electrical systems. Current projects include application of AI in DC microgrids, and AI-based control and design of smart electrical vehicle charging infrastructure. He has authored or co-authored more than 350 technical publications and filed for several patents. His research has been sponsored by various research foundations.

Chapter 1

Cyber-induced power system steady-state and dynamic issues

Jianzhe Liu[1], Subham Sahoo[2], and Bo Chen[1]

1.1 Introduction

It is well known that a power system is a cyber-physical system where the power assets not only interconnect through physical power flows but also commune with information streams.

The recent infrastructural development, including the increasing integration of emerging power electronics interfaced distributed energy resources (DERs) and communication technologies, has further strengthened the interplay between the physical and cyber layer components.

It is expected that the increasing coupling between the physical and cyber layer components can enhance power system resilience and reliability; nevertheless, the growing interdependence creates new challenges for power system operations.

First, many physical components work in response to operational commands from upstream operators. Failing to deliver such commands can create considerable service disruptions. For example, the reported cause responsible for the costly 2003 US Northeastern blackouts is the malfunction of alarm software in the operation center, which hinders protection devices to respond to a minor fault that should have been extinguished with negligible impact on the system. Other than software bugs, physical disruption on the information transmission can as well create interruptions to system operation.

Moreover, the cyber layer technologies are subject to an alarmingly increasing risk of man-made adversary attacks that might have an even greater impact than non-malicious software bugs or natural causes. In 2015, Ukrainian power grids were affected by an organized cyberattack that paralyzed power support for an entire region of tens of thousands of people. In March 2020 European Network of Transmission System Operators for Electricity that represents 42 European transmission system operators in 35 countries reported that it had experienced a successful

[1]Argonne National Laboratory, Lemont, IL, USA
[2]Department of Energy, Aalborg University, Aalborg Øst, Denmark

cyber intrusion. More operational measures targeting cyber threats are called for to prepare the system against cyber contingency events and to mitigate impacts from cyberattacks.

In general, a cyberattack disrupts data communication so that information obtained at the receiving end is different from what is delivered at the sending end. In terms of attack strategies, the adversary can either cause unavailability or integrity issues: denial of service (DoS) attacks are a commonly used attack strategy to cut off information flows, while attacks like man-in-the-middle (MITM) attacks modify transmitted information to achieve a variety of malicious objectives.

Discrepancies in the received and sent information create gaps in the recognized and actual system conditions. When these gaps are deliberated or are significant enough, especially in critical circumstances, the power system's responsiveness to changing conditions can be gravely impacted. As a result, power system operational issues associated with responsiveness to disturbances and contingencies might be aggravated.

Furthermore, when one takes into consideration of the inherent uncertainties in power electronics interfaced DERs that have already burdened system operations, the cyberattacks foster even more challenging operational risks to a cyber-physical power system with considerable DER penetrations.

Considering the opportunities and challenges in a cyber-physical power system, this chapter studies two fundamental power system issues that are potentially challenged by cyberattacks. First, we study a cyber-induced feasibility issue in a power system, where we are concerned with whether we can guarantee that the disrupted generation always matches a changing load. Second, we study the stability issue, that is, whether the system can maintain the generation-demand balance with a synchronized behavior.

1.2 System structure and problem description

In this section, we introduce the cyber-physical structure of a power system with a substantial DER penetration level; from the discussion, we show the vulnerability of the cyber-physical power system and describe two specific cyber-induced operational issues. These two issues show that power system operations across a broad range of time intervals are subject to cyber threats.

1.2.1 System operational structure

A traditional power system is operated under a hierarchical control scheme: the upper layer schedules economic dispatch for a handful of generating units, whereas the lower layer executes the dispatching command while rejecting all sorts of disturbances. Although this scheme worked well for decades, it is increasingly challenged by the rapid growth of power electronics interfaced DERs in today's power system.

The sheer number of DERs that are scattered in a power system and are heterogeneous in nature significantly increases the difficulty of exercising the traditional methods. On the one hand, the decision variable space of a centralized economic dispatch problem is overly enlarged, which makes it computationally forbidding to solve. On the other hand, controlling the DERs in a purely decentralized manner is ineffective in managing the potential regulation capacity. In addition, the lack of coordination can easily lead to a variety of new operational issues. For example, excessive power generation can lead to an unintentional reverse power flow and trigger protection relays.

Many have discussed a new cyber-physical DER operation scheme that can be efficiently utilized to manage many DERs. In such a scheme, the DERs are clustered and managed in multiple groups, such as microgrids, virtual power plants, and connected communities. Each group is operated by a single entity, such as a distributed energy resources management system, and is mathematically represented as one single generating unit in a power system economic dispatch problem as well. Subsequently, the computational complexity of the resulting problem is amenable.

When the dispatching signal for a single DER group is determined, the DERs in a group work under a distributed control scheme to fulfill the dispatched command. In distributed control, every agent has a distributed controller that processes data from the local and neighboring agents, as well as from other remote sites. These data are normally obtained by phasor measurement units (PMUs), which consist of dynamic-voltage phasors. Communication between the PMUs and the local controllers can be achieved in a centralized fashion, as shown in Figure 1.1 (a) and (b), where measurements from all the agents are collected centrally for processing and decision-making. The most prominent method of coordination between agents in the Supervisory Control and Data Acquisition (SCADA) system is usually employed [1] to alleviate monitoring in smart grid networks. If the number of agents is high, this approach not only requires significant communication resources but also is prone to potential cyberattacks. Another option, commonly referred to as decentralized control, refers to a scheme where only local measurements are used. While the communication infrastructure is completely avoided here, control capability is limited. As already explained above, a distributed control paradigm, as shown in Figure 1.1, introduces flexibility, since the computational resources are uniformly dispersed across the system to achieve coordination. Hence, low-bandwidth communication

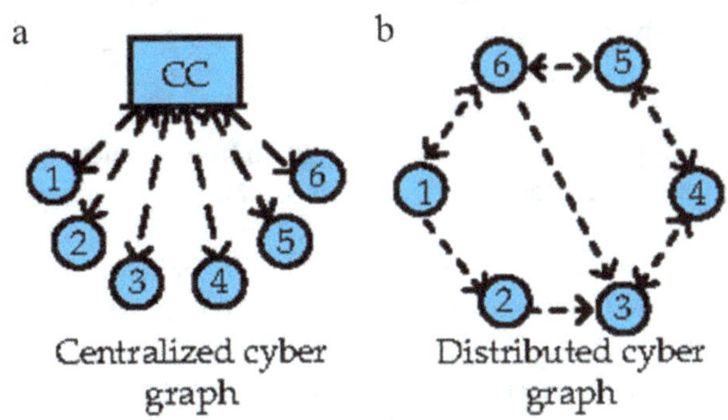

Figure 1.1 Communication topologies – (a) centralized and (b) distributed

channels can be employed to achieve the same function. Although it provides an obvious criterion for the assessment of intrusion attempts, vulnerability to cyberattacks cannot be necessarily guaranteed for coordinated attacks [2–4]. This can be explained owing to the insufficient information present in each node, which does not serve as adequate global information for the detection of cyberattacks.

Such a DER management scheme can effectively and efficiently manage a large number of DERs. Meanwhile, the performance of such a system in Figure 1.2 is dependent on the active information exchange. In the following sections, we show that dependence can potentially lead to multiple cyber-induced issues.

1.2.2 Feasibility in system level

In the aforementioned DER management scheme, the optimization layer decides the set points for the collective real and reactive power outputs of each DER group. Evidently, the successful delivery of such a dispatching signal is of critical concern for the performance of the entire system.

A power system operator usually needs to solve an optimal power flow (OPF) problem to determine the economic dispatch signals for each generating unit. For a power system with n buses included in a set defined as $\mathcal{N}$, a typical OPF problem is given conceptually as follows:

$$\min f(P_G) \text{ subject to:} \tag{1.1}$$

$$P_i = V_i \sum_{k=1}^{n} V_k \left(G_{ik}\cos\left(\theta_i - \theta_k\right) + B_{ik}\sin\left(\theta_i - \theta_k\right)\right), \quad \forall i \in \mathcal{N} \tag{1.2}$$

$$Q_i = V_i \sum_{k=1}^{n} V_k \left(G_{ik}\sin\left(\theta_i - \theta_k\right) - B_{ik}\cos\left(\theta_i - \theta_k\right)\right), \quad \forall i \in \mathcal{N} \tag{1.3}$$

$$P_G \in \mathcal{P}_G, Q_G \in \mathcal{Q}_G, \tag{1.4}$$

$$P_L = P_L^*, Q_L = Q_L^*, \tag{1.5}$$

where, P_i and Q_i are the real and reactive power injections at the ith bus; at each bus, V_i and θ_i represent the voltage magnitude and phase angles; G_{ik} and B_{ik} represent the conductance and susceptance of the power line between bus i and k; the real and reactive power injections at the generating buses form P_G and Q_G, respectively, and they need to lie in some operational constraints represented as P_G and Q_G; the real and reactive power injections at the load buses form P_L and Q_L, and they are modeled as constant power loads that equal to P_L^* and Q_L^*, respectively; the cost function $f(P_G)$ represents the generation costs.

Notice that OPF problem (1.1)– (1.5) seeks to find the most economic generation scheduling for the generation units that make the power flow equations (PFEs) (1.2) (1.3) feasible.

In fact, the PFEs are nonlinear equations and their inclusion in the OPF problem makes it a NP-hard non-convex problem, which is computationally demanding to find the global optimum.

Nevertheless, the PFEs are necessarily included, because when the OPF solution satisfies the PFEs, it represents that the law of conservation of energy is respected, meaning all the generated power is either consumed by the loads or line losses. In

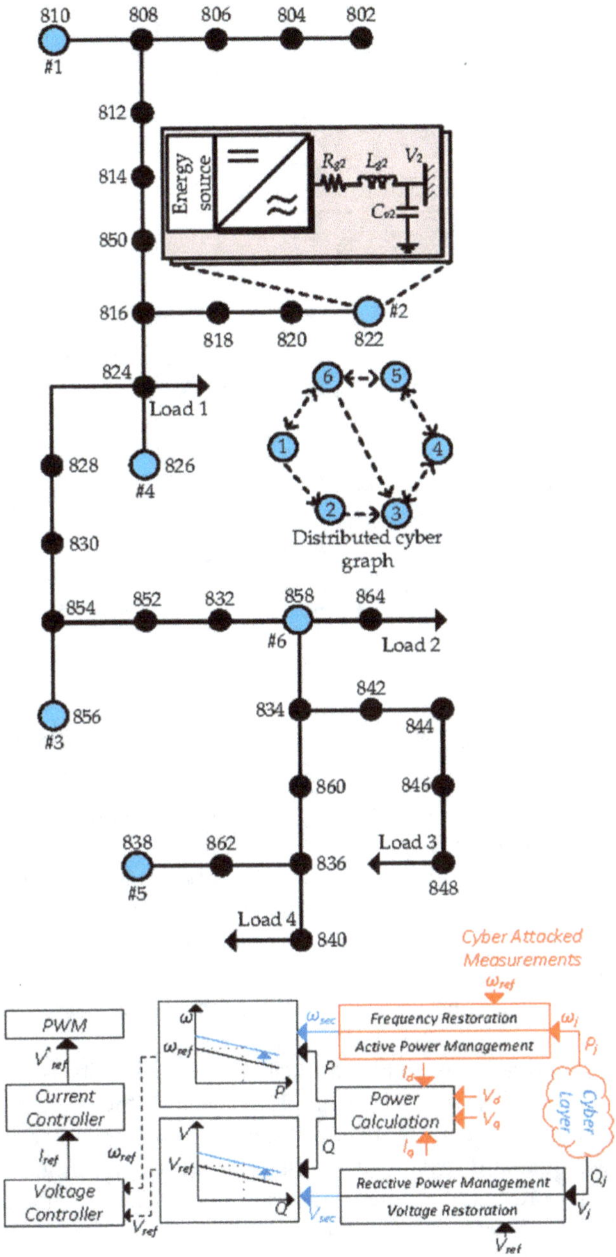

Figure 1.2 *Single-line diagram of an augmented IEEE 37-bus distribution network with* N=6 *power electronic converters – the local and secondary controller is embedded for each converter, which can potentially be attacked by adversaries.*

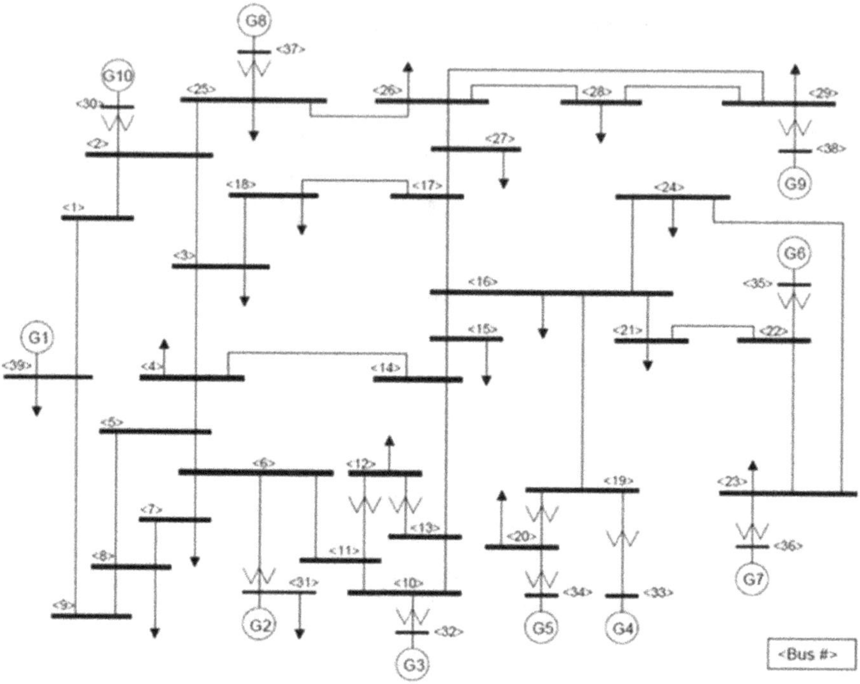

Figure 1.3 Configuration of IEEE 39-bus system [5, 6]

other words, the feasibility of the PFEs implies that the generation-demand balance has been achieved, and the system is long-term voltage stable.

It is because of this reason in spite of the cumbersome computational burden, the PFEs are still considered in the OPF problem. In practice, one often only needs to find a local optimum solution, but the solution must still satisfy the PFEs.

Recall that in our DER operation framework, each DER group is mathematically modeled as a generating unit in the OPF problem. After the OPF problem is solved, the group receives the generation schedule to execute.

We use one motivating example to illustrate the feasibility issue of PFEs and the importance of their research. The example is based on the standard IEEE 39-bus system, as shown in Figure 1.3. The test system describes the electric power flow within the New England district of the United States. It has ten generators, where Generator 1(G1) represents the interconnecting bus of the system to the external bulk power system.

Each generator of the test system has an operational constraint set for the real and reactive power outputs. Violating the constraints can cause the system operating points to locate in uneconomic regions, can lead to dissatisfaction with some engineering constraints for the rest of the grid, or can cause the operating points to cease to exist. For example, we can find a constraint set of the real and reactive power outputs from the generator to keep the PFEs feasible, as shown in Figure 1.4.

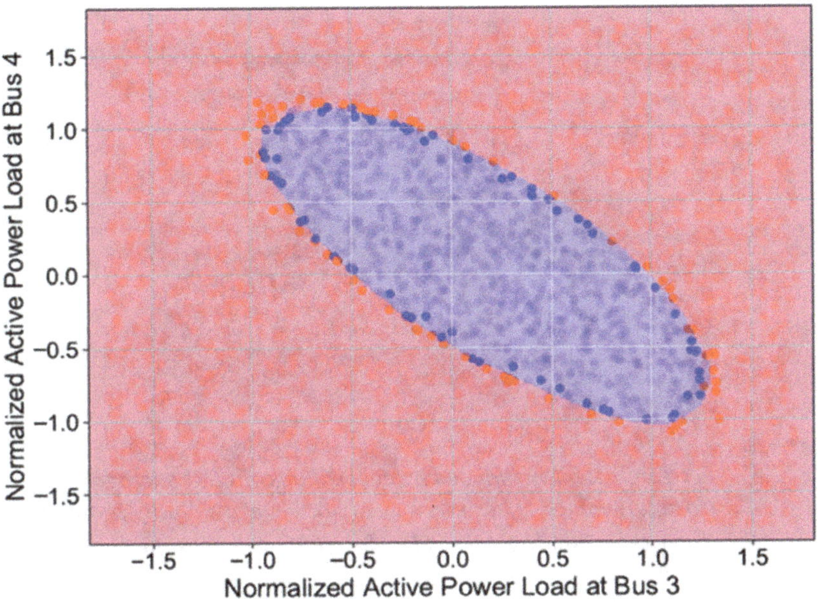

Figure 1.4 Feasibility region with respect to real and reactive power

The red shaded region is the infeasible region, that is, when the real and reactive power outputs are lying within the region, the PFEs of the power system do not have a solution; meanwhile, if the generation schedule lies within the blue region, the PFEs are feasible.

If the power generation is indeed lying outside the feasible region, the system can experience voltage stability issues, as shown in Figure 1.4. For traditional power systems, this usually happens when there is a physical contingency like a grounding fault that trips the tie line that connects a generator to the rest of the system. Usually, N-1 research is conducted for such circumstances if physical faults are the only concern, as physical faults are small-probability event and are unlikely to happen simultaneously.

New challenges surge with the emerging cyberattacks as cyberattacks can disrupt system operation in a similar way to a physical contingency, yet many aforementioned well-accepted engineering insights and contingency analysis methods no longer hold. On the one hand, cyberattacks can create similar impairing effects: for example, when the OPF dispatch signal is intercepted and replaced with malicious content, the generators will deviate from the schedule and the entire system can operate at incorrect or harming operating points. On the other hand, cyberattacks may happen simultaneously at multiple parts of a power system (e.g., the neighborhood attack that affects the entire fleet of assets in a region [7]) at a relatively low cost. If we consider the attacks that vary with the evolution of time, the attack surface is actually of infinite dimension, which is inapplicable to using traditional methods such as *N-k* analysis.

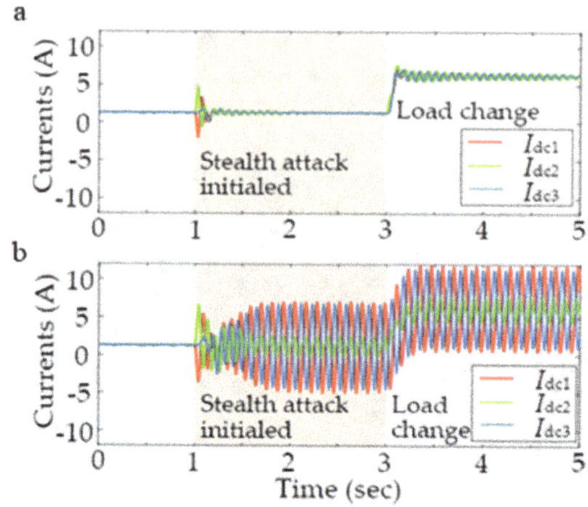

Figure 1.5 System instability induced by cyberattacks in DC microgrids – (a) small valued cyberattack does not incur a stability problem and (b) critical (higher) value of cyberattack can destabilize the same system

To prevent the potentially detrimental aftermath and to target the special features of cyberattacks, this chapter seeks to develop contingency analysis and defense methods that prepare and develop proactive measures for a cyber-physical power system to address the cyber-induced feasibility issue.

1.2.3 Stability issue in microgrid level

The definitions of stability for power electronics dominated power systems or microgrids have been revisited recently in the study of Sahoo et al. [2]. Although stability is primarily based on the physical manifestations, the penetration of threats and anomalies in the cyber layer has triggered a further assessment of the stability criterion for cyber-physical power electronics-based power systems [8]. This is also influenced by the attacker's resources, such as disruption resources and system information.

To explain this clearly, we consider an example of the impact of a stealth attack [10] injected into a DC microgrid, as shown in Figure 1.5. Here, we consider stealth voltage attacks of different magnitude in a microgrid with $N = 3$ DC/DC converters operating at a reference voltage of 315 V at $t = 1$ second. In Figure 1.5a, when an attack of magnitude 15 V is injected into agents I and II, the system is stable despite the presence of the cyberattack. However, when the magnitude of the attack is increased to 40 V in Figure 1.5(b)), the system immediately goes unstable with sustained oscillations. Hence, a sensitivity analysis is carried out in Figure 1.6 for different control parameters in the same DC microgrid with $N = 3$ agents. When increased from 0 to 80 V in a system with a reference voltage of 315 V, the stability boundary for the attack element changes differently for various system conditions. After conducting numerous simulations, it can be seen from Figure 1.6 that when the

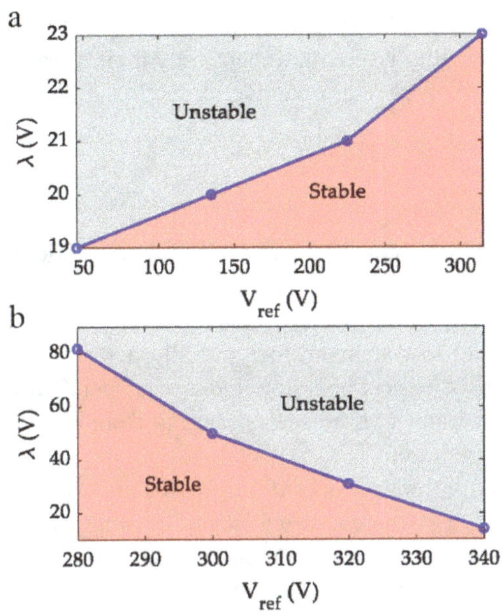

Figure 1.6 *Stability boundaries for DC microgrid with* N = 3 *agents for duty ratio (a)* D = 0.3 *and (b)* D *varies, under different values of cyberattacks* λ

global voltage reference is varied from 45 to 315 V with duty ratio $D = 0.3$, the difference among the values of λ, which makes the system unstable is quite small. On the other hand, when the global reference is increased such that D varies, the value of λ, which disturbs the system stability, will decrease, indicating that the stability margin of the system operating with a lower voltage reference is worse. Hence, the system with a relatively lower global voltage reference is more sensitive to stealth cyberattacks since even a small value of λ can disturb the system stability.

1.3 Cyber-induced power flow feasibility issue

In this section, we study the cyber-induced PFE feasibility issue, as mentioned in the previous section. Notice that PFE feasibility is associated with the steady-state generation-demand balance as well as the long-term voltage stability issue, hence it is one of the most fundamental issues in electric power system research. Due to the emerging impacts of cyber events, such feasibility issue is under new risks, and its research has had new thrusts and challenges. One such challenge is that the OPF problem, which normally incorporates PFEs as constraints, may become unsolvable. As OPF is at the heart of the operation of a power system, it is of critical significance to study the cyber-induced feasibility issue in its context.

In light of this, in the following, we analyze how MITM and DoS attacks can cause power flow feasibility issues, and then study their impacts on OPF problems. Targeting the issue, we propose a proactive protection strategy for resilient system hardening planning.

1.3.1 Introduction

The OPF problem is at the heart of the reliable and economic operation of a power system [1]. It aims to find the optimal real and reactive power generation schedules for the entire fleet of grid assets so that the system operates at a steady-state operating point with all load demand met and all operational constraints satisfied. It is solved continuously by grid operators like an independent system operator or a distribution system operator at time scales ranging from days to minutes to ensure normal steady-state operation.

Ever since its first introduction in 1963 [9, 10], the classic formulation of OPF has not changed much. Yet, there is still a lack of a numerically efficient and robust, reliable solver for its solution concerning power systems with all sorts of configurations and network parameters; one of the main computational challenges lies in its nonlinear equality constraints concerning PFEs [4].

PFEs for a power system admit the form of nonlinear quadratic equations. It describes the electric power transfer in an interconnected network: the existence of a PFE solution guarantees that the law of conservation of energy is respected such that all generated power is either sent to loads or lost during transmission, which is also associated with the long-term voltage stability issue. It is included in the OPF problem as constraints to provide such guarantees for a designed generation schedule. However, the nonlinear nature of the PFEs causes a nontrivial computational burden. For example, it is well-known that the number of solutions to a set of quadratic equations grows exponentially with respect to the number of equations. As a result, it is not only computationally demanding to select one optimal decision from the vast number of potential solutions; numerical solution algorithms can easily be trapped in a solution leading to limited economic or steady-state stability performance.

This long-standing power engineering issue commonly existing in almost all power systems is further exacerbated by cyber events, as shown in the subsequent discussions. We show the cause of the cyber-induced feasibility issue and discuss the main difficulty of the problem. A defense strategy is proposed to safeguard a power system under such risks.

1.3.2 System modeling and problem statement

In this section, we show the system model under study and the main problem considered in this chapter. We study a power system consisting of n load and generation buses; these buses form a set $\mathcal{N}$, where there is one slack bus denoted as bus 0, n^g generator buses* defined as buses 1 to n^g that are included in set $\mathcal{N}^g$, and there are n^l load buses† that are defined as buses $n^g + 1$ to n and are included in a set $\mathcal{N}$.

Conceivably, $n = n^g + n^l + 1$, and $\mathcal{N} = \{0\} \cup \mathcal{N}^g \cup \mathcal{N}^l$. Suppose these buses form a radial network; and let G_{ik} and B_{ik} represent the conductance and susceptance of the power line connecting bus i and bus k; let P_i and Q_i be the real and reactive power injections at bus i; let V_i and θ_i be the voltage magnitude and phase angle of bus i. We assume that the power lines in the system have uniform conductance–susceptance ratio. With the above discussion, the PFEs about bus i of the network are given as follows:

$$P_i = V_i \sum_{k=1}^{n} V_k \left(G_{ik} \cos\left(\theta_i - \theta_k\right) + B_{ik} \sin\left(\theta_i - \theta_k\right)\right), \quad \forall i \in \mathcal{N} \tag{1.6}$$

$$Q_i = V_i \sum_{k=1}^{n} V_k \left(G_{ik} \sin\left(\theta_i - \theta_k\right) - B_{ik} \cos\left(\theta_i - \theta_k\right)\right). \quad \forall i \in \mathcal{N} \tag{1.7}$$

There are four types of variables in the above two equations, i.e., the real and reactive power injections P_i and Q_i , and voltage magnitude and phase angle V_i and θ_i .

It is well known that a system of equations like the one discussed above is proper if there are equal numbers of equations and unknown variables‡. Hence, in the power engineering convention, one can usually fix two variables and solve the other two. For instance, when the voltage magnitude and phase angle of a bus are known, the bus is called a slack bus where only the real and reactive power is to be found; when the real and reactive power of a bus is known, it is called a PQ bus where only the voltage magnitude and phase angles are to be found.

In this section, we suppose that all the load and generator buses are PQ buses. In addition, recall that there is a slack bus in the system. It usually represents the interconnecting bus to the rest of the bulk power system, like the substation of a distribution system. As the capacity of the rest of the system is usually much larger than the system of concern, the voltage and phase angle of the bus are set as fixed reference values, such as $V_0 = 1$ and $\theta_0 = 0$ in per unit (p.u.) values.

Unlike load buses, the real and reactive power of a generator bus is determined in a centralized control center. An optimization problem like an OPF problem is usually solved for this purpose. An OPF problem is defined as follows by expanding the conceptual model (1.1)– (1.5), as shown in Section 1.2:

$$\min \sum_{i \in \mathcal{N}^g \cup \{0\}} f_i\left(P_i\right) \text{ subject to:} \tag{1.8}$$

$$P_i = V_i \sum_{k=1}^{n} V_k \left(G_{ik} \cos\left(\theta_i - \theta_k\right) + B_{ik} \sin\left(\theta_i - \theta_k\right)\right), \quad \forall i \in \mathcal{N} \tag{1.9}$$

*The point of interconnection (POI) real and reactive power outputs of a generator bus are controllable. It can be a microgrid or a synchronous generator.

†The POI real and reactive power outputs of a load bus are uncontrollable.

‡Otherwise, the system is overdetermined or underdetermined, meaning the number of equations are greater or smaller than the unknown variables. Such systems usually have no solution or has infinitely many solutions, respectively.

$$Q_i = V_i \sum_{k=1}^{n} V_k \left(G_{ik} \sin\left(\theta_i - \theta_k\right) - B_{ik} \cos\left(\theta_i - \theta_k\right)\right), \ \forall i \in \mathcal{N} \tag{1.10}$$

$$p_i \in [p_{i-}, p_{i+}], \ Q_i \in \left[Q_{i-}, Q_{i+}\right], \quad \forall i \in \mathcal{N}^g, \tag{1.11}$$

$$p_i = p_i^*, Q_i = Q_i^*, \quad \forall i \in \mathcal{N}^l, \tag{1.12}$$

$$V_0 = 1, \ \theta_0 = 0, \tag{1.13}$$

where f_i is a generation cost function; we let the designed generation power to lie in an operational constraint set defined as $\left[P_{i-}, P_{i+}\right]$ and $\left[Q_{i-}, Q_{i+}\right]$; the load buses power injections are known constants P_i^* and Q_i^* that are usually estimated from historical load patterns [11]; and the voltage magnitude and phase angle at the slack bus are considered as p.u. reference values.

We assume that the operational constraints included in the OPF for the generation schedules are known to the generators as well. This is to ensure the OPF decisions can be followed by the generators in practice so that the system operates as scheduled. Furthermore, when there is miscommunication about the generation schedules such that they are outside of the bounds, the generation sites can detect the difference and adjust the generation plans according to the operational constraints to ensure safe operation.

Suppose the designed power generation schedules for the generators are P_i^* and Q_i^*, $\forall i \in \mathcal{N}^g$. After solving the OPF problem, they will be sent to the generators to execute. However, when the data in transmission are corrupted, the generators may receive incorrect generation schedules to follow. Suppose that the received generation schedules are given as follows:

$$P_i^A = P_i^* + \widetilde{P}_i, \ Q_i^A = Q_i^* + \widetilde{Q}_i, \quad \forall i \in \mathcal{N}^g \tag{1.14}$$

where $\widetilde{P}_i$ and $\widetilde{Q}_i$ are malicious compromises made to the original signals, and P_i^A and Q_i^A are the alternated signals received by the generator i.

As discussed in Section 1.2, when P_i^A and Q_i^A are outside the feasibility region, the resulting network might not have a PFE solution. An adversary can take advantage of this trait and launch attacks to disrupt the system operation and endanger system steady-state stability. Defense strategies need to be developed to mitigate the impacts of cyberattacks.

In this section, we show a proactive grid planning layer defense strategy that modifies the optimization parameters in the OPF problem to harden a system against cyberattacks, aiming to enhance the survivability and fault ride-through capability during such an event.

Recall that we have discussed that normally operational constraints at the central control center are the same as those at the generator sites; a generator thus has the potential capability of correcting a compromised signal when it is outside the operational range.

Therefore, in general, a generator is capable of employing the following decentralized attack detection and correction strategy defined by a mapping $\boldsymbol{M_i}$ that projects $\{P_i^A, Q_i^A\}$ into a viable point $\left\{\hat{P}_i, \hat{Q}_i\right\}$ such that $\hat{P}_i \in \left[P_{i-}, P_{i+}\right]$ and $\hat{Q}_i \in \left[Q_{i-}, Q_{i+}\right]$, hoping that through this projection the system can restore feasibility with each generator using only local information.

We are concerned with the robust feasibility of a power system, which is defined as follows: a power system is called robustly feasible when the projected generation schedule $\{\hat{P}_i, \hat{Q}_i\}$ with $\boldsymbol{M}_i(\{P_i^A, Q_i^A\}) = \{\hat{P}_i, \hat{Q}_i\}$, for all $i \in \mathcal{N}^g$, always ensures the existence of solution for OPF (1.8) given any corrupted generation signal $\{P_i^A, Q_i^A\}$.

1.3.3 Defense strategy

From the definition of robust feasibility discussed above, we design the mapping $\boldsymbol{M}_i$ along with the associated generator operational bounds $[P_{i-}, P_{i+}]$ and $[Q_{i-}, Q_{i+}]$ to ensure the constant existence of PFE solution.

We propose the following application-amenable mapping $\boldsymbol{M}_i$ as follows:

$$\begin{cases} \hat{P}_i = P_{i-}, & \text{if } P_i^A < P_{i-} \\ \hat{P}_i = P_{i+}, & \text{if } P_i^A > P_{i+} \\ \hat{P}_i = P_i^A, & \text{otherwise} \end{cases} \tag{1.15}$$

$$\begin{cases} \hat{Q}_i = Q_{i-}, & \text{if } Q_i^A < Q_{i-} \\ \hat{Q}_i = Q_{i+}, & \text{if } Q_i^A > Q_{i+} \\ \hat{Q}_i = Q_i^A, & \text{otherwise} \end{cases} \tag{1.16}$$

which in essence applies the following rule: the generator applies the received generation schedule unless it lies outside the operational range, in which case the generator just generates the power at the nearest operational boundary.

In this way, the defense strategy can be easily and swiftly implemented in the generation sites without the need for sophisticated computational units or communication infrastructure connecting other parts of the system. This enables a quick response time that ensures better survivability and ride-through capability before the cyberattacks are cleared. With this strategy, the generated power always resides in $[P_{i-}, P_{i+}]$ and $[Q_{i-}, Q_{i+}]$.

1.4 Resiliency mechanism in microgrids

System resiliency against such extreme events is very crucial to ensure the security of supply. A tailor-made philosophy is to minimize the operation time interruption during such low-probability high-impact events, which defines the resiliency scale. However, a new self-healing scheme proposed in Refs. [12–14] ensures no interruption of supply, which operates on a trustworthy transaction mechanism. Basically, upon detection of cyberattacks, the mitigation philosophy allows system restoration and removal of cyberattacks from the system. As shown in Figure 1.7, upon detection, an authentication scheme labels the attacked measurement with “FALSE (F)” and vice versa. By doing so, the attacked measurements will be prevented from

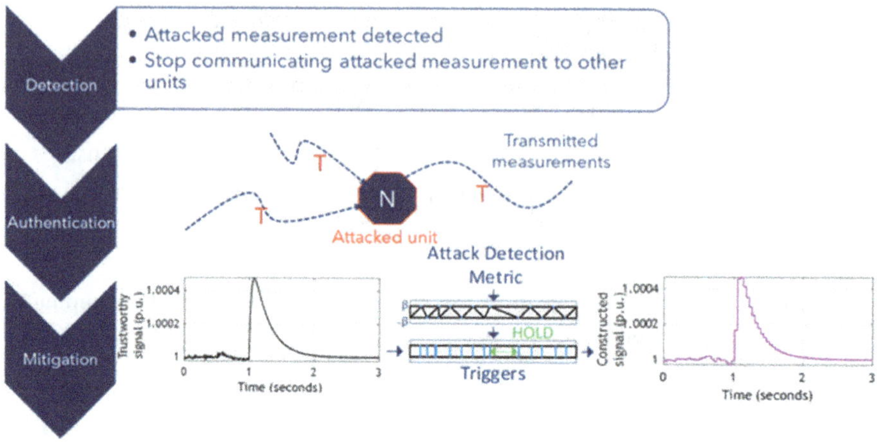

Figure 1.7 Self-healing mechanism for resiliency and prevention mechanism in microgrids [12]

being using anywhere. However, to continue seamless operation, we use the trustworthy communicated measurements (labeled with "TRUE (T)") from the neighboring converters and design a synthetic signal as shown in the right-hand half of Figure 1.7. The synthetic signal, is designed using a prediction policy considering the residuals of the cyberattack-detection mechanism and generating triggers once the threshold is hit. By using these triggers, a Sample and Hold operation is carried out to recreate the synthetic signal. Finally, this reconstructed signal is replaced with the attacked signal locally to account for system resiliency as soon as possible. Using this mechanism, the system can be brought back to normalcy only using a single converter, which has the capability of broadcasting trustworthy measurements. As a result, it provides the best resilience capability without using model information and data-driven indices. Hence, the development of such self-healing algorithms and reconfiguration plans will reinforce resilient power systems against critical natural and man-made events.

1.5 Conclusion

This chapter provides an overview of the steady-state and dynamic issues in cyber-physical power systems. Based on the steady-state resiliency requirements, the corrective measures can be accommodated based on the time scale of the steady-state updates. Moreover, investigating steady-state stability will also require firm thresholds and reconfiguration measures. However, the challenges with an assessment of cyber-induced threats in dynamic conditions naturally increase as it interferes with all the operational sectors, such as control, modeling, stability, and protection. Hence, we provide an overview of some of the commonly used resilient schemes and introduce some new self-healing innovations without reporting any interruption.

More details about these aspects will be discussed in the following chapters of this book.

References

[1] Boyer S.A. *SCADA: supervisory control data acquisition*. Pittsburgh, PA: International Society of Automation; 2009.

[2] Sahoo S., Mishra S., Peng J.C.-H., Dragičević T. 'A stealth cyberttack detection strategy for DC microgrids'. *IEEE Transactions on Power Electronics*. 2018;34(8):8162–74.

[3] Mazumder S.K., Kulkarni A., Sahoo S., *et al*. 'A review of current research trends in power-electronic innovations in cyber–physical systems'. *IEEE Journal of Emerging and Selected Topics in Power Electronics*. 2021;9(5):5146–63.

[4] Ye J., Giani A., Elasser A., *et al*. 'A review of cyber-physical security for photovoltaic systems'. *IEEE Journal of Emerging and Selected Topics in Power Electronics*. 2021:1–99.

[5] Hiskens I.A., Davy R.J. 'Exploring the power flow solution space boundary'. *IEEE Transactions on Power Systems*. 2001;16(3):389–95.

[6] IEEE 39-bus system. Available from http://sys.elec.kitami-it.ac.jp/ueda/demo/WebPF/39-New-England.pdf

[7] Donti P.L., Liu Y., Andreas J.S., Bernstein A., Yang R., Zhang Y. 'Matrix completion for low-observability voltage estimation'. *IEEE Transactions on Smart Grid*. 2019;3(11):2520–30.

[8] Leng M., Sahoo S., Blaabjerg F. 'Stability investigation of DC microgrids under stealth cyber attacks'. *IEEE Energy Conversion Congress and Exposition*. 2021:1427–32.

[9] Carpentier J. 'Contribution á l'étude du dispatching économique'. *Bulletin de la Société Française des Électriciens*. 1962;3(8):431–47.

[10] Carpentier J. 'Optimal power flows'. *International Journal of Electrical Power & Energy Systems*. 1979;1(1):3–15.

[11] Singh R., Pal B.C., Jabr R.A. 'Statistical representation of distribution system loads using Gaussian mixture model'. *IEEE Transactions on Power Systems*. 2009;25(1):29–37.

[12] Sahoo S., Dragičević T., Blaabjerg F. 'An event-driven resilient control strategy for DC microgrids'. *IEEE Transactions on Power Electronics*. 2020;35(12):13714–24.

[13] Sahoo S., Dragičević T., Blaabjerg F. 'Resilient operation of heterogeneous sources in cooperative DC microgrids'. *IEEE Transactions on Power Electronics*. 2020;35(12):12601–5.

[14] Sahoo S., Dragičević T., Blaabjerg F. 'Multilayer resilience paradigm against cyber attacks in DC microgrids'. *IEEE Transactions on Power Electronics*. 2021;36(3):2522–32.

Chapter 2

Modern power electronics in active distribution network

Minrui Leng[1], Subham Sahoo[2], and Frede Blaabjerg[2]

2.1 Introduction

With the advancement of power electronics and semiconductors technology, new faster devices with lower switching energy were developed, enabling faster switching frequency operation. As a result, the volume and size of the power converters were considerably reduced, bringing advantages for those applications in which the power density is very important [1]. As modern power electronics continue to transform the emerging challenges in the distribution networks, the controllability of interconnected power electronic converters becomes a severe problem. The advances in IT infrastructural developments have simplified this problem but have exposed the system to a larger threat of cyber attacks. This chapter overviews the different challenges faced by modern power electronics and the extent to which they are vulnerable to man-made threats, i.e, cyber attacks.

2.2 AC/DC distribution systems

As an innovative solution to increasing demand for power supply capability and power quality, AC/DC distribution systems can achieve more efficient conversion and more flexible regulation. A typical AC/DC distribution system includes different types of power converters for AC and DC distributed generations, loads and interfacing converters for the distribution networks, as shown in Figure 2.1. At the source side, large-scale distributed wind power and photovoltaic power are integrated. To implement a flexible regulation power supply, energy storage is connected, resulting in greater uncertainty at the source side. As for the networking, power interaction

[1]College of Electrical Engineering, Sichuan University, Chengdu, China
[2]Department of Energy, Aalborg University, Denmark

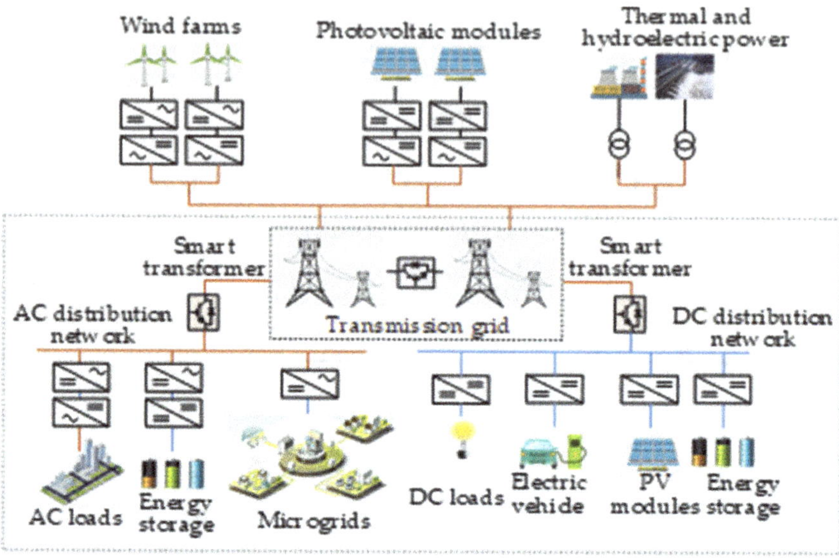

Figure 2.1 A typical structure of AC/DC distribution systems

coupling between not only AC and DC subgrids but also microgrids has diverse modes. Because of the diversified networks and the multipower conversion devices, the operation of the AC/DC distribution system is very complicated. On the load side, since a large number of source-containing loads featuring active response and bidirectional interaction capabilities have emerged, a strict request for user participation in energy management is needed and the uncertainty of the load side is added. To make the AC/DC distribution system more intelligent, new technologies such as DC line fault diagnosis, network power quality governance, distributed computing and autonomous control are applied using modern power electronic devices. The key equipment of distribution systems including distributed generation accessing converter and synchronization devices, flexible AC transmission system (FACTS) devices, power quality governance devices, DC protection, isolation devices and integrated distribution terminal units show the development trend of intelligence, modularity and high power density [2].

2.3 Modern power electronics – key enablers of intelligent grids

2.3.1 Renewable energy interfacing

2.3.1.1 Power conversion for energy storage system

A battery energy storage system can be used to regulate the load and the frequency of the grid side as well as easily connect the renewable energy generation to the grid side. With the rapid development of the energy storage capacity, the use of energy storage systems faces a series of technical challenges for high security and

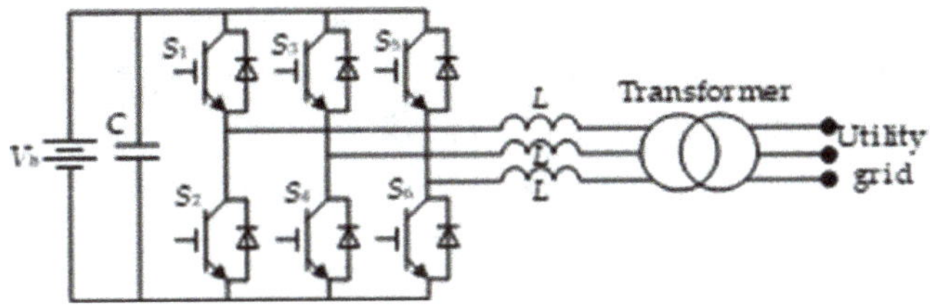

Figure 2.2 A typical single-stage power conversion system

high efficiency . The power conversion system plays a vital role in interfacing the battery energy storage and the grid. By investigating converter circuit topologies, combination modes and control strategies, the power conversion systems can effectively improve the efficiency, security and reliability, as well as reduce the cost of energy storage systems, promoting the popularization and application of energy storage systems. The energy storage power conversion system not only determines the power quality and dynamic characteristics of the energy storage system but also affects the lifetime of the batteries.

Low voltage power conversion systems include single-stage, two-stage and hybrid stage topologies. Figure 2.2 shows a typical single-stage power conversion system with a power frequency transformer, which is simple and widely used. To increase the power density and efficiency, the neutral point clamped converter and T-type three-level converter can be used in the power conversion system.

There are different types of batteries and their voltage level has no uniform standard. In addition, for different types of energy storage medium, the terminal voltage varies with the change of state of charge. Therefore, a DC–DC converter is necessary to provide a proper bus voltage for the next stage, as shown in Figure 2.3. The two-stage power conversion system is more flexible and has a strong matching ability as well as low requirements for batteries.

Compared to the power frequency transformer, the high frequency transformer is compact in size and light in weight, making it possible for reducing the cost and increasing the power density. An isolated DC–DC converter is used to transfer the power from batteries to the next stage, as shown in Figure 2.4. According to different gains and power requirements, the full-bridge structure can be replaced by the half bridge and push–pull structure.

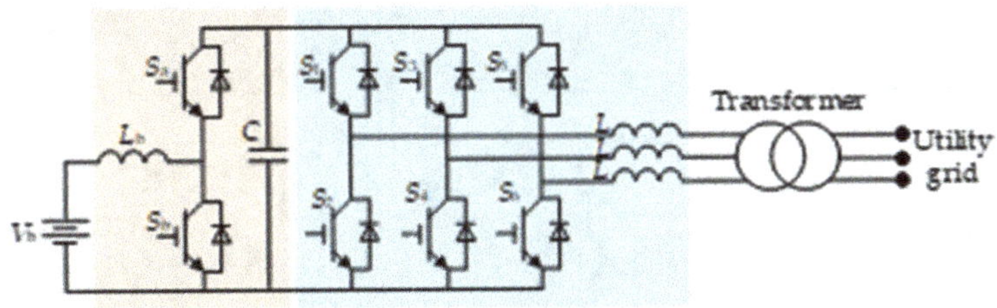

Figure 2.3 A typical two-stage power conversion system

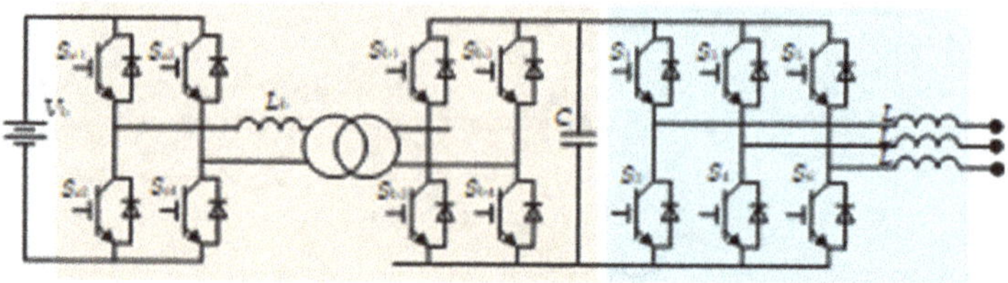

Figure 2.4 A two-stage power conversion system with an isolated DC-DC converter

To expand the capacity of power conversion systems, there are two general methods: expand from the AC side and expand from the DC side. Expanding from the AC side means connecting multiple power conversion systems on the AC side, as shown in Figure 2.5. By droop control and virtual synchronization control, the power among different power conversion systems can be distributed and the coordination can be realized. Expanding from the DC side means connecting multiple batteries to the DC bus by DC-DC converters, as shown in Figure 2.6. The energy distribution and the coordination of different cells are implemented by hierarchical control of multiple DC-DC converters. The power conversion system gathers energy through the DC bus, which is connected to the AC power grid by inverters, which makes it easy for a hybrid battery energy storage system to be used in high-power fields.

2.3.1.2 Power conversion for photovoltaic systems

Power conversion devices are the heart of PV systems that are known as the DC-DC converters and PV solar inverters, being responsible for interfacing with the grid.

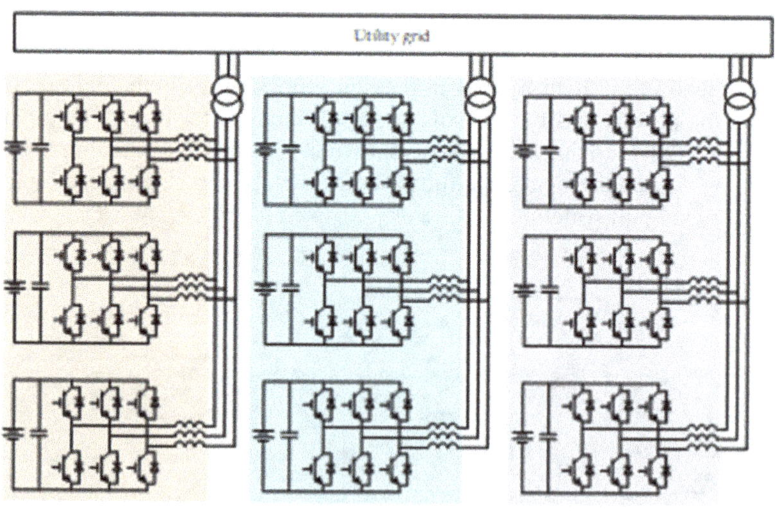

Figure 2.5 A power conversion system based on AC side expansion

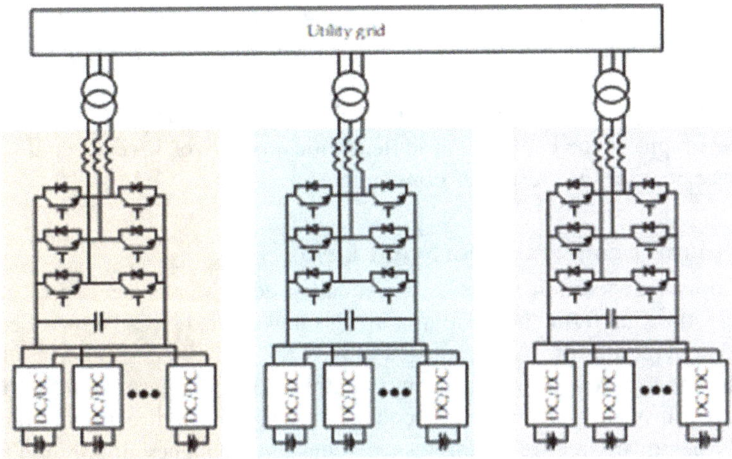

Figure 2.6 A power conversion system based on DC side expansion

The power conversion for PV systems is similar with that for the energy storage systems. Power conversion topologies for PV systems can be divided into PV topology with transformers and PV topology without transformers according to whether it is isolated.

The PV module is greatly affected by the external environment, and the output power changes with the change in light intensity and ambient temperature. For the independent PV system, when the light is strong in the daytime, the load cannot make full use of the power emitted by the PV module, resulting in waste. In case of insufficient light at night or on a cloudy day, the load cannot work normally because the PV module cannot provide sufficient energy. For large-scale grid-connected PV systems, if the energy injected into the power grid fluctuates greatly, the voltage of the subnetwork will increase or decrease, affecting the power supply quality. Moreover, the PV system injects more energy into the power grid during the daytime, and there is no energy injection at night, which has a great impact on the power grid and cannot alleviate the peak power consumption at night. In PV systems, the battery is necessary to regulate the power. PV systems with an energy storage unit is shown in Figure 2.7. If the DC link and inverter link in Figure 2.7 are taken as a whole, and the PV module port, energy storage port and load or grid port are connected externally, a three-port PV system is formed. In order to obtain better

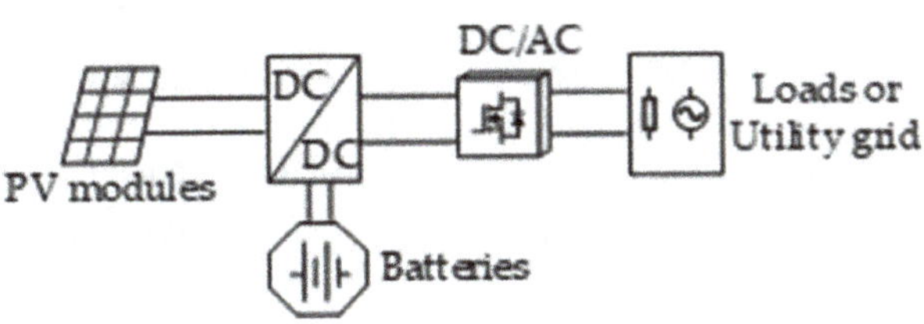

Figure 2.7 The PV system with energy storage unit

performance, the topology of the three-port PV system should have the following: less conversion order and fewer devices; compact structure and small volume; and low cost and high reliability. For the DC–DC converter, the buck-boost converter, boost converter, resonant converter and half-bridge converter are widely used to boost the output voltage of PV module, while multilevel inverter and Z source inverter are popular for DC to AC conversion.

2.3.1.3 Power conversion for wind farms

In wind power generation systems, a frequency converter is needed to complete the energy transfer from the generator to the power grid. AC–DC–AC converter effectively overcomes the disadvantages of AC–AC converters, such as high output voltage harmonics, low power factor at the input side and a large number of power components. It is easy to realize the control strategy and bidirectional conversion, especially being suitable for variable speed constant frequency doubly fed machine wind power generation system and brushless doubly fed machine wind power generation system. In addition, the use of power electronic frequency converters in offshore wind farms can realize the control of active and reactive power, making the wind turbine run in a variable speed state to capture the maximum wind energy and reduce mechanical stress and noise.

Moreover, wind farms need high-gain voltage and high-power DC–DC converters to boost the voltages generated by wind turbines. The three-phase dual active bridge (DAB) can be a good selection due to its high power capacity and low size filter requirements with soft-switching capability. Cascaded multilevel dual active bridge with galvanic isolation feature and parallel-input series-output configuration can ensure current sharing, step-up operation and better fault handling.

2.3.2 High-voltage DC transmission systems

A voltage source converter (VSC)–high-voltage DC (HVDC) system allows for the fully independent control of both the active and the reactive power flow within the operating range of the system. The use of VSC technology makes it possible to maintain voltage and power quality on the connecting AC networks. Over the last decade, several VSC–HVDC links have been installed to increase the power transfer capacity. The additional interconnection capacity will enable the better utilization of reserve and balancing capacities through power markets, potentially bringing down electricity prices while increasing the number of renewables in the energy mix.

A complete VSC–HVDC system consists of (a) AC transformers, AC filters and phase reactors on the AC side and (b) DC capacitors, DC cables, DC chopper and DC switchgear on the DC side. A typical topology of the VSC-HVDC converter station is shown in Figure 2.8. Generally, three types of converter configurations have been mainly utilized for VSC-based HVDC systems: two-level converter, three-level neutral point clamped (NPC) converter and modular multilevel converter (MMC).

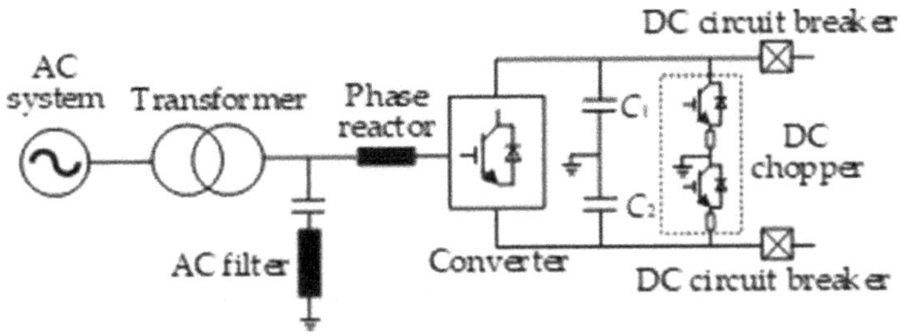

Figure 2.8 VSC-based HVDC system

2.3.3 Flexible AC transmission systems

2.3.3.1 Static synchronous compensator

A static synchronous compensator (STATCOM) is a flexible AC transmission system (FACTS) device of parallel reactive power compensation. It can emit or absorb reactive power, and its output can be changed to control specific parameters in the power system. Generally, it is a solid-state switching converter. When its input is connected with a power supply or energy storage device, its output can independently emit or absorb controllable active and reactive power. It can improve power system performances in the following aspects: dynamic voltage control, power oscillation damping, transient stability, voltage flicker control and so on. Compared with traditional reactive power compensation devices, STATCOM has the advantages of continuous regulation, small harmonic, low loss, wide operation range, high reliability and fast regulation speed, which has received extensive attention and rapid development.

2.3.3.2 Unified power flow controller

As the advanced and universal devices in FACTs, a unified power flow controller (UPFC) combines the advantages of parallel and series FACTs devices. It can provide voltage control, reactive or active power compensation, line impedance compensation, tidal current regulation and low-frequency oscillation damping for the power grid, and enhance the power angle and voltage stability of the system. Especially with the gradual development and maturity of MMC voltage source converter in recent years, MMC UPFC can be designed, manufactured and applied to practical projects. UPFC has advantages of STATCOM's strong reactive power compensation ability and node voltage maintenance ability, as well as static synchronous series compensator (SSSC)'s strong ability to compensate line voltage and control line power flow because of effectively combining the STATCOM and SSSC by back-to-back common DC bus connection to realize multifunctional integrated control of the line.

2.3.3.3 Interline power flow controller

With the expansion and complexity of power grid structure and scale, the UPFC cannot be fully suitable for different applications. Therefore, based on the structure

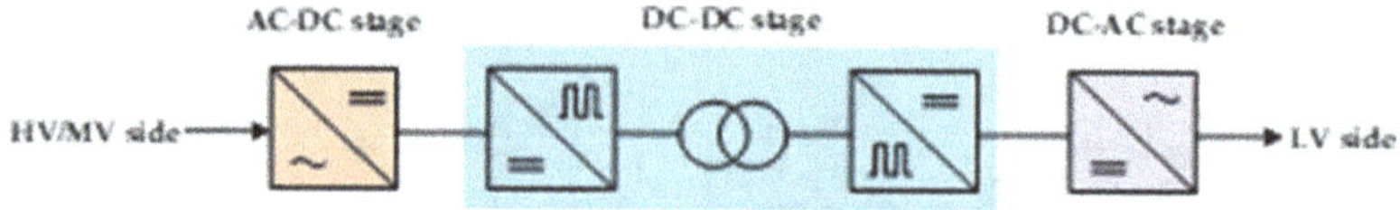

Figure 2.9 Structure of ST

of UPFC, many scholars have studied and developed a variety of variable structure power flow controllers through the improvement of converter combination and control mode to meet the needs of power grids with more different structures and application scenarios.

IPFC is an interline power flow controller with only two series converters. The two series side converters output series embed voltage to different transmission lines respectively. They are connected together through common DC lines and DC capacitors. In fact, it is equivalent to two SSSCs connected back-to-back. Generally, the active power loss between the converter of IPFC and DC bus is very small and negligible. Therefore, approximately lossless active power exchange can be carried out between different transmission lines through IPFC, so as to realize the active power transfer between different transmission lines.

2.3.4 Smart transformer

The smart transformer (ST) is a solid-state transformer (SST) that provides ancillary services to the distribution and transmission grids to optimize their performance. The ST possesses a wide range of functions, such as reactive power support in medium-voltage (MV) or high-voltage (HV) grids, DC connectivity at the HV, MV and LV levels, and load control on the LV side. For ST in Figure 2.9, there are three-stage structure, four-stage structure and five-stage structure in different applications, among which, the three-stage solution with the isolation stage in the DC/DC converter is basic and widely used. This three-stage structure enables galvanic isolation between the two grids, guaranteeing the appliances' safety during abnormal conditions (e.g., faults or lightning strikes). The most suitable topologies to implement the ST from the power electronics viewpoint are VSC four wires, based on the full-bridge (FB), T-type or NPC topologies [3].

In the three-stage structure, the formation of local or regional DC grids is possible because of the availability of the DC stage. This feature enables controlling the two grids independently, with only the constraint of the active power link. The requirements for this stage, such as high-rated power and high current capability on the LV side and high-voltage capability, high-frequency isolation and high efficiency on the HV or MV side. The basic configuration includes series-connected modules on the HV or MV side and parallel-connected ones on the LV side. The DC/DC stage is more likely to be an isolated DC/DC converter, implemented normally using the DAB converter [4] or the series-resonant converter [5]. An additional approach is the use of multiwinding-based topologies, such as the quadruple-active bridge (QAB) [6] as a basic cell, which shares the same advantages as the DAB converter but with a lower number of high/medium frequency transformers.

At the HV or MV level, the HV or MV DC link voltage is maintained constant at the nominal value by controlling the active current. Moreover, the reactive power can be controlled separately from the active power, representing a degree of freedom for the system control. With the introduction of communication, the ST can perform power factor correction at the HV or MV substation, reducing the reactive power request to the transmission grid. The active power is the only link between the AC HV or MV and LV ST stages. Oscillating power components can be controlled separately from the active power. Thus, harmonic voltage and current compensation are possible. The most promising topologies to implement the HV or MV stage of the ST are the cascaded H-bridge (CHB) converter and MMC. Both converters share the same features: modularity, possibility of fault-tolerance implementation, multilevel operation, reduced dv/dt and filter size. The NPC converter may provide a possible topology to implement the MV stage, and it represents a standard solution for MV applications [7].

Power electronics has played an important role in AC/DC distribution system, which possesses the following features: can provide AC and DC interfaces for distributed renewable energy generation and energy storage; can realize the comprehensive regulation ability of voltage conversion, voltage isolation, improving power quality and realizing bidirectional energy flow; can be independently separated from the main system according to fault conditions, so as to improve the self-healing of the power grid. The development of modern power electronics in AC/DC distribution systems is moving toward multiports, multilevels, cascaded connections and multicommunication.

2.4 Controllability

There are three layers for the hierarchy of control in AC/DC distribution systems, including primary, secondary and tertiary control layers. The main targets of the primary control are accurate active and reactive power sharing among the distributed generation units by incorporating the power controllers, maintaining VSC terminal voltage magnitude and frequency by using the voltage controllers and achieving the required filter inductor current by employing the current controllers. As a result, the primary control can provide internal control of the converter output voltage and current, as well as power sharing using centralized and decentralized control methods. The islanding detection, power sharing, output control and subsequent change of controller modes need a fast response, which is implemented by the primary control because it is the fastest among all the control levels. The primary control can be classified as linear and nonlinear types due to its characteristics. The linear controllers include synchronously rotating reference frame-based proportional–integral (PI) controllers, state feedback, stationary reference-based proportional-resonant (PR) and deadbeat controllers. PI controller is preferable for DC command following as it tracks the reference with zero steady-state error while the PR controller is appropriate for AC command following. The nonlinear controllers include hysteresis, predictive, sliding-mode, delta-modulation, online optimized, neural network and fuzzy logic controllers.

The secondary control plays an important role in the energy management system of the AC/DC distribution systems [8]. It contributes to power quality improvement, restoring frequency and voltage in the AC/DC distribution systems due to primary level droop control actions. It is helpful for resynchronization of AC/DC distribution systems to the utility grid and ensuring optimal and coordinated operation of distributed energy resources (DER) units within the AC/DC distribution systems. The secondary control processes the frequency and voltage error deviations to produce control signals, for the error correction that is sent to the primary controller in islanded mode, which can be centralized, decentralized or distributed in nature of its operation. There are optimization-based supervisory control, coordinated control, intelligent supervisory control, consensus-based control techniques and multiagent-based techniques.

The goal of tertiary control is to deal with the power flow between the AC/DC distribution systems and the utility grid and to optimize the operation of AC/DC distribution systems. The tertiary control coordinates the operation of multiple AC/DC distribution systems and communicates the requirements such as voltage and frequency regulation from the large-scale utility grid.

2.5 Artificial intelligence in modern power systems

Passive and unidirectional power systems are transformed into active, distributed and transactive power systems due to the massive adoption of heterogeneous and spatially distributed electrical components in power systems. In order to ensure the good performance and management of power systems, fast and reliable communication and computational networks are necessary for efficiently processing measurements and transferring control signals to relevant electrical components for quick actions. There is a large amount of data generated and exchanged among different entities in these modern power systems. Artificial intelligence is an attractive candidate to realize the generalization of the model based on large amounts of collected data that is the potential to be used in power systems. As a result, the artificial intelligence techniques can offer fast and accurate data-driven solutions for different power system applications [9], including energy forecasting, scheduling and electricity markets, system optimization and stability control, fault detection and protection, and cybersecurity, as shown in Figure 2.10.

2.5.1 Energy forecasting

Taking advantage of artificial intelligence in regression, research on power prediction of multiple forms of energy generation at the source end and research on energy load prediction at the load end will better support the planning, operation and service of power systems.

As the penetration of intermittent renewable energy increases, the impact of intermittent and fluctuating power generation on the power systems becomes more

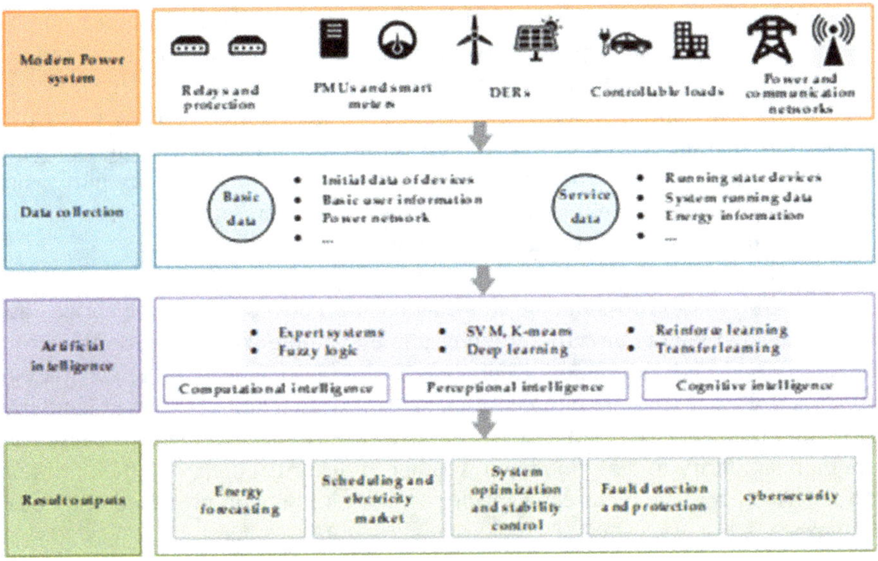

Figure 2.10 Applications of artificial intelligence in modern power systems

and more obvious, and accurate renewable energy power prediction is particularly important for system stability and economic operation. The key to improving the accuracy of intermittent renewable energy power prediction is to build a prediction model with data processing and feature extraction capabilities, and good self-learning correction capability. For example, a deep long-short-term memory recurrent neural network model can be constructed to predict PV system output power with the help of the recurrent structure in the long-short-term memory network and the ability of the memory unit to capture temporal change features. The deep convolutional neural network is used to extract the advantages of nonlinear features and invariant structure of daily light data sequences to improve the prediction accuracy of system output power. The wind speed prediction is performed with the advantage that the deep confidence network can effectively characterize the internal structure and features of complex wind speed series, and then the wind power is evolved.

Except for source prediction, accurate load forecasting is crucial for secure and reliable power system operations, which is becoming complicated and difficult due to the variability and nonstationarity load features caused by seasonal effects, weather conditions and socioeconomic dynamics. Artificial intelligence is an effective tool to predict loads, sharing a similar method with that in source-end forecasting.

2.5.2 Scheduling and electricity market

The relationship among electricity customers, electricity suppliers and system operators becomes more complicated considering the increasing penetration of smart grid applications and the restructuring of the electricity market worldwide. The interactive patterns between suppliers and customers make electricity prices more unpredictable. The accurate electricity price forecasting is extremely crucial for electricity market participants who engage in optimal bidding strategies and risk management. Moreover, with the wider ownership of behind-the-meter DERs and advanced metering infrastructures, electricity customers can proactively participate in electricity market decisions. Accordingly, electricity price forecasting can guide customers to adjust their consumption profiles for optimizing electricity purchasing costs. Extensive research efforts have been vested on addressing the important but challenging task of electricity price forecasting.

Artificial intelligence uses historical electricity prices, socioeconomic factors and other information to simulate the relationship between electricity prices and their influencing factors directly through sample learning, with strong nonlinear fitting ability and high prediction accuracy. At present, to improve the accuracy of prediction for electricity price forecasting, data mining techniques such as similarity search or fuzzy classifier and association analysis are used to identify data series with high similarity to daily load and weather in historical data, and then artificial neural network and support vector machine are used to obtain predicted daily electricity price.

In terms of market trading and bidding strategy, due to the existence of market operation mode, energy supply, demand and other factors, it is also necessary to consider the game between the operation of production, transmission, transfer, consumption, storage and other aspects. The balance of electricity, gas, heat, oil and other multienergy transactions and pricing strategies in a perfectly competitive market further complicates the operation of energy systems. Therefore, the complexity, incomplete and asymmetric information of this problem can be effectively decomposed by using multi-intelligent body technology, and the self-learning capability of the intelligent body can be improved by combining Q-learning method of reinforcement learning.

2.5.3 System optimization and stability control

The power system is not a simple superposition of several independent energy supply systems, but through the synergistic regulation of multiple energy supply units and the synergistic optimization of both supply and demand, it can meet the system energy demand while obtaining higher efficiency than each energy system operating independently.

Stability control of power systems mainly considers whether the system can operate normally after being disturbed. The development of transmission and distribution networks and the increased participation in the power market have increased the uncertainty and difficulty of power system stability control. Traditional control methods are mostly based on the analysis of the physical characteristics of the power grid. However, physical modeling needs to be done based on different topologies, operation modes and fault types, so it is not very adaptable to changes in grid structure and new power equipment and devices to meet the needs of grid development. Therefore, a data-driven approach can be attempted to replace the process simulation and obtain the stability control strategy directly by fully mining the system environment information with the autonomous decision-making capability of reinforcement learning. It is also noted that reinforcement learning is weak in information perception and acquisition. When analyzing the grid environment information in the early stage, we can first extract the grid operation characteristics with the advantage of deep learning in feature extraction and then provide high-value density information as the input data for reinforcement learning to improve the correctness of decision-making and control efficiency.

2.5.4 Fault detection and protection

Fault diagnosis of electrical equipment is essentially a complex pattern recognition problem. The traditional pattern recognition method for fault diagnosis suffers from the problem that fault features rely on subjective selection, while the strong recognition ability of artificial intelligence is bound to enhance the fault diagnosis capability. Deep learning of artificial intelligence can identify the coupled features inherent in big data and integrate the acquired feature information into the modeling process, thus eliminating the shortage of human-selected features and the complexity caused by traditional feature extraction.

As an example, in order to solve the difficulties of asynchronous motor fault diagnosis methods caused by complex motor structure, nonsmooth signal and mechanical big data, stacked noise reduction self-coding is employed to extract signal features and combine with Softmax classifier to achieve efficient and accurate motor fault diagnosis; the deep autoencoder network model and layer-by-layer encoder network model are used for fault detection of gearboxes and main bearings of new energy wind turbines, overcoming the reliance of traditional methods on diagnostic experience and the lack of generalization ability of model diagnosis under large data input; combining the characteristics of oil chromatography data and fault types of power transformer online monitoring, a new method of transformer fault diagnosis based on classification deep autoencoder network can be used. There is also a study online tripping faults that often occur in power systems, using the advantage of long- and short-term memory networks that are good at extracting time-series features to capture faults from transmission and distribution processes.

2.5.5 Cybersecurity

The modern power system and integrated energy system are no longer a network of power equipment in the traditional sense but have evolved into a cyber-physical system with deep coupling of information and physics [10]. The weak link of the system is no longer limited to the physical subsystem, and the information link may become the new weak point. As a result, cybersecurity and the protection of power information and physical systems have become more and more important. The Iranian nuclear power grid attack in 2010 and the Ukrainian blackout in 2013 are reminders of the importance of cybersecurity in power systems.

Attack detection is the primary protection technology in network security. The known intrusion attacks against the power grid and power information system mainly include virus attacks, vulnerability attacks, false data injection, eavesdropping attacks and denial of service attacks, which are developing toward the trend of team-based combination attacks.

Currently, some studies introduce artificial intelligence into the attack detection process by applying conditional Gaussian-Bernoulli restricted Boltzmann machines to extract high-dimensional features of false data attack behaviors or using Convolutional neural networks to learn intrusion attack features from the time-space dependence of network data messages. Then, they detect false data and malicious attack messages in real time. Some scholars have also taken advantage of deep learning in dimensionality reduction by using auto-encoders to first reduce the dimensionality of data and then perform attack feature extraction. Also, multi-intelligence techniques have been used for cybersecurity protection of power systems [11].

In addition, security situational awareness assesses system security from a network-wide perspective, identifies potential weak areas and highly concealed attacks and then predicts upcoming security events by retracing historical attacks [12]. With the powerful ability of artificial intelligence in big data analysis and situational awareness, network security situational awareness can be performed. Based on the self-learning capability of reinforcement learning, the Q-learning-based sequence approach can identify the vulnerable sequences that lead to severe power outages in the power system; a modified deep confidence network is used to build a situational prediction model with multiple inputs and outputs for grid security situational prediction.

2.6 Cyber-physical security in active distribution network

2.6.1 Potential vulnerability

An active distribution network is potentially vulnerable to attacks, including attacks on hardware, attacks on controllers, attacks propagating throughout the supply chain, attacks targeting the monitoring and diagnostics platform, and attacks directed at the grid and impacting the plant operation and its overall safety. The potential points that are vulnerable to attacks are shown in Figure 2.11. They are classified as physical attacks and cyber attacks. The actual physical attack is to tamper with the hardware by stealing and removing the devices in the systems. The cyber attacks are intruded in an active distribution network, such as in sensors, smart meters, load aggregators, faulty electronic components, subpar analog, digital parts and plant supervisory systems to wreck system performance such as voltage and frequency regulation-based ancillary services, energy management, falsifying energy demand and disconnecting the grid from the plant [13]. With the intelligence and informatization of the power system, cyber security is paid increasing attention.

2.6.2 Vulnerability analysis of cyber attacks on control of VSCs

VSCs play a key role in renewable energy sources such as wind and photo-voltaic, energy storage systems, electrical vehicles, ST and HVDC transmission systems, because they serve as the most common energy conversion interfaces between these

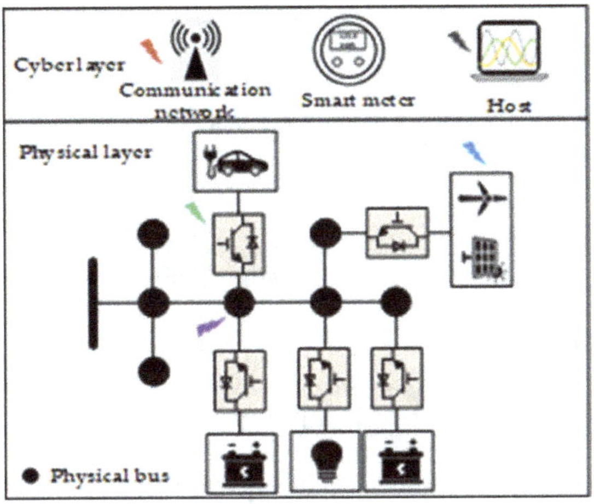

Figure 2.11 *Potential points vulnerable to attacks*

technologies and the power grid. As a result, vulnerability analysis of cyber attacks on control of VSCs is taken as an example and presented to show how cyber attacks influence the active distribution network. It should be noted that other power converters in active distribution networks share a similar vulnerable analysis of cyber attacks with that of VSCs.

2.6.2.1 Grid-forming control for VSCs

Grid-forming VSCs are responsible to hold voltage and frequency. When it is synchronized with other AC sources, primary control can be aligned locally by the available measurements, which is considered safe from the perspective of cyber security because it is difficult for the attackers to access the physical layer. Moreover, suitable physical layer security alternatives such as beamforming are commonly used recently [14]. Nevertheless, decentralized control philosophies suffer from an operational point of view in matching the commercial regulatory standards [15]. This drawback has been made up by the secondary controller, which compensates the primary control by the offsets generated from the information of other VSCs, called distributed or centralized secondary control. The application of secondary control will introduce the communication, making it vulnerable to the attackers to inject various attacks into the sensors, communication link or the controller.

By penetration of the adversary inside the control platform, which can be easily achieved by Trojan Horse [16] to use remote systems as hosts, the data of sensors can be manipulated. The acquisition gains of sensors based on a linear plotting theory are used to correct the actual measurement, which can be easily changed by attackers, leading to errors in the reported measurements.

The manipulation of data for communication exchanging can happen in the communication part involving a router/encoder/decoder. The transmitted data can be changed by authorizing violation, interrupting the transmitted signals, the illegitimate opening of information logs, replaying the transmitted information from the past and so on.

As for the controller, the reference inputs used in either the outer control loop or secondary controller for control of VSCs can be illegitimately changed by the Trojan Horse, destroying the objectives of the controller.

2.6.2.2 Grid-feeding and supporting control for VSCs

In grid-connected applications for integrating renewable energy sources, grid-feeding controls for VSCs are popular, which are used to inject active and reactive power into the grid-forming units [17]. Moreover, desired control inputs are added to the overlaying grid-forming controller to improve grid supportive services. For the grid-feeding controlled VSCs, attacks are likely to inject attacks on sensors and controllers to manipulate the reference input v_{dcref} and the measured information, leading to a stability/coordination issue in the network due to limiting or increasing the power flow from VSCs. The grid-supportive services provide P_{gss} and Q_{gss}, which can be sneaked with false data, resulting in misguiding the controlled units. In brief, the upper-level control responsible either for grid-supportive services or for secondary control objectives provides the measurements/references for the lower control, being vulnerable to cyber attacks. Moreover, some challenges such as project system outage, nonoptimal operation, economic feasibility, instability and consumer discomfort caused by cyber attacks should be considered.

2.6.3 Cyber attacks in active distribution network

2.6.3.1 Distributed power sharing in AC and DC microgrids

Cyber attacks on AC microgrid: Taking a setup of an AC microgrid with three grid-forming VSCs as an example, a false data attack is conducted into the outgoing communication links from unit II, shown in Figure 2.12. The AC microgrid needs improved voltage regulation and load sharing to have a reliable and economical operation, so the distributed control is applied. The primary control is used to control the grid-forming VSCs, and the active power among the distributed generations is shared equally for equal active power droop m_p. However, the objective of the

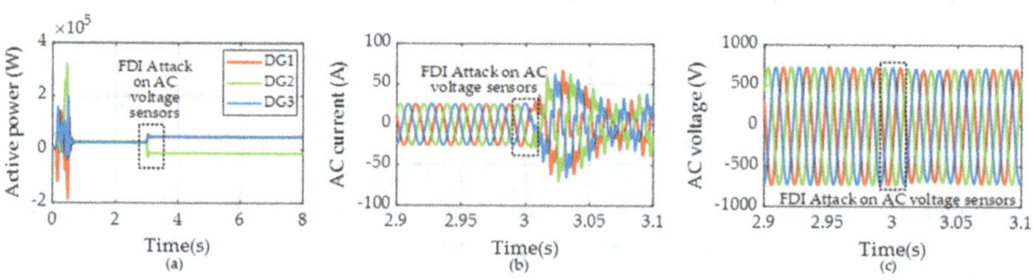

Figure 2.12 Cyber attacks on AC microgrid: (a) active power, (b) DC currents, (c) DC voltages – FDIA to cause derivation and unstable AC outputs

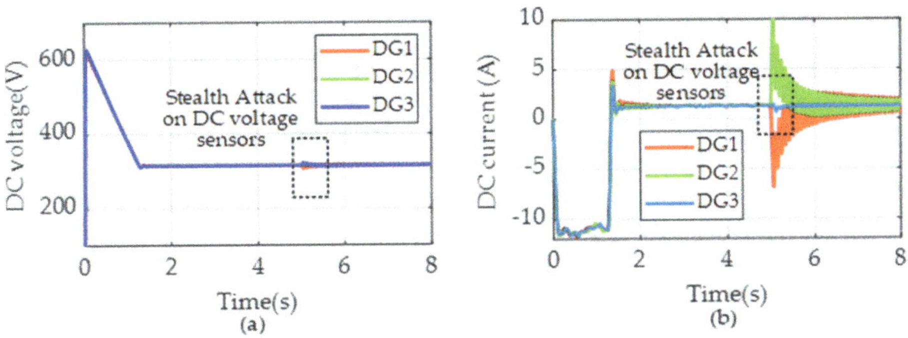

Figure 2.13 Cyber attacks on DC microgrid: (a) DC voltages, (b) DC currents – stealth attacks to cause oscillations

distributed control failed due to the injection of false data. As a result, the operation of microgrids can be seriously affected and shut down.

Cyber attacks on DC microgrid: Figure 2.13 shows the results of attacks on DC voltage sensors in unit II. Before initiating the attack, the DC microgrid currents and average voltages are converged and the system operates stably. Introducing the stealth attack, which is a well-planned set of balanced attacks, the DC microgrid currents and average voltages can maintain consensus even with large balanced values. However, instability may occur due to the stealth attack. With a relatively small value of attack, the currents and average voltages are in good performance while they oscillate slightly after the load changing. When relatively large attacks are injected into the voltage sensors, severe instabilities appear even though the currents and average voltages are in good convergence.

2.6.3.2 Cyber attack on VSC-based HVDC stations

The rising share of renewable energy sources in the electric mix requires an increase in long-distance transmission capacities. The recourse to HVDC reduces the transmission losses and is required for nonsynchronous zone interconnections or long-distance underground links. Furthermore, the development of meshed HVDC grids to replace existing point-to-point links will increase the grid reliability and flexibility while decreasing the investment costs, where two-level VSCs-based HVDC is popular and sophisticated. The control law of VSC-HVDC is quite commonly a set of grid-feeding VSCs with one station following P–Q control whereas the other station following DC voltage regulation. Cyber attacks are possible to penetrate VSC–HVDC through sensors or grid-supportive services. Taking the 200 MVA, ±100 kV VSC-based HVDC as an example, Figure 2.14 shows the results when an False data injection (FDI) attack is injected into the DC voltage sensor in station II at $t = 1.5$ s. It can be seen that there are large oscillations in the VSC-based HVDC as the attacks initiated, arising the concerns of stability and reliability of the systems.

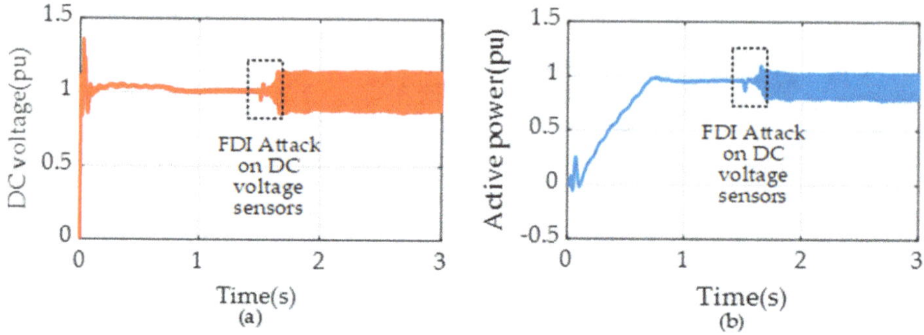

Figure 2.14 Cyber attack on station II in VSC-HVDC: (a) DC voltage, (b) Active power – FDIA to cause oscillatory instability

2.6.3.3 Impact of cyber attack on wind farms

Attacks on STATCOM: With large installations of wind farms, it is necessary to extract maximum output by using many robust and reliable control strategies [18]. The traditional wind farms adopt squirrel-cage induction generators that need large capacitor banks for reactive power to be absorbed by the Induction generator (IG) because its stator is directly connected to the grid. If the reactive power requirement increases, it is withdrawn from the grid. Since the wind farms are usually connected to a 25-kV distribution network, excess withdrawal of reactive power deteriorates the voltage profile. To prevent this, STATCOMs are usually connected at the Point of common coupling (PCC) to provide reactive power support to the wind farm [19]. The STATCOM can maintain the constant voltage by continuously exchanging the reactive power with the system to improve the voltage fluctuation. Therefore, it is efficient to use STATCOM to compensate for the variation in the reactive power of the system. In the analysis, a false data injection attack is initiated in the AC voltage sensor at t = 12 s, as shown in Figure 2.15. The results show that there is an unnormal increase in reactive power when the attack is injected, which needs the STATCOM to absorb the reactive power [20].

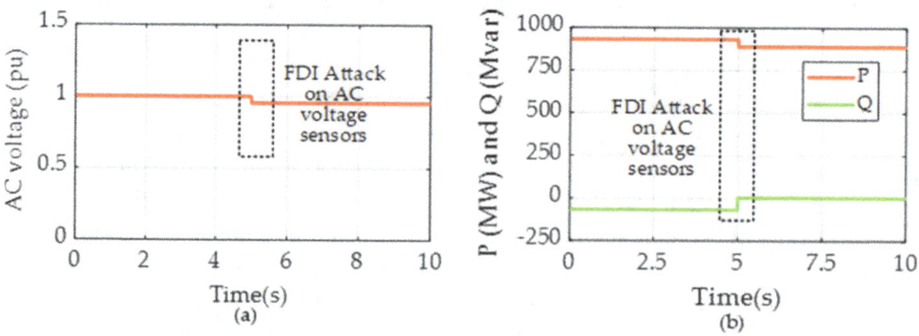

Figure 2.15 Attacks on STATCOM: (a) AC voltage, (b) Active and reactive power – FDIA to cause derivation

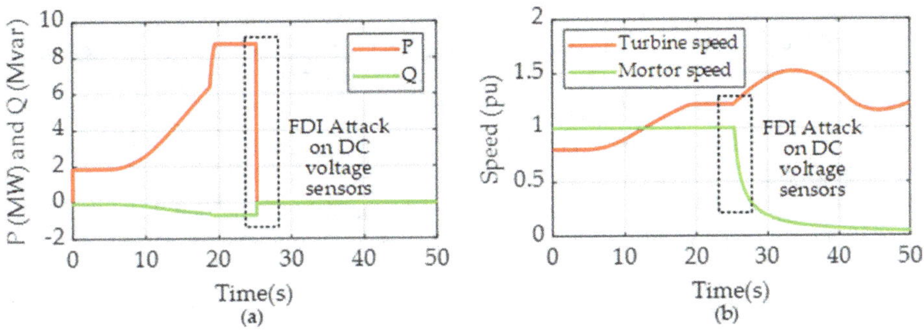

Figure 2.16 Impact of FDIA on DC voltage sensor in DFIG: (a) Active and reactive power, (b) speeds – FDIA to cause shut down or tripping

Attacks on doubly-fed IG (DFIG): The DFIG is largely used in wind energy conversion systems, with the intent of having a relatively cheap power electronics interface while allowing high-performance torque control and avoiding machine oversizing when operating with constant DC voltage and variable speed. The DFIG consists of a wound-rotor induction machine interfaced with a single downsized VSC. The VSC provides the required magnetizing current and control action, which is usually on the rotor side, which is to extract maximum power from the IG based on the power-tip speed ratio graph. In addition, the DC voltage is regulated by the grid side converter to transfer the power from VSC into the grid. To reveal the influence of cyber attacks on DFIG, an FDIA is injected into the AC voltage sensor, as shown in Figure 2.16. The protection is activated easily due to the overvoltage threshold, which affects the operation of the VSC. The DFIG is vulnerable to cyber attacks, which can shut down or trip a large renewable generating unit, failing to provide reliability for system operation as a result.

References

[1] Alhurayyis I., Elkhateb A., Morrow J.D. 'Isolated and non-isolated DC-to-DC converters for medium voltage DC networks: a review'. *IEEE Journal of Emerging and Selected*. 2020;9(6):7486–7500.

[2] Liu X., Liu Y., Liu J., Xiang Y., Yuan X. 'Optimal planning of AC-DC hybrid transmission and distributed energy resource system: review and prospects'. *CSEE Journal of Power and Energy Systems*. 2019;5(3):409–22.

[3] Liserre M., Buticchi G., Andresen M., De Carne G., Costa L.F., Zou Z.-X. 'The smart transformer: impact on the electric grid and technology challenges'. *IEEE Industrial Electronics Magazine*. 2016;10(2):46–58.

[4] She X., Yu X., Wang F., Huang A.Q. 'Design and demonstration of a 3.6-kV–120-V/10-kVA solid-state transformer for smart grid application'. *IEEE Transactions on Power Electronics*. 2014;29(8):3982–96.

[5] Zhao C., Dujic D., Mester A., *et al*. 'Power electronic traction transformer—medium voltage prototype'. *IEEE Transactions on Industrial Electronics*. 2014;61(7):3257–68.

[6] Callegati F., Cerroni W., Ramilli M. 'Man-in-the-middle attack to the https protocol'. *IEEE Security & Privacy Magazine*. 2009;7(1):78–81.

[7] Ferreira Costa L., De Carne G., Buticchi G., Liserre M. 'The smart transformer: a solid-state transformer tailored to provide ancillary services to the distribution grid'. *IEEE Power Electronics Magazine*. 2017;4(2):56–67.

[8] Sahoo S., Mishra S. 'A distributed finite-time secondary average voltage regulation and current sharing controller for DC microgrids'. *IEEE Transactions on Smart Grid*. 2019;10(1):282–92.

[9] Bose B.K. 'Artificial intelligence techniques in smart grid and renewable energy systems—some example applications'. *Proceedings of the IEEE*. 2017;105(11):2262–73.

[10] Sahoo S., Dragičević T., Blaabjerg F. 'Cyber security in control of grid-tied power electronic converters—challenges and vulnerabilities'. *IEEE Journal of Emerging and Selected Topics in Power Electronics*.;9(5):5326–40.

[11] Boyer S.A. *SCADA: Supervisory control and data acquisition*. International Society of Automation; 2009.

[12] Liu Y., Ning P., Reiter M.K. 'False data injection attacks against state estimation in electric power grids'. *ACM Transactions on Information and System Security*. 2011;14(1):1–33.

[13] Wood A.D., Stankovic J.A. 'Denial of service in sensor networks'. *Computer*. 2002;35(10):54–62.

[14] Shiu Y.-S., Chang S., Wu H.-C., Huang S., Chen H.-H. 'Physical layer security in wireless networks: a tutorial'. *IEEE Wireless Communications*. 2011;18(2):66–74.

[15] Ovalle A., Ramos G., Bacha S., Hably A., Rumeau A. 'Decentralized control of voltage source converters in microgrids based on the application of instantaneous power theory'. *IEEE Transactions on Industrial Electronics*. 2014;62(2):1152–62.

[16] Stefanov A., Liu C.-C. 'Cyber-power system security in a smart grid environment'. *IEEE PES Innovative Smart Grid Technologies*. 2012:1–3.

[17] Rocabert J., Luna A., Blaabjerg F., Rodríguez P. 'Control of power converters in AC microgrids'. *IEEE Transactions on Power Electronics*. 2012;27(11):4734–49.

[18] Saad-Saoud Z., Lisboa M.L., Ekanayake J.B., Jenkins N., Strbac G. 'Application of STATCOMs to wind farms'. *IEE Proceedings - Generation, Transmission and Distribution*. 1998;145(5):511–6.

[19] Ahsan S., Siddiqui A.S. 'Dynamic compensation of real and reactive power in wind farms using STATCOM'. *Perspectives in Science*. 2016;8(1):519–21.

[20] Molinas M., Jon Are Suul., Undeland T. 'Low voltage ride through of wind farms with cage generators: STATCOM versus SVC'. *IEEE Transactions on Power Electronics*. 2008;23(3):1104–17.

Chapter 3
Microgrids in mission-critical applications

Suman Rath[1], Charalambos Konstantinou[2], Behnaz Papari[3], Christopher Shannon Edrington[4], Pudong Ge[5], and Fei Teng[5]

Similar to utility grid and other terrestrial microgrids, aircraft electric power systems and shipboard microgrids have their own power generation, distribution, utilization and generation storage. The rapid development of power electronics technology has allowed the converters to operate at DC voltage levels required for transmission, distribution and consumption. However, the coordination among power electronic converters can lead to malicious intrusions that aim to manipulate microgrids' operation and cause deviation from their mission-critical objectives. Such manipulations may lead to disturbances during motion, thus posing serious risks to passenger lives and cargo. The disturbances can also affect operations in critical sectors such as commercial transportation and defence, which may directly influence the global economy and national security. This chapter overviews the industrial security measures considered for microgrids in mission-critical applications, such as those found in electric aircrafts and shipboard power systems, with further insights on vulnerable areas, which can be exploited by stealthy attackers.

3.1 Introduction

Depleting reserves and rising prices of fossil fuels (e.g., petroleum) are some of the primary concerns for the entire mankind. These energy sources are instrumental in supporting major forms of transportation including aerial, terrestrial and water-based movement. Despite fully electric (battery-driven) terrestrial vehicles have been in use [1], pre-charged batteries cannot be considered as viable energy sources for long-distance aircrafts and ships [2]. Hence, engineers and researchers have been

[1]Oklahoma State University, Stillwater, OK, USA
[2]KAUST, Thuwal, Saudi Arabia
[3]Texas A&M University, Galveston, TX, USA
[4]Clemson University, Clemson, SC, USA
[5]Imperial College London, London, UK

trying to develop strategies for better usage of fossil fuels in such vehicular units. One of the techniques to achieve this is by increasing the level of electrification in those vehicles, which enables a more efficient utilization of the input fuel [3, 4]. The input fossil fuel can be used to facilitate production of electrical energy necessary to run the electrified components. Thus, a setup for in-vehicle power generation can be devised, which involves the use of rotating turbine units (powered by fossil fuels such as petroleum) in aircrafts and ships to generate electric power for its consumption into other subsystems. An intermediate stage may also be present where the generated electric power is stored before consumption. A completely autonomous microgrid is designed on-board the vehicular system consisting of local generation, loads, storage and distribution units.

Power electronic devices are essential components of aircraft and shipboard microgrids since they facilitate regulated power conversion [5] and can act as interfacing elements between the turbines, storage units and the common AC/DC buses. As interfacing elements in the modern power system, power electronic converters are responsible for stability and regulation of system parameters (e.g., frequency and voltage) as per the specific ratings at various bus-levels, for storage at the battery-level [6] and consumption at the load-level [7, 8]. Hence, stabilization and synchronization of output parameters among such conversion-level devices [9] are important control objectives, which have to be fulfilled in order to eliminate potentially disruptive issues such as voltage and frequency instabilities and unauthorized load sharing. in the autonomous microgrid system [10–12]. Such disruptions may lead to malfunction or shutdown of various critical subsystem loads in the aircraft/ship and affect overall functioning of the vehicular unit [13, 14]. Achievement of the above-mentioned control targets requires the use of information and communication technology (ICT) for enabling exchange of information between the controllers associated with the converters. More specifically, the communication network is responsible for transmitting inputs received from sensors and controllers to the appropriate recipient units, which will then process the received data to generate control commands for the conversion-level devices [15].

Clearly, trustworthiness of the communication network is of paramount importance for reliable operation of the control framework present in the microgrid system. However, this may not be the case as inadequate security measures in ICT expose the microgrid to additional cyber vulnerabilities, which may be exploited by malicious actors to jeopardize reliability of the control system and disturb working of the overall vehicular body [16]. These vulnerabilities can be at the sensor, the communication and/or the controller level. Considering, the human lives that might be at stake, any rogue attempt to manipulate the aircraft or shipboard microgrid can have devastating consequences. Elimination of cyber vulnerabilities is the only way to fortify the system against such malicious intrusion and manipulation attempts. This chapter presents an insight into the control structure of aircraft and shipboard microgrids along with a review of the industry-standard security protocols that are generally adopted to safeguard such power systems and highlights their loopholes and vulnerabilities. An understanding of such vulnerabilities is an essential first step towards their elimination.

Remaining parts of this chapter are organized as follows. Section 3.2 presents a detailed insight into the architecture of aircraft microgrids. This section also highlights the role played by power electronics in enhancing the electrification level in modern aircrafts. Furthermore, some of the latest trends in aircraft power electronics have been discussed here. Section 3.3 elaborates the architecture of shipboard microgrids specifically focusing on the crucial role of conversion-level power electronics. Similar to the preceding section, this section also reviews some of the latest trends in shipboard power electronics. Section 3.4 provides a detailed analysis of how coordinated control is achieved among power electronic converters in such vehicular units. Section 3.5 presents an extensive analysis of vulnerabilities in the discussed control structures after reviewing the industry-level security fortifications guarding them. Section 3.6 provides simulation results to demonstrate the level of damage that cyberattacks can cause in medium voltage DC (MVDC) shipboard microgrids. Finally, Section 3.7 presents a summary of the chapter along with some guidelines for improving the existing security measures.

3.2 Electric aircrafts

Records indicate that more than 4.5 billion people travelled through air in 2019 [17]. Despite a slump due to travel curbs in the COVID-19 era [18], air travel is set to regain shape in the coming years. A major deterrent that dissuades people from less-privileged sections of the society to travel through air is the high ticket price. High ticket costs are partly due to the rising prices of aviation turbine fuel (or jet fuel), which is the primary source of energy for aircrafts. Hence, it is essential to increase the efficiency of aircrafts so that the necessary power requirements can be derived from significantly lower quantities of jet fuel. As mentioned in the preceding section, this can be realized through electrification of various aircraft subsystems. Electrification necessitates the building of a complete power system on-board the vehicle, where electrical energy is generated through the conversion of mechanical energy, which is obtained through combustion of the input jet fuel. Figure 3.1 shows a schematic diagram depicting the process of electricity generation in a typical aircraft microgrid.

Aircraft power systems are autonomously operating microgrids, which derive power for the connected loads from local generator units. The generator units can be attached to rotating parts of the main engine(s), the auxiliary power unit, etc. After generation, the power is fed to one or more AC/DC buses, each of which has to be maintained at a constant value of voltage. Since there are multiple generator units, the output power from each of them has to be appropriately conditioned before being fed to the common buses. Similarly, power conversion has to be done before the power from various buses can be stored in batteries or consumed by subsystem loads. Power conditioning (/conversion) is usually performed through the use of controllable power converters, which form the interfacing layer between various elements of the aircraft microgrid. Thus, the role of conversion-level power electronics expands across all levels of power flow (i.e., generation, storage, and

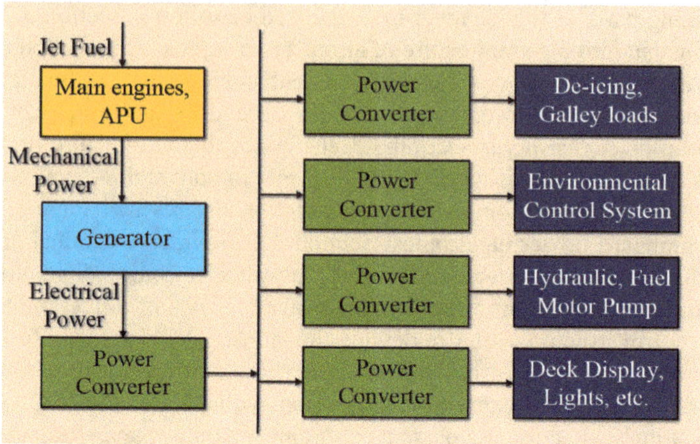

Figure 3.1 Generation and consumption of electrical energy in the aircraft power system

distribution). Recent technological advancements in the domain of power electronics have enabled even the traditionally non-electrified components (/subsystems), which were conventionally powered through mechanical, hydraulic or pneumatic arrangements to be run using electricity [4].

Figure 3.2 shows an example depicting the architecture of a typical aircraft power system. Output parameters of the generated AC power are modified to generate pre-fixed values of frequency and voltage using AC–DC and DC–AC converters before being connected to the common AC bus. Similarly, power from the common bus is appropriately conditioned using DC–DC converters and inverters before

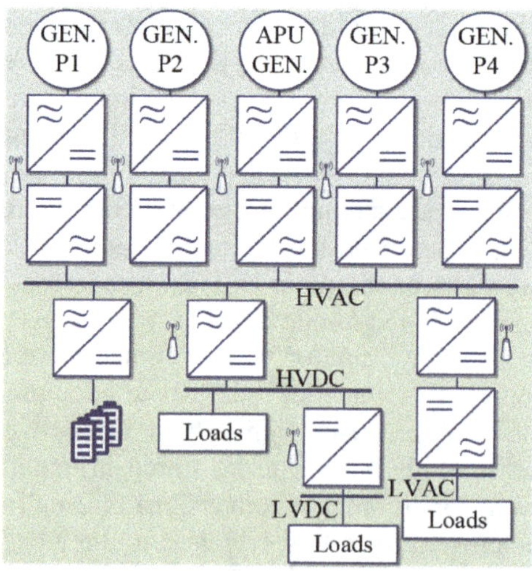

Figure 3.2 Layout of a typical aircraft microgrid

being provided to the loads through DC and AC buses (of corresponding voltage and frequency levels), respectively. Bus voltage levels in a typical aircraft power system may include 230 V AC (large loads such as de-icing systems, galley loads and AC motor pumps may be connected to this bus), 270 V DC (loads operating at this level may include variable speed motors controlled using appropriate motor controllers), 115 V AC (connected loads may include fans, equipment coolers, window heating systems, etc.) and 28 V DC (associated loads may include deck display unit, DC fuel pump, etc.) [19]. Typical storage devices such as batteries are also present in the aircraft microgrid. Bidirectional DC–DC converters are employed to feed the DC storage batteries with unused, excess power, which can be utilized as and when the need arises. An aircraft power system generally uses transformer rectifier unit (TRU) and auto-TRU (ATRU) [20] for AC–DC power conversion. These are specialized devices that can integrate the abilities of both transformers and rectifiers into a single device and are useful for obtaining DC power at varied voltage levels.

The modern aircraft power system is a cyber-physical system in which the physical devices mentioned in the preceding paragraph form the physical layer. This layer also consists of numerous sensors that record vital data measurements and send them through a communication network to enable the control layer for operation of the power electronic converters [21]. The communication elements along with the controllers and other units involved in the handling of data form the cyber layer in the cyber-physical aircraft power system. More details about the control of power electronic converters in the aircraft microgrid have been provided in Section 3.4.

3.3 MVDC shipboards

The shipping trade industry has an annual financial value of more than 14 trillion US dollars [22], which translates to almost 70% of the total value of world trade and accounts for nearly 80% of the world's overall goods' volume [23]. Apart from trade, ships are also used for luxury cruising and ferrying of passengers. Clearly, these marine vehicles are crucial elements for the well-being of the global economy and their efficiency can directly affect the world GDP. One of the ways to improve the efficiency of ships is by increasing the proportion of their electrified subsystems. Electrification allows better utilization of fossil fuels, reduces overall weight of the ship and helps lower the production of (air and water) polluting elements [24]. Power to run the electrified subsystems is generated through local generation techniques, which mostly involve the use of alternators or other generating units connected to rotating parts (e.g., engine). Through this technique, electricity can be produced in the ship through the conversion of mechanical energy. The electrical power finally obtained is transferred to the appropriate loads using proper distribution channels. Thus, an autonomously operating power system with local generation facilities is built on the shipboard. There have been many advances in the concept of ship power systems over the last 30 years. The following paragraph briefly tracks down the development timeline of shipboard power systems.

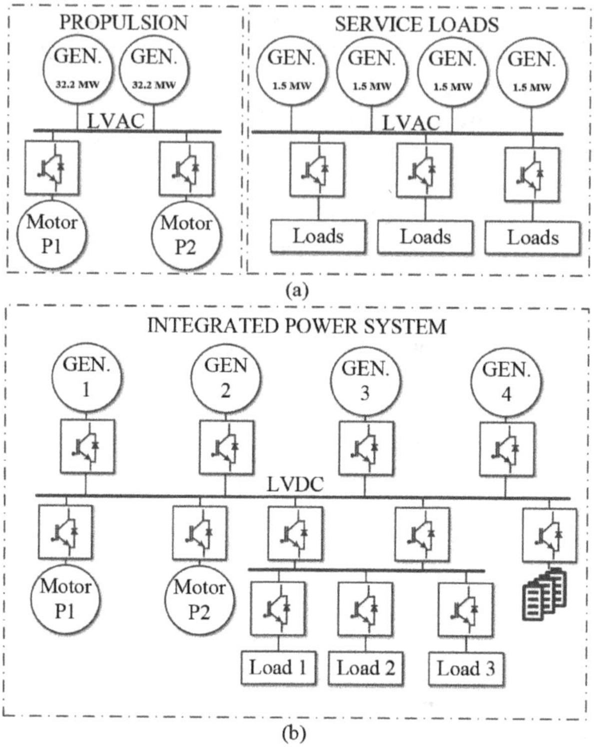

Figure 3.3 Architecture of (a) a segregated AC shipboard power system and (b) an integrated DC shipboard power system

In the past, the logic was to have separate generation for various parts of the ship such as hotel loads, radar and propulsion [25–28]. This was mostly due to the variation in power ratings at different stages, which did not allow them to be connected together. However, with the advent of power electronics as an energy routing and shaping enabler along with advances in control, optimization and computational capability, new integrated system architectures have emerged that facilitate a higher degree of resiliency, functionality and performance [29, 30]. Modern power electronics have also allowed shipboard power systems, which were traditionally maintained at low voltage AC (LVAC) distribution levels [31] to be maintained at both medium voltage AC and MVDC. However, MVDC is preferable for power distribution due to higher efficiency and non-requirement of heavy and space-consuming transformers for voltage modification [31]. Figure 3.3(a) shows the power system architecture of SS *Canberra* [32], which followed the traditional LVAC power distribution system layout [31]. It has dedicated generation units to power the propulsion subsystem and separate units to power the other loads. Figure 3.3(b) shows the possible architecture of a DC shipboard power system, which uses a common set of generators to power all of its subsystems. Both the examples shown in the figure

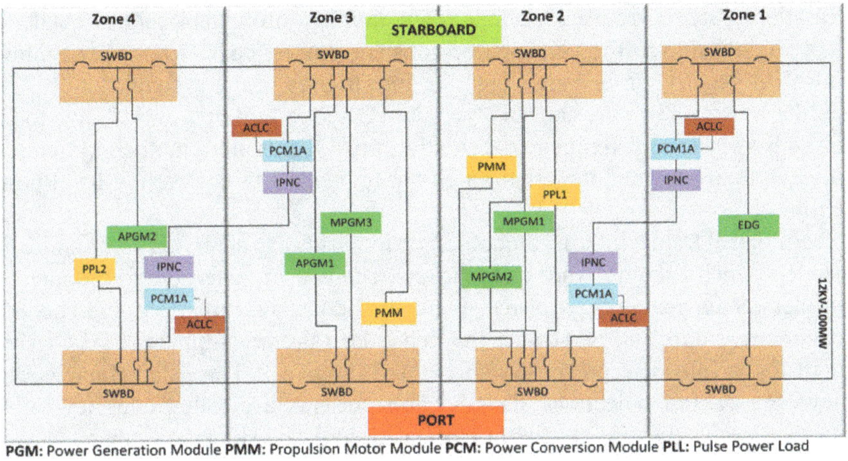

Figure 3.4 Notional MVDC 4-zone ship power system architecture [35]

have radial distribution systems. However, modern MVDC shipboards generally do not use radial power distribution systems due to lower efficiency and reliability levels. The state-of-the-art technology in MVDC shipboard power systems is discussed in the next paragraph.

The state of the art is to utilize the concept of zonal power distribution architectures. This type of architecture subdivides the distribution system into zones that are interconnected to the port and starboard buses of the main distribution and subsequently interconnect adjacent zones. Such an architecture is enabled by power electronics that allow energy to be routed via distributed energy resources (storage and generation) to any load centre in the ship. Thus, it provides a complete capability to reconfigure the ship's power distribution to meet mission demand in the presence of faults and failures. This highly reconfigurable architecture is leveraged by the application of advanced control methodologies and optimization techniques [33, 34]. An example of a four-zone system (one of many potential zonal architectures) is shown in Figure 3.4.

Clearly, the ability of controllable power electronics to perform regulated power conversion has revolutionized shipboard power systems by allowing zonal distribution systems and operation at MVDC levels. Power electronic converters in such systems are controlled through a communication-assisted framework. A detailed discussion on the control structure of shipboard power electronics is provided in Section 3.4.

3.4 Coordinated control in aircrafts and shipboards

Aircraft and shipboard power systems require the use of dedicated control structures for the regulation of power flow arrangements and achievement of coordination among system parameters at various levels. In vehicular microgrids where DC distribution systems are used, voltage regulation is a prominent control challenge. This is mainly due to the presence of constant power loads, whose behaviour is identical to negative impedance, creating instabilities in the system [36–38]. Due to similar reasons [39], the control layers in AC and hybrid microgrids have to perform both voltage and frequency regulation [40]. Controllers in such systems try to achieve their assigned targets through the regulation of conversion-level power electronics.

The control framework in autonomous microgrids has to rely on communication networks, which carry forward data packets received from sensors to the intended controllers. Such networks are also responsible for exchange of information between two or more control units. Hence, the vehicular (aircraft/shipboard) microgrid is a smart, cyber-physical power system, which uses an ICT-assisted framework to achieve the control objectives. This section presents a detailed overview of the control structures that are present in electric aircraft and shipboard power systems. A brief insight into the associated communication networks has also been provided focusing on the information-exchange protocols and the possible network topologies.

3.4.1 Aircraft power systems

There are two main control challenges in aircraft microgrids. The first one is the management of load sharing between the generating units, and the second is the regulation of system parameters (voltage and frequency) as per the requisite specifications at various bus levels. These challenges are generally met through the use of a multi-layered hierarchical control structure, which improves reliability and allows a more independent level of operation for each individual layer [41]. A detailed overview of each of the layers along with their functions is provided below.

The primary controller forms the first layer of hierarchical control and is installed at the load level [42]. It is responsible for providing local control of generator-level power electronics. At the distribution level, it allows the adjustment of bus parameters through the provision of current and voltage controls. It generally operates through a droop-based control action, which also allows proportional load sharing among various generating units. The droop-based controller has a number of limitations, the most significant of which is its inability to restore system parameters (voltage and frequency) to their nominal steady-state values.

The secondary controller forms the second layer of hierarchical control and aims to remove limitations of primary control in order to improve system stability for effective operation. It utilizes a communication-assisted framework and facilitates the restoration of voltage and frequency to their nominal steady-state values. The communication network enables information exchange between various nodes of the system to improve coordination and synchronization. This network can either have a centralized (involving a single central controller that commands all the nodes) or distributed (involving a master-slave configuration where all other nodes follow a reference node) architecture. However, the distributed framework is preferable to centralized one as it stays unaffected during single-point failures.

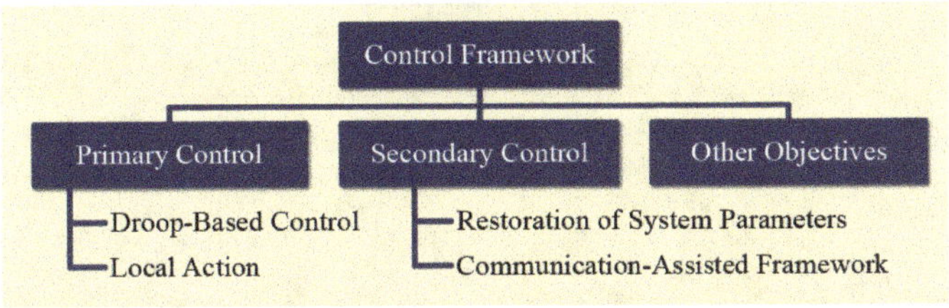

Figure 3.5 Control framework in a typical aircraft microgrid

Other levels of control in this structure may aim to fulfil multiple singular objectives. An example of this is the storage-level control framework. Regulation of the storage-level bidirectional DC–DC converter is an important element in the control framework of the aircraft microgrid as it allows regulation of battery state of charge (SOC) and voltage at this node. The battery SOC has a significant impact on the range and overall weight of the system. A brief overview of the discussion in this subsection is presented in Figure 3.5.

3.4.2 Shipboard power systems

There are essentially three layers of control that exist in shipboard power systems. The device layer control, which enables local control for generators, storage, converters and loads (also known as primary layer control); power management (PM) layer, which ensures that bus voltage/frequency stability is maintained and that converters appropriately share current/power demand in the system (also known as secondary layer control); and energy management (EM) layer, which provides optimal resource allocation, reconfiguration and power set points for the PM layer (also known as tertiary layer control). Figure 3.6 illustrates a potential layered distributed controller architecture.

In particular, PM and EM layers are intended to be distributed in order to maximize architectural benefits and performance and therefore as networked controller structures must synergize with cybersecurity considerations. Contingencies can be addressed with distributed control to operate in decentralized hierarchical structures if necessary, in emergency situations when there is an absence of or compromise in the communications network.

In general, most control structures depend on specific models of electrical apparatus integrated into the system as well as anticipated load profiles. However, models of systems are often only known in the abstract and thus behavioural models that capture overall input/output relationships and simple dynamics are typically available or can be created [44]. Therefore, an essential gap that needs to be addressed is the addition of data-driven-based modelling and control structures that allow base models of systems and subsystems to evolve over time to further capture non-linearities and other operational nuances [45]. As such, given the significant reliance of data-driven techniques on sensor observation and associated communication, cyber-physical security is essential [46–48].

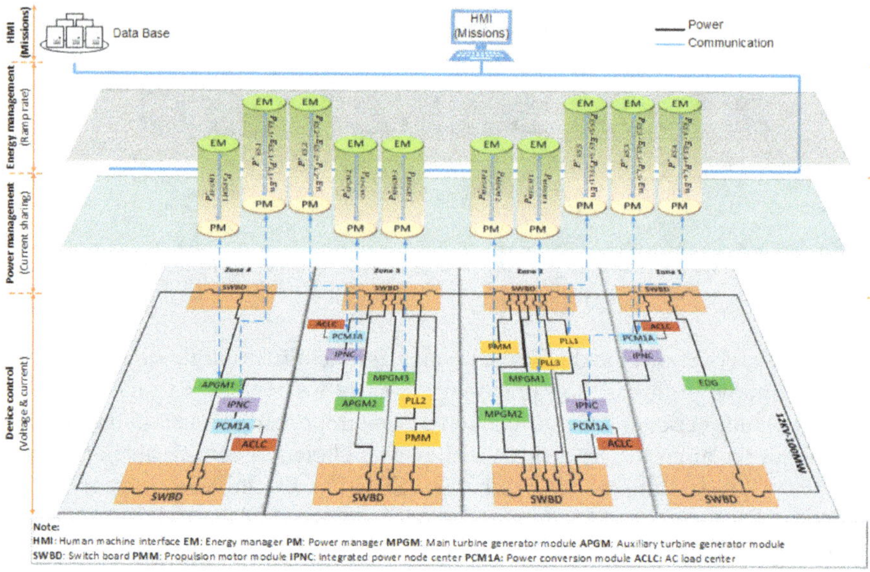

Figure 3.6 Notional hierarchical distributed control structure [43]

Section 3.5 presents a detailed assessment of the risk to aircraft and shipboard power systems due to cyber-attacks. It also presents a discussion on the vulnerabilities, types of cyberattacks that can exploit them and the possible consequences of such attacks.

3.5 Threat analysis against cyber attacks

The function of mission-critical applications such as shipboards and aircraft highly rely on their power supply systems, i.e., microgrids. The digitalization and modularization are the main characteristics of mission-critical cyber-physical microgrids that are regulated through either centralized or distributed control systems, as shown in Figure 3.7. The islanded and localized energy supply system of such critical applications leads to vital security requirements [21]. Digitalized data or information exchange, that is designed for optimality, among power electronic-based modules may impose cyber threat to the cyber-physical energy supply systems, even causing cascading failures and tremendous damages [49–51].

The sources of cyber threats can be generalized as two aspects: random communication failures and malicious attacks. The random communication failures of the microgrids could occur due to the vulnerability of the sophisticated communication scheme, while mission-critical applications require accurate and real-time operations [52]. Delayed or unsuccessful communication attempts possibly lead to an ineffective or inefficient response to emergency and alarm events, which may further trigger the response of the microgrid protection scheme, and cause severe cascading failures. For example, cyber threats such as communication failures on

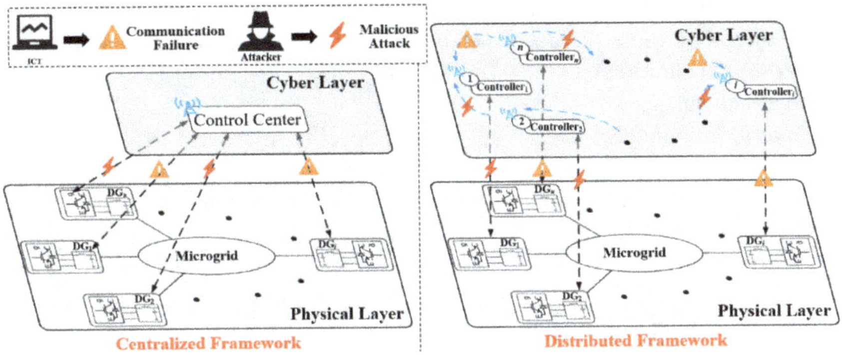

Figure 3.7 Diagram of microgrid control including cyber threats

shipboards or aircraft are more vulnerable than other applications due to the nature of maritime or aviation systems, many of which are far away from land and depend on long-range communication [53]. Hence, a single-point communication failure of mission-critical cyber-physical microgrids potentially leads to the failures of power electronic devices, damaging the partial and even absolute mission-critical power supply systems.

The malicious attacks can be classified into three types: attacks on availability, integrity and confidentiality [54, 55]. Malicious attacks on availability, generally referred to as denial-of-service (DoS) attacks attempt to limit or delay access to data. Attackers may optimally elaborate the occurrences of communication "failures" to maximize the consequences in terms of mission-critical power supplies. In a microgrid, a single-point data unavailability attack may be extended to the unavailability of power supply owing to the inevitable coupling between the cyber layer and physical power electronic devices. In islanded microgrids of mission-critical applications, DoS attacks can block modular information access and delay the sequences of critical message transmissions.

Malicious attacks on integrity interfere with the accuracy or trustworthiness of data. For mission-critical microgrids, attacks on integrity, i.e., false data injection (FDI) attacks may spoof or tamper with the data or critical messages in order to negatively affect the operation [10], thereby destroying the whole power supply of microgrids [56]. FDI attacks could occur anywhere the data exchange exists, if attackers have the corresponding authorities or access. Misleading one or more controller(s) in the microgrid through manipulation of sensing devices may lead to faults of power electronic devices or modules [57], which could be detected if appropriate detection schemes are designed. Moreover, a class of FDI attacks, which distributes false commands to the controller [58], may cause incorrect action of protection schemes, causing unwanted faults. Furthermore, adjusting the setting parameters of protection relays may cause the relays to trip erroneously [55].

Malicious attacks on confidentiality are attempts to eavesdrop on privileged information. Any unauthorized disclosure may uncover sensitive information of mission-critical microgrids, either operational data or consumer behaviours. The attacks on confidentiality are not prioritized as those on availability and integrity that may directly impose threats on secure operations of mission-critical microgrids. However, it still should be emphasized to avoid the leakage of critical information.

The attack surfaces of all malicious attacks are from the potential interfaces of cyber layers. The attack surfaces vary in the specific microgrid designs of critical missions. The generalized attack surfaces are summarized into six categories [3]: wired links, wireless links, unencrypted communications, unauthenticated communications, exposed endpoints and humans. In brief, signals from wired or wireless links can be jammed, tampered and eavesdropped. Unencrypted communications make operational data leaked or modified, while unauthenticated communications can be utilized by attackers pretending to be legitimate data sources or users. Some of the commonly used (microgrid) communication protocols, which are particularly vulnerable to such attacks include controller area network bus [59], IEC 61850 [60], Distributed Network Protocol (DNP3) [61] and Modbus/TCP [62]. Exposed endpoints and human are physical and social interface surfaces, respectively, which can both be illegally used. These attack surfaces can be utilized for conducting three types of malicious attacks without specific corresponding relationships, and it should be analysed case by case.

Confronted with the aforementioned cyber threats, i.e., communication failures and malicious attacks, mitigation measures for cyber threats should consider cyber-system hardening and operational strategy updating to enhance the resilience of microgrids [63–65]. Cyber-system hardening focuses on predictive and protective methods that make both cyber infrastructure and software less vulnerable, while operational strategy updating strengthens defensive capability in response to increasingly complicated cyber events. Elimination of vulnerabilities and strengthening of defence mechanisms in both software and hardware units are equally important as modern microgrids are increasingly reliant on software-defined control [66], where formulation and implementation of a significant portion of the control logic are performed using software tools.

3.6 Case study of MVDC shipboard microgrid

As evident from the discussion in section 1.5, there exist numerous potential vulnerabilities, through which cyberattacks can be launched to manipulate conversion-level power electronic devices in aircraft and shipboard microgrids. This chapter presents a case study on the MVDC two-zone shipboard microgrid along with simulation results to help the reader visualize the extensive damage that such attacks can cause in a mission-critical application-based power system.

A notional two-zone MVDC shipboard power system microgrid model, as described in [67], is designed as a virtual test model in MATLAB/Simulink®. Figure 3.8 shows the schematic diagram of the designed model. Each of the zones consists of one power generation module, including a synchronous generator (having a generation capacity of 36 MW), which is interfaced with the distribution system using a rectifier. In the nominal state, the generators share the system load

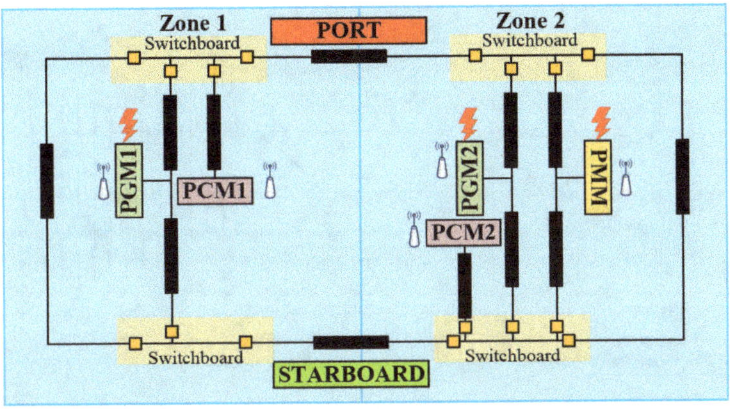

Figure 3.8 *Test system model of 2-zone MVDC shipboard microgrid*

equally. The MVDC distribution system is maintained at 12 kV. The zones also consist of power conversion modules, which are utilized for converting MVDC to low-voltage DC [68]. The propulsion motor module (consisting of two sub-modules, each of which is an 18 MW resistive load connected to the rest of the system using an inverter) is connected in Zone 2. Values of all other system parameters may be obtained from [67]. In order to demonstrate the impact of cyberattacks on the system, two different attack scenarios are considered.

3.6.1 Attack Scenario 1

Let us consider a scenario where the attacker has partial access, i.e., only to a single vulnerable node in the system. Consider the power generation module 1 (PGM1) to be the target node. The adversary realizes an attack vector able to modify the voltage output of the rectifier at the generator side in Zone 1. The attack vector aims to achieve its target objective by manipulating the controller module associated with the rectifier to track a periodically changing value of voltage. Figure 3.9(a) shows the value of the voltage at the attacked node after the attack is initiated. Clearly, the attacker successfully forces the voltage to follow an oscillating trajectory, affecting the stability of the MVDC microgrid system.

3.6.2 Attack Scenario 2

Consider a scenario where the attacker has complete access to the system. In addition to the manipulation at PGM1, the attacker also exploits the vulnerabilities at the generator side in Zone 2 and the propulsion motor module to initiate a coordinated manipulation. At the second generator, the attacker periodically manipulates the voltage output of the rectifier while simultaneously spoofing the sensor measurements at the load side to mislead controllers at the generation side. Figure 3.9(b) depicts the trajectory of voltage as measured at PGM1 after the attack is initiated.

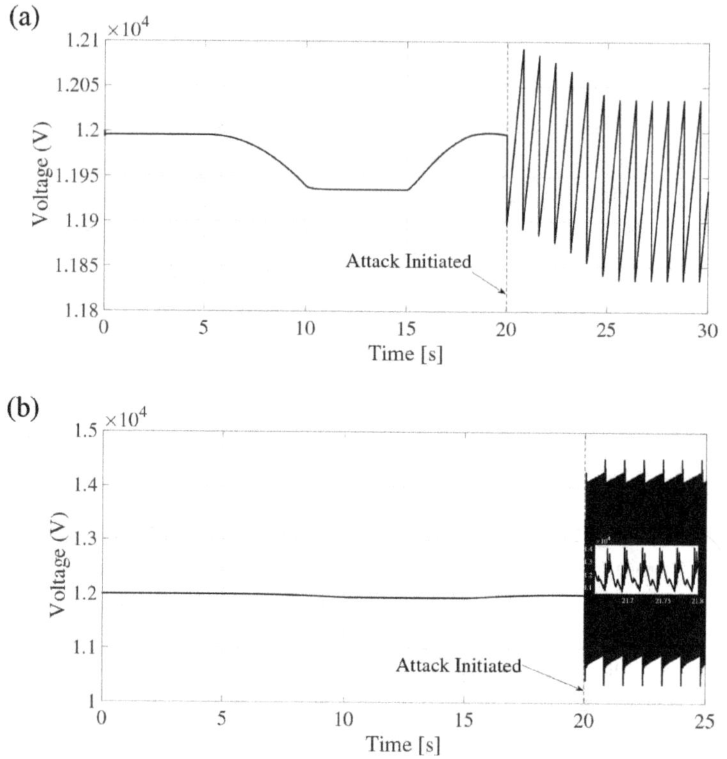

Figure 3.9 Voltage at PGM1 in (a) Scenario 1 and (b) Scenario 2

Numerous oscillations are seen in the parameter trajectory. The voltage deviation is more than ±16.67% as compared to the nominal value.

As shown in the first scenario, an attacker can manipulate voltage at any particular bus against the permissible limits. Moreover, simultaneous manipulation of voltage at multiple nodes may also be possible based on the attacker's capability and level of access [21]. Voltage instability may also occur in the system when multiple power electronic converters feeding the same bus have different output parameter values. Cyberattacks can have a devastating impact on microgrids, which can hamper the achievement of mission-critical objectives.

3.7 Conclusion

This chapter focuses on aircraft and shipboard microgrids, which are autonomous power systems with in-vehicle power generation facilities. Power electronic devices in such systems enable power conversion for operation at multiple voltage and frequency values as required at various generation, distribution, storage and utilization levels. Coordinated control among the power electronic converters is generally achieved through a hierarchical control framework where one or more levels are dependent on communication networks for the exchange of real-time sensor measurements between various nodes.

Dependence on communication networks gives rise to a number of cyber vulnerabilities, which may be exploited by malicious agents (through cyber-attacks) to manipulate power electronics in the vehicular microgrid. Attacks can be aimed at the communication channels to disrupt the availability, confidentiality or integrity of data [69]. Some of the examples in this regard include – DoS attacks, which aim to reduce availability and are capable of completely blocking information flow to critical units of the vehicular system, and FDI attacks, which aim to manipulate critical information flowing to controllers associated with specific power electronic devices. Attacks on confidentiality include eavesdropping attempts, which are often aimed towards learning critical information of the system to design more sophisticated and lethal manipulation attacks.

Attack surfaces can include wired or wireless links, unencrypted messages, unauthorized data packets, outwardly exposed (and hence, vulnerable) points and even, humans. Links can be spied upon through eavesdropping tools and/or subjected to FDI and DoS attacks. Unencrypted (/weakly encrypted) data packets can be easily monitored to learn their contents. Unauthorized data packets can help miscreants mask themselves as authentic, legitimate users. Human elements in the system can also be attacked/manipulated through social and psychological tools. All these forms of cyber threats can jeopardize the working of the aircraft and shipboard microgrid causing economic losses and putting human lives at stake. Hence, these vulnerabilities should be eliminated through fortification of ICT for obtaining a higher degree of resilience.

References

[1] Zakaria H., Hamid M., Abdellatif E.M., Imane A. 'Recent advancements and developments for electric vehicle technology'. *International Conference of Computer Science and Renewable Energies (ICCSRE)*; IEEE, Agadir, Morocco, 22–24 Jul; 2019. pp. 1–6.

[2] Adams E. 'The age of electric aviation is just 30 years away'. *Wired.* 2017.

[3] Kumar D., Zare F. 'A comprehensive review of maritime microgrids: system architectures, energy efficiency, power quality, and regulations'. *IEEE Access*. 2019;7:67249–77.

[4] Yang T., Gao F., Bozhko S. 'Power electronic systems for aircraft'. *Control of Power Electronic Converters and Systems*. Elsevier; 2018. pp. 333–68.

[5] Enslin J.H., Whisenant S.G., Hadidi R. 'Third eGrid workshop maps the grid of the future: attendees engage to examine the role of power electronic applications in modern electric power systems'. *IEEE Power Electronics Magazine*. 2019;6(1):48–55.

[6] Singh B.S. 'Control of three-phase four wire grid integrated multiple solar PV arrays-battery with bidirectional converter based microgrid'. *IEEE*

International Conference on Power Electronics, Drives and Energy Systems; PEDES (ed.); 2020. pp. 1–6.

[7] Emadi A., Williamson S.S., Khaligh A. 'Power electronics intensive solutions for advanced electric, hybrid electric, and fuel cell vehicular power systems'. *IEEE Transactions on Power Electronics*. 2006;21(3):567–77.

[8] Nguyen B.L.H., Vu T., Ogilvie C. 'Advanced load shedding for integrated power and energy systems'. *IEEE Electric Ship Technologies Symposium (ESTS)*, 3-6 Aug; 2021. pp. 1–6 pp..

[9] Effler S., Halton M., Mooney J., Rinne K. 'Scalable digital power controller with phase alignment and frequency synchronization'. *IEEE Transactions on Circuits and Systems I: Regular Papers*. 2013;61(4):1289–97.

[10] Rath S., Pal D., Sharma P.S., Panigrahi B.K. 'A cyber-secure distributed control architecture for autonomous AC microgrid'. *IEEE Systems Journal*. 2021;15(3):3324–35.

[11] Yue X., Boroyevich D., Lee F.C., Chen F., Burgos R., Zhuo F. 'Beat frequency oscillation analysis for power electronic converters in DC nanogrid based on crossed frequency output impedance matrix model'. *IEEE Transactions on Power Electronics*. 2017;33(4):3052–64.

[12] Yoon C., Bai H., Beres R.N., Wang X., Bak C.L., Blaabjerg F. 'Harmonic stability assessment for multiparalleled, grid-connected inverters'. *IEEE Transactions on Sustainable Energy*. 2016;7(4):1388–97.

[13] Liu W., Tarasiuk T., Gorniak M., *et al.* 'Power quality assessment in shipboard microgrids under unbalanced and harmonic AC bus voltage'. *IEEE Transactions on Industry Applications*. 2018;55(1):765–75.

[14] Gao F., Bozhko S., Costabeber A., Asher G., Wheeler P. 'Control design and voltage stability analysis of a droop-controlled electrical power system for more electric aircraft'. *IEEE Transactions on Industrial Electronics*; 64, 12; 2017. pp. 9271–81.

[15] Ospina J., Konstantinou C., Stanovich M., *et al.* 'Evaluation of communication network models for shipboard power systems' in ESTS, (ed.)'. *IEEE Electric Ship Technologies Symposium*; 2021. pp. 1–9.

[16] Rath S., Zografopoulos I., Konstantinou C. 'Stealthy rootkit attacks on cyber-physical microgrids: poster'. *Proceedings of the Twelfth ACM International Conference on Future Energy Systems*, Jun; 2021. pp. 294–5.

[17] Mazareanu E. 'Number of scheduled passengers boarded by the global airline industry from 2004 to 2021'. 2020.

[18] '*Air travel down 60 per cent, as airline industry losses top $370 billion: ICAO*'. 2021. Available from https://news.un.org/en/story/2021/01/1082302 [Accessed 21 Jul].

[19] Venkataramanan V., Mallikeswaran A., Srivastava A. 'Analysis of aircraft electric microgrid system with auxiliary power unit using real time simulation'. *"IEEE 24th International Symposium on Industrial Electronics (ISIE)*. IEEE; Buzios, Brazil, 3-5 Jun; 2015. pp. 245–50.

[20] Tao Y., Bozhko S., Wheeler P., *et al.* 'Generic functional modelling of multi-pulse auto-transformer rectifier units for more-electric aircraft applications'. *Chinese Journal of Aeronautics*. 2018;31(5):883–91.

[21] Zografopoulos I., Ospina J., Liu X., Konstantinou C. 'Cyber-physical energy systems security: threat modeling, risk assessment, resources, metrics, and case studies'. *IEEE Access*. 2021;9:29775–818.

[22] *Shipping and world trade: driving prosperity*. Available from https://www.ics-shipping.org/shipping-fact/shipping-and-world-trade-driving-prosperity/ [Accessed 23 Jul].

[23] Kalouptsidi M. *The role of shipping in world trade*. 2021. Available from https://econofact.org/the-role-of-shipping-in-world-trade [Accessed 23 Jul].

[24] Ghimire P., Park D., Zadeh M.K., Thorstensen J., Pedersen E. "Shipboard electric power conversion: system architecture, applications, control, and challenges [technology leaders]". *IEEE Electrification Magazine*. 2019;7(4):6–20.

[25] Pan P., Sun Y., Yuan C., Yan X., Tang X. 'Research progress on SHIP power systems integrated with new energy sources: a review'. *Renewable and Sustainable Energy Reviews*. 2021;144:111048.

[26] Edrington C.S., Ozkan G., Papari B., Perkins D. 'Distributed adaptive power management for medium voltage SHIP power systems'. *Journal of Marine Engineering & Technology*. 2021;5(1):1–16.

[27] Taheri S.I., Vieira G.G.T.T., Salles M.B.C., Avila S.L. 'A trip-ahead strategy for optimal energy dispatch in SHIP power systems'. *Electric Power Systems Research*. 2021;192:106917.

[28] Ozkan G., Hoang P.H., Ramezani Badr P., Edrington C.S., Papari B. 'Real-time thermal management for two-level active rectifier with finite control set model predictive control'. *International Journal of Electrical Power & Energy Systems*. 2021;131(2):107057.

[29] Amy J.V., Doerry N. "Design considerations for a reference MVDC power system'. Paper presented at the SNAME Maritime Convention'. *Bellevue, WA,*. 2016.

[30] Doerry N., Amy J. The road to MVDC. *ASNE Intelligent Ships Symposium*, May; 2015.

[31] Jayasinghe S., Meegahapola L., Fernando N., Jin Z., Guerrero J. 'Review of SHIP microgrids: system architectures, storage technologies and power quality aspects'. *Inventions*. 2017;2(1):4.

[32] Skjong E., Rødskar E., Molinas M., Johansen T.A., Cunningham J. 'The marine vessel's electrical power system: from its birth to present day'. *Proceedings of the IEEE*. 2015;103(12):2410–24.

[33] Steinke J.K., Maibach P., Ortiz G. 'MVDC applications and technology'. *International Exhibition and Conference for Power Electronics, Intelligent Motion, Renewable Energy and Energy Management*; Nuremberg, Germany, 7–9 May; 2019. pp. 1–8.

[34] Javaid U., Freijedo F.D., Dujic D., van der Merwe W. 'Dynamic assessment of source–load interactions in marine MVDC distribution'. *IEEE Transactions on Industrial Electronics*. 2017;64(6):4372–81.
[35] Edrington C., Vu T., Vahedi H. 'Demonstration and evaluation of largescale distributed control for integrated power and energy MVDC systems on ships'. *Advanced Machinery Technology Symposium Philadelphia, PA:*; Drexel University - Philadelphia, 27–29 Mar; 2018.
[36] Jin Z., Meng L., Guerrero J.M. 'Constant power load instability mitigation in DC shipboard power systems using negative series virtual inductor method'. *IECON 2017-43rd Annual Conference of the IEEE Industrial Electronics Society*; IEEE, Beijing, China, 29 Oct–1 Nov; 2017. pp. 6789–94.
[37] Khaligh A., Rahimi Am., Emadi A. 'Modified pulse-adjustment technique to control DC/DC converters driving variable constant-power loads'. *IEEE Transactions on Industrial Electronics*. 2008;55(3):1133–46.
[38] AL-Nussairi M.K., Bayindir R., Padmanaban S., Mihet-Popa L., Siano P. 'Constant power loads (CPL) with microgrids: problem definition, stability analysis and compensation techniques'. *Energies*. 2017;10(10):1656.
[39] Dillon J.A., Weaver W.W., Robinett R.D., Wilson D.G. 'Electro-mechanical-thermal performance and stability of aircraft energy networks with pulse power loads'. *IEEE Transactions on Aerospace and Electronic Systems*. 2019;56(4):2537–47.
[40] Zhaoxia X., Tianli Z., Huaimin L., Guerrero J.M., Su C.-L., Vasquez J.C. 'Coordinated control of a hybrid-electric-ferry shipboard microgrid'. *IEEE Transactions on Transportation Electrification*. 2019;5(3):828–39.
[41] Gao F., Kang R., Cao J., Yang T. 'Primary and secondary control in DC microgrids: a review'. *Journal of Modern Power Systems and Clean Energy*. 2019;7(2):227–42.
[42] Buticchi G., Bozhko S., Liserre M., Wheeler P., Al-Haddad K. 'On-board microgrids for the more electric aircraft—technology review'. *IEEE Transactions on Industrial Electronics*. 2018;66(7):5588–99.
[43] Edrington C.S., Ozkan G., Papari B., *et al.* 'Distributed energy management for SHIP power systems with distributed energy storage'. *Journal of Marine Engineering & Technology*. 2020;19(sup1):31–44.
[44] Cupelli L., Cupelli M., Monti A. 'Data-driven control of converters in DC microgrids for bus voltage regulation'. *IECON 2018-44th Annual Conference of the IEEE Industrial Electronics Society*; IEEE, Washington, DC, 20–23 Oct; 2018. pp. 3389–94.
[45] Fang S., Xu Y., Wen S., Zhao T., Wang H., Liu L. "Data-driven robust coordination of generation and demand-side in photovoltaic integrated all-electric SHIP microgrids". *IEEE Transactions on Power Systems*. 2019;35(3):1783–95.
[46] Gilligan B.K. *Electric ship digital twin: framework for cyber-physical system security*. Boston: Massachusetts Institute of Technology; 2019.

[47] Mustafa A., Bidram A., Modares H., Poudel B.P. 'Detection and mitigation of cyberthreats in the DC microgrid distributed control system'. *International Journal of Electrical Power & Energy Systems*. 2020;120(7):105968.

[48] Mazumder S.K., Kulkarni A., Sahoo S., *et al.* 'A review of current research trends in power-electronic innovations in cyber–physical systems'. *IEEE Journal of Emerging and Selected Topics in Power Electronics*. 2021;9(5):5146–63.

[49] Xenofontos C., Zografopoulos I., Konstantinou C., *et al.* 'Consumer, Commercial and Industrial IoT (In)Security: Attack Taxonomy and Case Studies'. *IEEE internet of things journal*. 2021:1–1.

[50] Ospina J., Zografopoulos I., Liu X., Konstantinou C. 'Demo: Trustworthy cyberphysical energy systems: time-delay attacks in a real-time co-simulation environment'. *CPSIOTSEC'20: Proceedings of the 2020 Joint Workshop on CPS&IoT Security and Privacy. Association for Computing Machinery*; New York, NY, 9 Nov; 2020. p. 69.

[51] Zografopoulos I., Konstantinou C., Tsoutsos N.G., *et al.* 'Security assessment and impact analysis of cyberattacks in integrated T&D power systems'. *arXiv preprint arXiv*. 2021;210203215.

[52] Ogilvie C., Ospina J., Konstantinou C., *et al.* 'Modeling communication networks in a real-time simulation environment for evaluating controls of shipboard power systems'. *IEEE CyberPELS (CyberPELS)*"; Miami, FL, 13 Oct; 2020. pp. 1–7.

[53] Kushal T.R.B., Lai K., Illindala M.S. 'Risk-based mitigation of load curtailment cyber attack using intelligent agents in a shipboard power system'. *IEEE Transactions on Smart Grid*. 2018;10(5):4741–50.

[54] Li Z., Shahidehpour M., Aminifar F. 'Cybersecurity in distributed power systems'. *Proceedings of the IEEE*. 2017;105(7):1367–88.

[55] Mishra S., Anderson K., Miller B., Boyer K., Warren A. 'Microgrid resilience: a holistic approach for assessing threats, identifying vulnerabilities, and designing corresponding mitigation strategies'. *Applied Energy*. 2020;264(7):114726.

[56] Konstantinou C., Sazos M., Musleh A.S., Keliris A., Al-Durra A., Maniatakos M. 'GPS spoofing effect on phase angle monitoring and control in a real-time digital simulator-based hardware-in-the-loop environment'. *IET Cyber-Physical Systems: Theory & Applications*. 2017;2(4):180–7.

[57] Yang B., Guo L., Li F., Ye J., Song W. 'Impact analysis of data integrity attacks on power electronics and electric drives'. *IEEE Transportation Electrification Conference and Expo (ITEC)*; IEEE, Detroit, MI, 19–21 Jun; 2019. pp. 1–6.

[58] Sargolzaei A. 'A secure control design for networked control system with nonlinear dynamics under false-data-injection attacks'. *American Control Conference (ACC)*; IEEE, New Orleans, LA, 25–28 May; 2021. pp. 2693–9.

[59] Hyo. J J., Choi W. 'A survey of attacks on controller area networks and corresponding countermeasures'. *IEEE Transactions on Intelligent Transportation Systems*. 2021:1–19.

[60] Rajkumar V.S., Tealane M., Ştefanov A., Presekal A., Palensky P. 'Cyber attacks on power system automation and protection and impact analysis'. *IEEE PES Innovative Smart Grid Technologies Europe (ISGT-Europe)*; IEEE, Virtual/online, Delft, Netherlands, 26–28 Oct; 2020. pp. 247–54.

[61] Singh C., Nivangune A., Patwardhan M. Function code based vulnerability analysis of DNP3. *IEEE International Conference on Advanced Networks and Telecommunications Systems (ANTS)*; IEEE, Bangalore, India, 6-9 Nov; 2016. pp. 1–6.

[62] Chen B., Pattanaik N., Goulart A., Butler-purry K.L., Kundur D. Implementing attacks for modbus/TCP protocol in a real-time cyber physical system test bed. *2015 IEEE International Workshop Technical Committee on Communications Quality and Reliability (CQR)*; IEEE, Charleston, SC, 11–14 May; 2015. pp. 1–6.

[63] Ge P., Teng F., Konstantinou C., Hu S. 'A resilience-oriented centralised-to-decentralised framework for networked microgrids management'. *arXiv preprint arXiv*. 2021;210900245.

[64] Konstantinou C., Anubi O.M. 'Resilient cyber-physical energy systems using prior information based on Gaussian process'. *IEEE Transactions on Industrial Informatics*. 2021;18(3):2160–8.

[65] Konstantinou C., Stergiopoulos G., Parvania M., Esteves-Verissimo P. 'Chaos engineering for enhanced resilience of cyber-physical systems'. *arXiv preprint arXiv*. 2021:2106.14962:.

[66] Danzi P., Angjelichinoski M., Stefanovic C., Dragicevic T., Popovski P. 'Software-defined microgrid control for resilience against denial-of-service attacks'. *IEEE Transactions on Smart Grid*. 2018;10(5):5258–68.

[67] Ravindra H., Stanovich M., Steurer M. 'Documentation for a notional two zone medium voltage DC shipboard power system model implemented on the rtds'. *ESRDC*. 2016.

[68] Xie R., Mo R., Shi Y., Li H. 'Comparative study of the DC-DC power conversion module based on dual active bridge converter and modular multilevel converter for shipboard MVDC system'. *IEEE Electric Ship Technologies Symposium (ESTS)*; IEEE, Arlington, VA, 14–17 Aug; 2017. pp. 36–43.

[69] McLaughlin S., Konstantinou C., Wang X., *et al.* 'The cybersecurity landscape in industrial control systems'. *Proceedings of the IEEE*. 2016;104(5):1039–57.

Chapter 4

Situational awareness of cyber attacks in smart grids

Vivek Kumar Singh[1], Philip Hart[2], Pengyuan Wang[2], Honggang Wang[2], and Manimaran Govindarasu[3]

Development of a secure critical infrastructure, such as the power system, necessitates addressing the associated cybersecurity challenges at the network, physical, and application layers. With the continuously evolving cyberattacks in grid networks, including advanced persistent threats (APTs), the development of cybersecurity situational awareness (CSA) is an emerging trend that facilitates precise and detailed command and control to enhance the security and resilience of power systems. The comprehensive and timely CSA is necessary to predict cyberattacks at an early stage and provide an intelligent incident response to minimize their impacts on the power system. In this chapter, we present a high-level conceptual architecture of CSA against cyber threats in the smart grid. The proposed conceptual architecture integrates data-driven anomaly detection algorithms, heterogeneous database, and event visualization dashboard to monitor grid network and analyze and predict cybersecurity threats that can affect the normal grid operation. We then discuss the several components of Iowa State University PowerCyber (ISU PC) testbed that can be utilized to emulate substation and control center networks and further test and validate anomaly detection methods and intrusion detection system (IDS) tools in a cyber-physical testbed environment. As a proof of concept, we present a data-driven anomaly detection algorithm for wide-area monitoring system (WAMS) using synchrophasors and evaluated its performance by computing true positive rates for line fault and cyberattack events. Furthermore, we present the prototype demonstration at the commercial scale by integrating the proposed detection algorithm with the General Electric (GE)-based WAMS platform to support CSA in power system. Finally, we conclude this chapter by outlining how a cybersecurity training can enhance the overall CSA while emphasizing the need to develop robust anomaly detectors to secure the grid network.

[1]Idaho National Laboratory, Idaho Falls, ID, USA
[2]General Electric Global Research, Niskayuna, NY, USA
[3]Iowa State University, Ames, IA, USA

4.1 Introduction

Today's energy infrastructure is continuously evolving into an interconnected cyber-physical system (CPS) that is integrated with a high-speed communication network, Internet of things technology, and sophisticated wide-area controls to make the grid smarter, reliable, and efficient. In recent years, several wide-area monitoring, protection, and control (WAMPAC) applications, including wide-area monitoring system (WAMS), automatic generation control (AGC), oscillation damping, wide-area voltage control (WAVC), etc. are developed and widely applied in the energy management system (EMS) [1]. These control center-based applications rely on insecure data-sharing devices and unencrypted supervisory control and data acquisition (SCADA) communication to support real-time monitoring and control operations [2]. Furthermore, with the recent development in synchrophasor technology and its increment in the industry deployment, there has been a strong urge in incorporating phasor measurement units (PMUs) measurements to develop mission-critical applications, including WAMPAC [3]. The increasing reliance on information technology (IT) and unauthenticated data-sharing devices in the legacy grid infrastructure to support WAMPAC applications has exposed power grid to a multitude of cyberattacks [4, 5]. Since the main objective for developing these WAMPAC applications is to mitigate power system disturbances, any unusual malfunction or operational delays, triggered through malicious events, can compromise the regular grid operation. In fact, it has been observed that the attacks are evolving with high frequency while operating in stealth mode through various methods, including social engineering, phishing, and malware installation-based attacks [5]. In addition, the APTs can bypass the existing conventional security solutions, which are based on limited attack surfaces, and inject severe disturbances to destabilize the normal grid operation [6]. Real-world incidents include Stuxnet Worm [7] and Ukraine's grid hack in 2015 and 2016 [8], which showed how the normal grid operation could be compromised through stealthy and sophisticated cyberattacks. Therefore, CSA is a necessary and mandatory practice to secure the grid against the evolving threats. It provides efficient and secure ways to monitor grid network, manage ongoing events, anticipate, detect, and analyze possible attacks at an initial stage, and prepare system operators to initiate intelligent mitigation strategies against attacks.

CSA is one step closer toward developing the defense-in-depth architecture for critical infrastructures, such as the power system. The MITRE Corporation [9] has classified CSA into three processes: network awareness, threat awareness, and mission awareness. The network awareness includes configuration and patch management, vulnerability auditing, and incident recognition and understanding. The threat awareness consists of identifying and analyzing internal and external threats, and mission awareness includes forensic analysis, risk assessments, task planning, and task execution. Most of the existing cybersecurity solutions focused on detecting anomalies using network or system logs and applies rule-based preventive strategies against known attacks [10, 11]; however, developing situational awareness against different classes of cyberattacks is still not comprehensively addressed in the smart grid. The accurate and faster situational awareness can predict possible attacks before they can affect the grid and notify control center operators to take necessary corrective or preventive actions to maintain the regular operation of power system. With the high integration of communication technology and deployment of smart sensors, it is possible to generate security-related big data that can be applied in developing cybersecurity situational awareness. The North American Electric Reliability Corporation (NERC) discussed how synchrophasor can be applied in developing wide-area situational awareness (WASA) though the improved grid visibility, high quality and robust measurements, and smart decision-making using the extracted phasor measurements [12]. Therefore, the synchrophasor data can be leveraged to form security-related big data and eventually develop a robust and efficient CSA for power grid.

4.1.1 Related work

In general, the situational awareness, as mentioned by the United States (US) Army [13], is defined as "Knowledge and understanding of current situation which promotes timely, relevant, and accurate assessment of friendly, enemy and other operations within the battle space in order to facilitate decision-making." The original definition of situation awareness is introduced by Endsley [14]. Figure 4.1 presents a conceptual model that includes extraction of environmental factors in time and space (level 1), comprehension of the current situation (level 2), detection of the existing situation (level 3), and prediction of future trends (level 4). Most of the past research works were focused on developing network-based CSA by collecting multi-dimensional data from distributed sensors [14, 15]. The authors of [15] presented the foundation for developing heterogeneous data fusion system and designing a network-based cyberspace for situational awareness. The white paper [16] discussed how the SPLUNK platform can be utilized to provide improved network visibility, quickly detect abnormality in the grid network, and develop holistic and data-analysis-based security metrics while supporting traditional security information and event management (SIEM). Although this foundation is mandatory, most of these research efforts are focused on network-based CSA [14–18] that only captures limited attack surfaces as the smart grid network that compromises various communication protocols and different network architectures. Furthermore, the previously

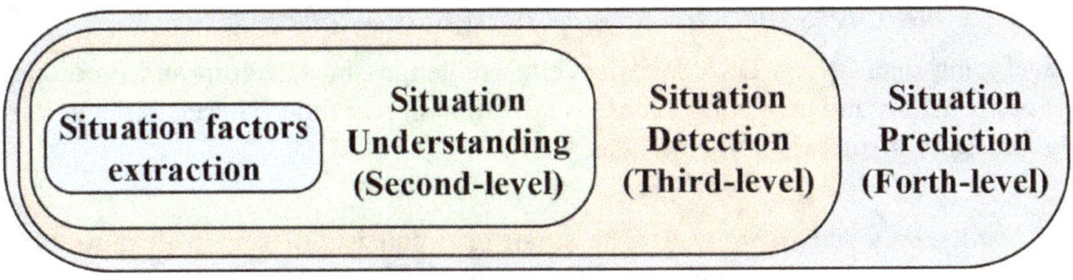

Figure 4.1 Conceptual model of CSA

proposed network-based CSA relies on distributed method that certainly lacks centralized architecture for fusing multi-dimensional datasets from multiple sensors.

As the number of PMUs installed in substations is continuously increasing, the synchrophasor-based applications in the EMS are also gaining more popularity. The PMU provides high-resolution measurements at a sampling rate of 30–120 samples/second, and hence it can be applied to form big data as required for CSA in smart grid.

Previous research efforts have shown the application of synchrophasor in developing situational awareness for the power system operation. The authors of [19] proposed an integrated system of linear state estimation (LSS) and advanced synchrophasor applications to provide better situational awareness and later tested the proposed architecture on simulated data from Electric Reliability Council of Texas and Los Angeles Department of Water and Power. The proposed approach assists system operators to analyze and diagnose current grid conditions and take necessary corrective actions in real time. In [20], the authors discussed the synchrophasor-based WASA, installed at the Hydro-Québec Research Institute (IREQ), by inferring knowledge from PMU measurements and developing cognitive models for grid operators, engineers, and data analyst. The advanced capabilities of WASA include faults localization, prediction of geomagnetic disturbances, and summarizing grid events.

In the context of cybersecurity, the recent research efforts showed how PMU measurements could be applied in forming security-related big data and developing anomaly detectors for WAMPAC. The cyberattacks and physical disturbances inject different flavors of transient phenomena that can be identified and distinguished using PMU measurements and their derived features. Besides, due to the high availability of PMUs, data mining technique and machine learning algorithms have gained high popularity in developing multi-classification models for the WAMPAC cybersecurity. The authors of [21, 22] presented how the derived features of PMU measurements can be applied to develop decision tree (DT)-based classification models that can detect cyberattacks and physical disturbances in real time. In [23], the authors discussed the Hoeffding adaptive trees (HAT)-based classification model for detecting cyberattacks and distinguishing them from the traditional power system contingencies in the context of WAMS. In particular, network (cyber) properties (logs, alerts) and system parameters (voltage, frequency, etc.) can be utilized in developing this detection model in real time.

4.1.2 Challenges for wide-area situational awareness

Developing data-driven CSA faces several challenges in detecting and predicting cyber intrusions and providing event visualization in real time. These challenges are elaborated in greater detail below.

1. *Legacy grid infrastructures*: The smart grid consists of several legacy infrastructures, such as programmable logic controllers (PLC), remote terminal units (RTU), etc., which generally have a life span of 10–15 years and are equipped

with an obsolete technology. Furthermore, the complex grid topology and interconnected multi-level networks have made difficult for the control center operator to receive an accurate situational awareness in a small time frame.

2. *Network monitoring*: The network monitoring is one of the critical steps toward developing a defense-in-depth architecture by efficiently managing the network and providing a real-time situational awareness to the operator. The efficient network monitoring requires a collection, detection, network analysis, and escalation of indications and warnings to detect and respond to intrusions at an early stage. Therefore, several network and data sensors have to be deployed at substation and communication networks to provide the bird's eye view of the grid network to the system operator.
3. *Data management*: The current power grid relies on different communication protocols and several hardware and software resources to receive multidimensional data at different sampling rate. The enhancement in volume, velocity, and veracity of data in power grid leads to the challenge of big data that is difficult to handle using conventional data management solutions. Furthermore, there is a compelling urge to develop a real-time sensing platform that can collect grid information, network packets, and other necessary information while supporting necessary configurations for event visualization in real time.
4. *Human vulnerabilities and ignorance*: Smart attackers can exploit human vulnerabilities and leverage their social engineering skills, including installing malware, phishing emails, etc. to perform stealthy cyberattacks in the grid network. Furthermore, due to the lack of cybersecurity training and their ignorance of security landscape, certain employees are exposed to security breaches and exhibit incompetence in handling unexpected events, including cyberattacks.

4.2 Wide-area cybersecurity situational awareness

In this section, we present a high-level conceptual architecture of wide-area CSA that is seamlessly integrated with the control center-based EMS while supporting the existing WAMPAC applications.

4.2.1 Conceptual architecture

The proposed architecture in Figure 4.2 receives power system measurements and network logs to provide an accurate situational awareness to the control center operator and also coordinates with these WAMPAC applications to provide necessary corrective actions, if necessary, to maintain the regular grid operation. It consists of several components that are discussed here.

1. ***Data-driven anomaly detection algorithms***: It includes several data-centric algorithms that leverage grid measurements, network properties, system topology, and geographical information to detect different classes of cyberattacks

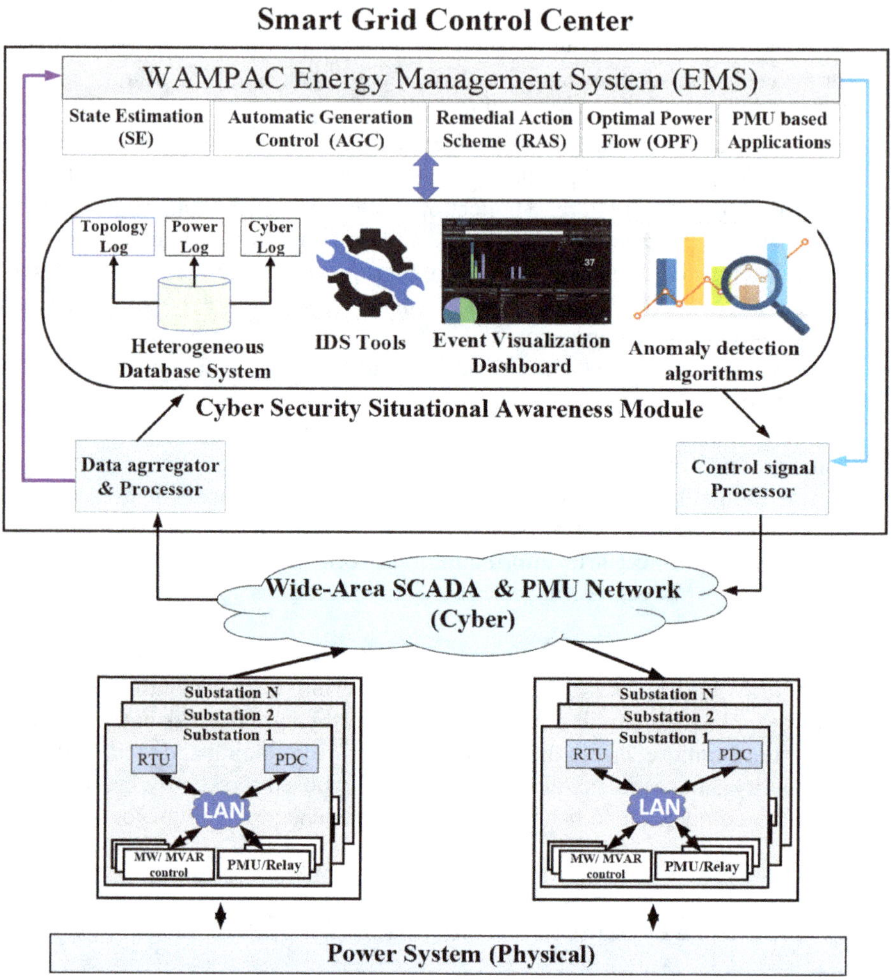

Figure 4.2 Conceptual architecture of wide-area CSA in smart grid

and physical disturbances. Furthermore, these algorithms can be applied to predict system behavior during attacks and initiate necessary mitigation strategies to restore the regular grid operation. We have classified the data-driven algorithms into three categories that are discussed here.

(a) ***Model-based algorithms*:** These algorithms include developing predictive models by leveraging redundant information, such as generation forecasts, load forecasts, wind power forecast, etc., and cyberattacks are detected based on deviations in the observed data. The authors of [10-13] presented how the forecast information can be utilized in detecting data integrity attacks on measurement signals pertinent to the WAMPAC cybersecurity, including state estimation and automatic generation control (AGC), and wind farm SCADA system. Note that the proper tuning of the threshold parameters are required during offline analysis to minimum false positives and false negatives that happen due to the existing error in forecast information and during physical disturbances, such as line faults.

(b) ***Machine learning-based algorithms*:** These algorithms include various types of supervised algorithms, such as support vector machine, DT, random forest, etc., and unsupervised algorithms, including K-means clustering, k-nearest neighbors, principal component analysis, etc. to detect stealthy and coordinated cyberattacks using SCADA and synchrophasor information. Furthermore, several signal decomposition methods, such as empirical mode decomposition, variational mode decomposition, etc. can be applied to compute input features. These input features are filtered using features selection techniques, such as filter (Pearson correlation, chi-square, etc.), wrapper (best-first search method, backward elimination, etc.), and embedded methods (DT, L1 (LASSO) regularization) to optimize the performance of classification models. The selected input features are utilized for training and developing classification models during the offline process. The trained classification model is applied later for online testing and performing multi-events classification in real time.

2. ***Intrusion detection system (IDS) Tools*:** Developing IDS tools, which can be deployed in the control center network, is one of the key steps in addressing cybersecurity challenges in smart grid. Snort, Zeek (BRO), and Suricata IDS are the most popular and open source network-based IDS tools that identify intrusions based on the known attack signatures and defined spatio-temporal behaviors of power system. However, these existing IDS tools are protocol-specific and not applicable to unknown or new attacks that are not specified in the signature database. Recently, two state-of-the-art IDS tools are proposed that detect different classes of attacks in SCADA and synchrophasor networks, which are discussed here.

   (a) ***Hybrid intrusion detector for energy system*:** This IDS tool presents a hybrid IDS by integrating network-based IDS, model-based IDS, and machine learning-based IDS to identify known and stealthy cyberattacks in the SCADA network. This tool presents a real-time sensing platform that supports SCADA measurements and cyber logs for detecting intrusions while presenting real-time visualization [22].

   (b) ***Platform for self-healing and attack-resilient PMU network (SHARP-Net)*:** This tool has been developed on minimega and Python-based software platforms that detect attacks on data aggregators in the synchrophasor network. Furthermore, it supports mitigation actions through network re-configurations and software-based dynamic routing [24].

3. ***Event visualization dashboard*:** The control center receives various types of data, including SCADA and phasor measurements, network logs, grid topology, and geographical data information at a different sampling rate. Therefore, developing the event visualization dashboard assists the control center operator to actively monitor network traffic, analyze the incoming communication and

alert logs in real time, and hence provide a better situational awareness of the grid network in real time.

4. ***Heterogeneous database system (HDS)*:** The EMS receives multi-dimensional grid measurements from the deployed distributed sensors to perform threat analysis, attack detection, and events prediction. Furthermore, it also receives grid security data from various sources, including IDS alerts, firewall alerts, system logs, etc. Therefore, HDS is required to store these datasets and later perform data pre-processing for several applications, including anomaly detection, events prediction, forensic analysis, etc. Furthermore, it also supports real-time events visualization using several software products and tools.

4.3 Smart grid cybersecurity testbed

Figure 4.3 presents a high-level architecture of hardware-in-the-loop (HIL) testbed for cybersecurity research in smart grid. We have leveraged resources available at Iowa State University's PowerCyber Laboratory to develop the smart grid cybersecurity testbed that is utilized later to perform different classes of attack-defense experiments in smart grid [25]. This testbed consists of a hybrid combination of hardware, software, emulators, and simulators that are interconnected through a wide-area communication network to support the bi-directional information exchange and emulate the grid network as close to real world. The physical components include simulators, emulators, physical relays, smart meters, PMUs, etc., while the software and emulated components include SCADA and synchrophasor communication protocols, network emulators, IDS tools, and information infrastructure. We have categorized testbed components into four major layers: physical layer, substation layer, wide-area communication layer, and control center layer. These layers are elaborated upon in the remainder of this section.

1. *Physical layer*: This layer presents commercial off-the-shelf real-time simulators like RTDS and OPAL-RT that have advanced computational capabilities to understand and capture the dynamic and transient behaviors of power system. Real-time simulators solve characteristic equations based on physics-based models, simulate complex power system at the same rate as "wall-clock time" in customized time-steps, and support HIL testing using high-speed input/output ports. Furthermore, it also supports industry-grade communication protocols to closely represent the real-grid infrastructure. In particular, the RTDS simulates power system models in an electromagnetic transient (EMT) domain at smaller time-steps of 50 microseconds. The OPAL-RT simulates power system models at smaller time-steps of 50 microseconds during an EMT simulation, and also supports phasor domain simulation at a few-millisecond time-steps (10 milliseconds) to simulate comparatively large systems (1,000 bus system) in real time. Apart from various communication protocols, such as ModBus, Distributed Network Protocol 3, IEC 61850, and IEEE C37.118, these

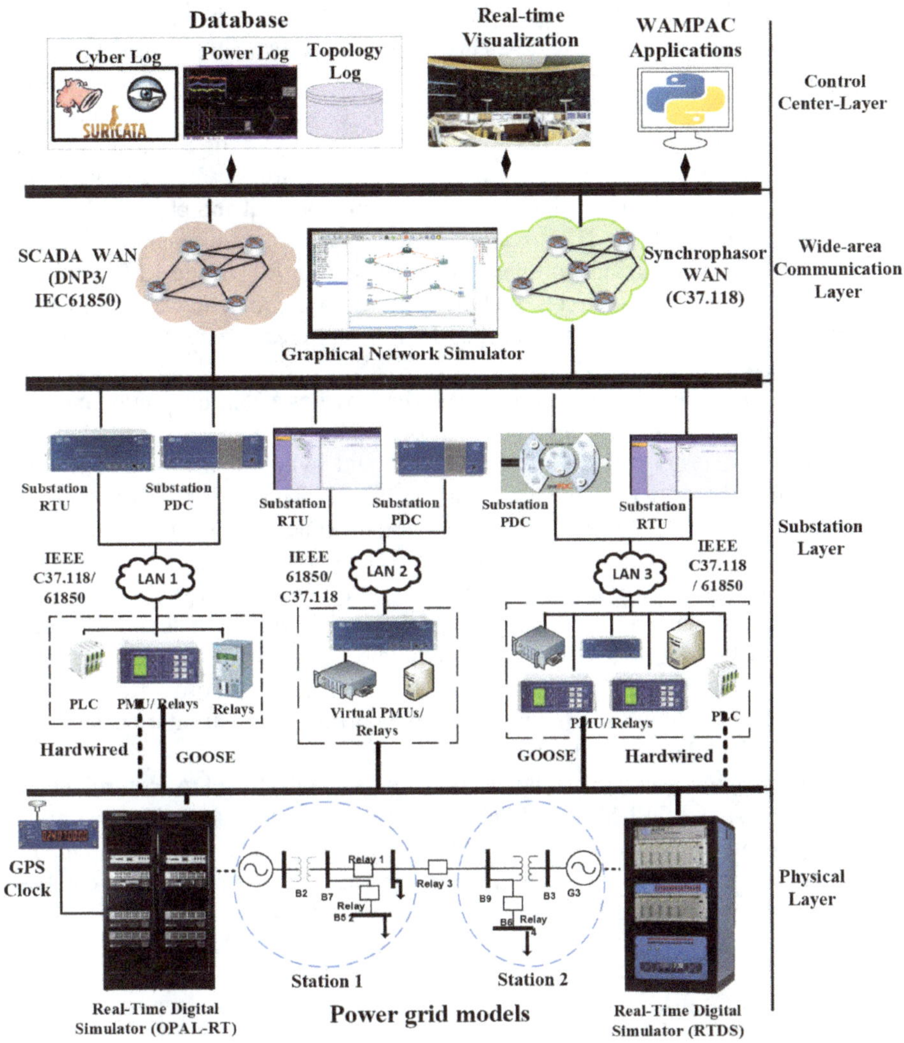

Figure 4.3 Smart grid CPS testbed at Iowa State University (ISU)

simulators are also physically integrated with relays and PMUs using analog and digital input/output cards to capture physical response characteristics when subjected to disturbances.

2. *Substation layer*: This layer consists of various field devices, such as multi-functional relays (7SJ610, 7SJ82, and SEL-421), PLC (P1-550), and PMUs (SEL-421) that are interfaced with real-time simulators and are communicating with substation-based local phasor data concentrators (LPDC (SEL-3573)) and RTU (SICAM PAS, real-time automation controller) through a local-area network, as shown in Figure 4.3. This substation layer is designed within the

testbed in three different ways: (1) using a combination of hardware devices where physical RTUs and LPDCs are communicating with physical IEDs that are mapped to real-time simulators, (2) using software resources where software programs, emulating RTU and LPDC, are interacting with virtual IEDs that are modeled in simulators to support scalability, and (3) using a hybrid combination of hardware devices and software resources, where software- and hardware-based LPDCs and RTUs are mapped to virtual and physical IEDs. In all three scenarios, the IEDs receive power system measurements from real-time simulators and transmit them to the substation RTUs and LPDCs, which perform necessary functions, including close-loop protection and human-in-the-loop operations while forwarding the relevant measurements to the control center layer.

3. *Wide-area communication layer*: This layer provides a communication medium to support the bi-directional information exchange between substation and control center networks for the SCADA and synchrophasor-based control operations. In particular, we have utilized the Internet-Scale Event and Attack Generation (ISEAGE) environment and graphical network simulation environment, using graphical network simulator (GNS3) tool, to develop a scalable cyber infrastructure, log-network interfaces, and cybersecurity attack-defense experimental platform in smart grid. GNS3 is free and open source software, written in Python, that is utilized to emulate complex network and supports emulated and simulated devices. In a similar way, ISEAGE provides an emulation of desired network topology while supporting physical interfaces to integrate physical devices and networks [26].
4. *Control center layer*: This layer includes Siemens SCADA control center (Power TG) and GEEMS, SCADA and synchrophasor-based wide-area monitoring, protection, and control (WAMPAC) applications, including remedial action scheme (RAS), that are written in Python script, heterogeneous database capturing cyber logs, grid measurements, and topology logs, and real-time visualization platform using several software products and tools.

4.4 Case study: anomaly detection for WAMS

4.4.1 Synchrophasor-based WAMS

System operators are responsible to balance the cost of system operation against the risks to system reliability. Managing this trade-off is becoming more complex as the ever-increasing share of intermittent generation has to be managed to meet more variable system demand. Therefore, enhanced system visibility is critical to effectively manage power system stability and respond to system disturbances. WAMS provides this visibility based on real-time data streamed from PMUs. A host of WAMS software applications use this data to provide context-specific information, presented through an intuitive user interface. Early warning of dynamic issues is not readily observed in the EMS. Commercially available WAMS solutions complement traditional SCADA-based EMS to deliver fast and proactive grid stability management to meet the complex challenges of modern power systems.

4.4.2 Problem formulation

Since PMUs are increasingly deployed in several substations to provide control room operators with critical WASA, they are also exposed to growing cybersecurity vulnerability. It is conceivable that an adversary with malicious intent will seek to compromise these wide-area voltage and current measurements in order to affect control decisions that unnecessarily disrupt typical grid operations. In this section, we describe the development and validation of a near-real-time anomaly detection algorithm for synchrophasor data that reduces the attack surface of the control center and its synchrophasor-dependent software applications.

An anomaly on the physical layer of WAMS network can have a number of manifestations. More than one PMU may be compromised by the attacker. The synthetic synchrophasor data injected into a compromised PMU channel may appear as zero values, dropped packets, or sudden changes in magnitude and phase angle. Depending upon the sophistication of attack, the signals may or may not follow the underlying physics (Kirchhoff's voltage and current laws) of the power system. It is acknowledged that in one of the extreme cases where all PMUs are compromised by the attacker, it would be nearly impossible to detect attacks. Nonetheless, anomaly detection algorithms should be developed to detect all kinds of attacks, including the sophisticated attacks that can impact control decision-making. From this standpoint, a particularly stealthy attack vector is one that spoofs or replays the transient that occurred during a fault event by compromising one or more PMUs, also known as fault replay attack. Once a PMU or the synchrophasor-based communication channel is compromised, a hacker may compel this unit or channel to transmit prerecorded phasor measurements to the phasor data concentrator (PDC)

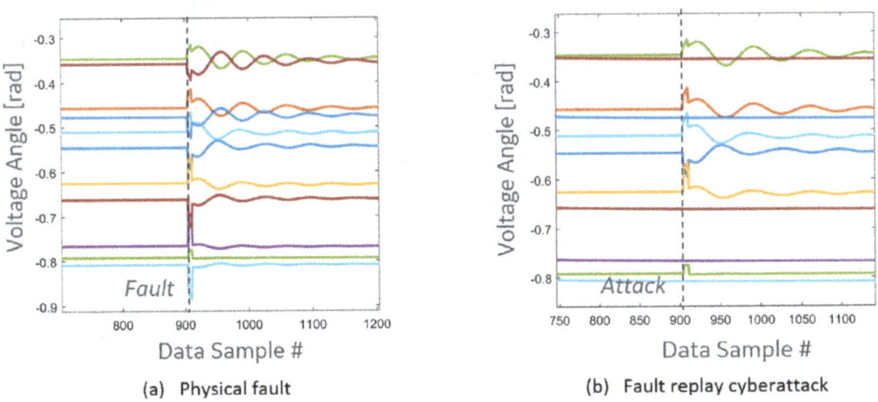

Figure 4.4 Simulated PMU node angles in the aftermath of a fault transient (a) and after a fault replay cyberattack (b) for 11 buses within a power system (angles are unwrapped and detrended)

while resembling legitimate measurement that would have been observed during an actual fault. This particular attack vector could easily trick a human operator and EMS-based wide-area controller to initiate protective actions that would potentially result in loss of load or generation. In the case of a fault replay attack on a large number of PMU channels, the injected signals may replicate the physics of the system, making it difficult for a system operator to detect this attack.

Figure 4.4(a) shows simulations of filtered and unwrapped synchrophasor angle measurements for a faulty event on the IEEE 11-bus system where PMUs are deployed at each bus of the system. Figure 4.4(b) shows a simulated fault replay attack where half of the PMUs are compromised. In this case, if a small time period is provided to recognize the event and take a control decision, an operator might get easily convinced that an actual fault event has occurred and might take wrong decisions. Furthermore, several automated decision-making tools are susceptible to this replay attack.

4.4.2.1 Proposed anomaly detection algorithm

There exists a significant research gap associated with developing robust and accurate anomaly detection algorithms for WAMS that can accurately detect fault replay attacks or "spoofed" fault signals by leveraging data-driven and model-based approaches. To achieve high classification accuracy, autonomy, and ease of implementation, a novel anomaly detection algorithm is proposed in this section for WAMS cybersecurity that integrates data-driven and model-based approaches for rapid and convenient training. The proposed algorithm is capable of performing online training and can function entirely in a purely data-driven manner, if necessary. Alternatively, online training can be supplemented or replaced entirely by model-based training, using the physics-based analysis of the power system. During the model-based training, the grid topological data are required and should be available in the EMS. Because of this flexibility, the algorithm can be implemented as a component of the substation-based LPDC, as a part of data quality reporting service. It can also be implemented as a preconditioning step within the EMS platform at the periphery of the control center. The proposed algorithm is capable of discerning

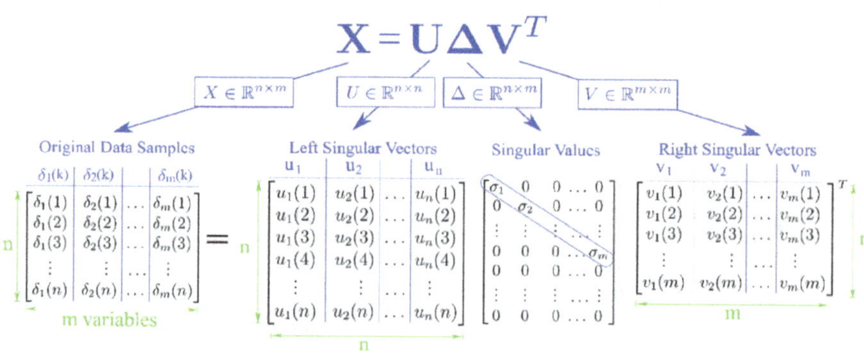

Figure 4.5 Annotated overview of the singular value decomposition

actual physical faults (e.g., short circuits, line faults) from replay attack that mimics fault-like behavior by replaying faulty measurements.

The proposed algorithm applies the singular value decomposition (SVD) to pre-process synchrophasor data and later develop classification model. It is based on the notion that the SVD of a moving window of synchrophasor measurements will betray a set of spatiotemporal correlations that are uniquely associated with the incoming PMU measurements. The computed singular values and singular vectors of an SVD serve as critical inputs to anomaly detection and event classification subroutines. Figure 4.5 provides an annotated overview of the SVD transformation using notation and conventions, such as the orientation of the data matrix, that are adopted by this algorithm.

Spatiotemporal correlations are expected to exist within PMU measurements taken from three-phase, sparsely connected, reactive power grids of arbitrary size and interconnection under various states of time-varying load and generation. These correlations are clearly visible after the application of SVD and are difficult to spoof unless every single PMU, deployed in the system, is compromised while assuming that the adversary has access to the detailed information of the grid topology. The correlations would likely to be less apparent if only a subset of the synchrophasor measurements are spoofed by the adversary. However, the exact nature of each correlation can be more deeply understood using a physics-based partitioning algorithm that utilizes a prior knowledge of the exact power system topology. Since the exact power system topology may not be always known to grid operators, use of physical parameters and other relevant physical information from the power system should ideally be optional and contingent upon availability. However, when topological information is available, the physics-based partitioning algorithm can be used to supplement the data-driven approach to achieve better detection performance.

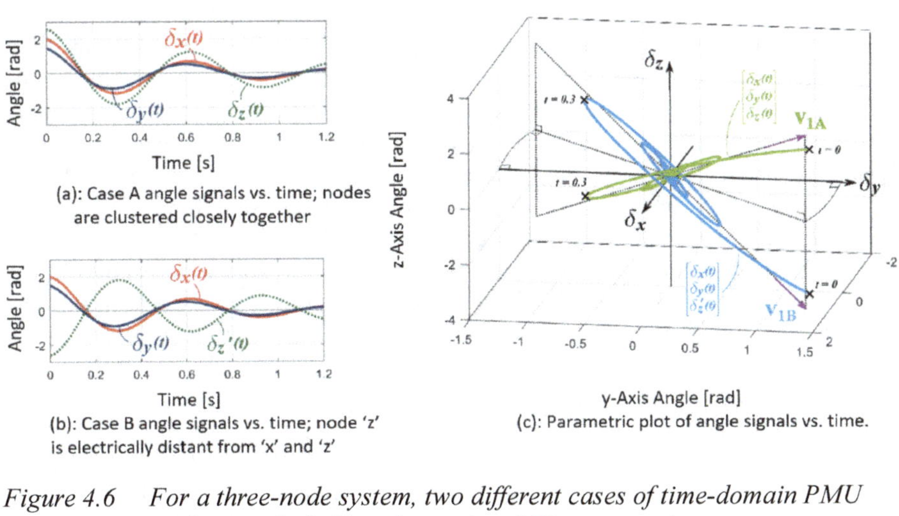

Figure 4.6 For a three-node system, two different cases of time-domain PMU angle signals (left) and associated 3D parametric curve

In electromechanical timescale, the transient behavior of voltages and currents at each node of power system is contingent upon a set of well-known nonlinear differential equations and algebraic constraints [26, 27]. Over the decades, the network structure that underpins this differential-algebraic equation set has been closely studied, and tools have been developed that leverage this underlying structure to partition the network into clusters of nodes [27–29]. Such partitioning strategies have historically been applied in the development of reduced-order models that represent clusters of nodes as a single aggregate node, drastically reducing the number of differential and algebraic equations, and thereby improving computational efficiency of the model.

For a simple three-node power system, Figure 4.6 demonstrates the impact of electrical distance between nodes on the behavior of voltage angles due to a significant event. In the aftermath of any significant event, unique signatures are observed in the right singular vectors (RSVs) that are more-consistently associated with the network topology and structure compared to the location or severity of a particular fault. In [30], the intent is to apply the SVD of measured sensor data as a data-driven means to inform an aggregation strategy and ultimately resulting in the optimized construction of a (reduced-order) physics-based model, similar to the objective in previous works [27–29]. The cybersecurity algorithm, described in this work, is not concerned with reduced-order modeling; however, it leverages (i) "real" RSVs obtained from the SVD of a moving-time-window of historical fault events, (ii) "synthetic" RSVs derived from partitioning analysis, and (iii) a combination of both real and synthetic RSVs for purposes of signature identification and algorithm training. The signatures extracted from the RSVs ultimately drive the algorithm's main event of detection and classification subroutines.

While a variety of different partitioning analysis techniques could be employed to train the proposed cyberattack detection algorithm, the Generalized Eigenvalue Perturbation (GEP) method [27] has several desirable qualities, including expediency and scalability. For a power system of arbitrary size and interconnection, a vector-valued set of nonlinear differential equations and algebraic constraints that dictate the dynamic behavior of a power system in the electromechanical timescale is described in [27]. This vector-valued equation set is most appropriate and widely applied for classical transient stability analysis as prime mover dynamics, exciter dynamics, and power system stabilizers are not well represented in the model. Loads are assumed to be constant power, the network is assumed to be symmetric and balanced, and the transfer conductances of the network are neglected. While this partitioning method is most appropriate for synchronous machine dynamics in the transmission system, the authors of [31] have demonstrated its applicability to the dynamics of grid-forming, droop-controlled distributed generation resources within a microgrid-based distribution system. The proposed use of the GEP partitioning algorithm [27] to train the proposed WAMS anomaly detection algorithm is reviewed below.

1. **Step 1:** The reduced generalized eigenvalue problem is defined by $\mathbf{E}$ and $\mathbf{R}_S$ matrices, as defined in [27]. These matrices are closely related to a linearized

version of the aforementioned set of nonlinear differential equations describing the electromechanical network dynamics. As a first step, these two matrices are computed while assuming a particular equilibrium operating point. For the differential-algebraic system comprised by $(\mathbf{E}, \mathbf{R}_S)$, the smallest generalized eigenvalue and its associated eigenvector are identified, which are denoted by λ and v, respectively.

2. **Step 2:** The gradient $\vec{g}_i$ of the generalized eigenvalue λ with respect to the vector of network susceptances, $\overrightarrow{bl}$, is then computed. Gradient $\vec{g}_i$ is used to identify the element of the network chosen for deletion.
3. **Step 3:** Repeat previous steps for the new system with the missing branch, until a sufficient number of branches have been eliminated such that the nodes of the original network are fully partitioned into two disjoint subsets. A single iteration of the GEP algorithm has now been completed.
4. **Step 4:** Once the network has been partitioned into two disjoint subsets of nodes (Subset 1 and Subset 2) using the GEP algorithm, a "synthetic RSV" is created, as discussed in [32]. This synthetic RSV constitutes a prediction of a vector within the **V** matrix obtained from the SVD. This synthetic RSV is first instantiated as a zero vector with length equal to the number of PMUs associated with the synchrophasor angle data matrix. In this matrix, the first element of the synthetic RSV corresponds to the first column of data (i.e., the first PMU), the second element of the synthetic RSV corresponds to the second PMU, and so on. The elements of this synthetic RSV are populated with a "1" or "–1" depending on whether the node that hosts each PMU is located in Subset 1 or Subset 2. If PMUs are not included within the boundaries of the portion of the network based on the above GEP partitioning analysis, the element associated with that PMU remains zero. Once all elements in the matrix are appropriately populated, the synthesized RSV is then normalized to the magnitude of 1.
5. **Step 5:** To generate the additional synthetic RSVs for the network corresponding to the faster inter-area system modes across smaller areas of the network, steps 1–4 can be iteratively repeated for the sub-networks of the power system associated with Subset 1 and Subset 2.

In the proposed anomaly detection algorithm for WAMS, the computed features and signatures, which are obtained from the application of the SVD using synchrophasor data, are post-processed and utilized within event detection and classification subroutines. The anomaly detector runs continuously to examine the SVD features for evidence of an event. Upon detection of an event, the event classification subroutine uses the features and system signatures obtained from the SVD of the time window of data corresponding to the new event. For performing multi-classification events, including faults and cyberattack, the classification subroutine also examines relevant information obtained from partitioning analysis and from historical events. The use of parameters and other relevant information from the partitioning analysis is optional and is contingent upon availability of recent system topological information. Additional information regarding the proposed anomaly detection algorithm for WAMS cybersecurity is discussed in [32].

Since PMUs have a sampling rate varying from 30 samples/second to 120 samples/second, these measurements can quickly accumulate and grow to unmanageable proportions in the phasor aggregators. The sheer volume of sensor data from PMUs, SCADA sensors, and smart meters can be a hindrance to the grid operators who require timely and accurate information to take corrective actions. Over 2,000 PMUs have been installed in the United States [33] and Midcontinent Independent System Operator (MISO) has access to over 3,500 synchrophasor measurements from 489 PMUs. Using conventional sampling rates, the synchrophasor network in the MISO generates over 100 megabytes (MB) of synchrophasor data every minute and multiple terabytes over the course of a month. However, advanced and improved techniques are needed to distill the sensor data and reduce these measurements to timely and actionable information in the context of WAMS cybersecurity in real time. Regardless of the type of intrusion detection algorithm, different features, obtained from the PMU data, must be down-sampled to develop an efficient classification model.

4.4.2.2 Experimental setup and case study

A CPS testbed is required to facilitate the development and rigorous evaluation of the proposed WAMS-related cyberattack detection algorithm. The previously discussed testbed environment is capable of simulating a large number of randomized transient events, including actual fault events and spoofed data injection attacks, using the two-area Kundur bus system [26]. For the studied power system, the PMU data are collected in the testbed environment throughout the duration of each event while resembling the typical behavior of WAMS. The PMU data comprise time-stamped and synchronized samples of voltage and current phasor magnitude and phase angles that are collected at a user-defined sampling rate under the realistic off-nominal frequency conditions. We have developed automated scripts that perform different functionality as discussed here.

1. Specification of parameters relevant to the physical characteristics of the power network, including fault events within the network.
2. Specification of parameters relevant to the construction and testing of a detection algorithm for data spoofing attacks.
3. Construction of a nested cell variable of sliding-window PMU data matrices and attack events using simulation results that are obtained from running the power system model based on Simulink® or importing the existing simulation results.
4. Training of attack detection algorithm using a model-driven GEP partitioning script.
5. Training of the cyberattack detection algorithm using SVD features and signatures calculated from previous events.

6. Application of the cyberattack spoofing detection algorithm to the nested cell variable containing PMU data matrices.
7. Three-dimensional visualization of the test matrix results.

Regarding specification of parameters relevant to the physical characteristics of the network, the simulation environment is highly customizable. It provides users a high degree of control over a large number of relevant parameters, including:

1. The nature, placement, and severity of physical fault events, including short-circuit and line faults.
2. The number of PMUs within the system.
3. Placement of PMUs within the system.
4. Specification of the compromised PMUs.
5. Specification of attack type on each PMU unit. Currently, the choices are limited to the following: (1) a compromised PMU replays historical fault data, (2) a compromised PMU erroneously reports a user-defined constant magnitude and phase angle, and (3) a compromised PMU erroneously reports a magnitude and phase angle of zero values.
6. Specification of line impedance and generator properties, including inertia and damping factors.
7. Specification of different features of simulation data and experiment results that are to be stored for future analysis.

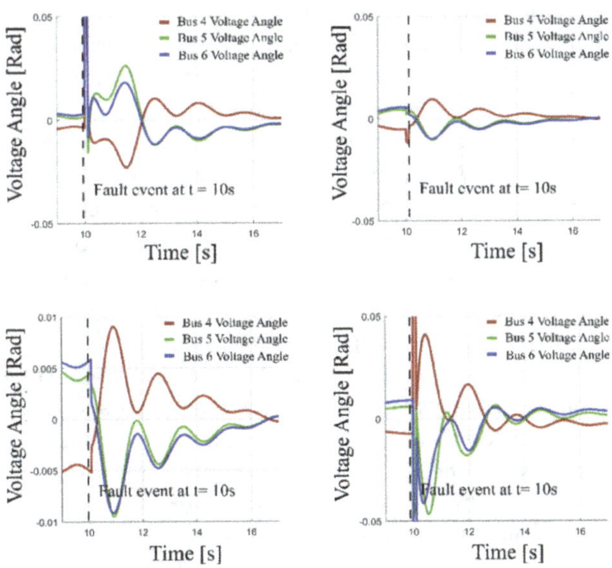

Figure 4.7 Waveforms of post-processed synchrophasor voltage angle for three PMUs and four random fault events based on nature (e.g., three-phase short-circuit or line fault) and location

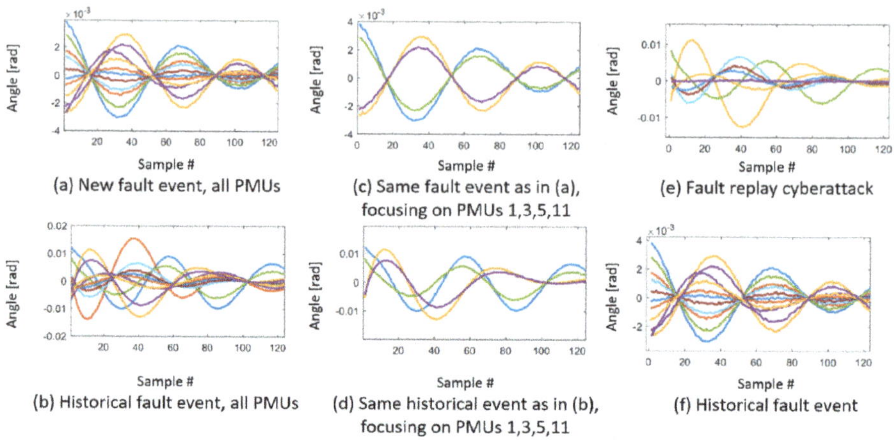

Figure 4.8 Simulated voltage angles for all nodes in a 4-machine, 11-bus network

Many of the above environmental parameters (e.g., 1–5) can be randomized to facilitate large-scale studies across a range of potential environmental conditions. Additionally, other parameters, such as number of PMUs in the network and number of compromised PMUs, can be swept across a range of values, which allows a greater insight into the impact of parameter on the performance of algorithm. Furthermore, the software-based CPS testbed supports the specification of parameters relevant to the operation of algorithm, including all relevant decision thresholds and size of the data window in SVD. The testbed environment permits most of the parameters of algorithm. These parameters can be varied across a range of values in large-scale test matrices to empirically evaluate the performance of classifier in real time.

For example, Figure 4.7 demonstrates the post-event synchrophasor waveform results that are derived from the CPS testbed for four different combinations of environmental parameters. In this case, severity and location of a physical fault is varied between four cases, and waveforms of post-processed synchrophasor voltage angle are plotted for three PMUs.

Figure 4.8 represents an additional example where windows-based simulated voltage angle waveforms are collected using the software-based data aggregator in the testbed environment. The waveforms are collected after several events, including a new fault event, historical fault event, and fault replay attack. Figure 4.8(a) and (c) corresponds to the same "new" fault event. Figure 4.8(c) presents the behavior of voltage angle for nodes 1, 3, 5, and 11. Likewise, Figure 4.8(b) and (d) corresponds to the same historical fault event. In this case, the SVD was applied to time windows of data and considered as a part of the detection algorithm. Using the post-processed features and signature information, obtained from new and historical events, the algorithm has correctly identified the real fault, as shown in Figure 4.8(a), and fault replay attack, as shown in Figure 4.8(e).

For each tested condition and combination of parameters, the anomaly detection algorithm is applied to 600 datasets of randomized fault events and fault-replay attacks. Random short- circuit faults and line faults are applied, and their transients are recorded for developing a fault-replay attack. For a given fault-replay attack, the transients replayed at the compromised PMUs are recorded from the same event, and the fault transient is replayed at the same PMU from which it was recorded. A matrix of different algorithm parameter selections, number of PMUs, and number of compromised PMUs are studied. The data window length used for the SVD is varied from 10 to 200 samples, and a particularly important decision threshold (referred to here as "rotation threshold angle") is varied from 0.5 degrees to 60 degrees.

4.4.2.3 Performance evaluation

This section demonstrates results based on the performance of anomaly detection algorithm for a wide range of power system conditions and across multiple parameters of the proposed algorithm. In this work, outputs of a classification subroutine are: (i) true-positive, which is a correct classification of an abnormal event as abnormal; (ii) false-positive, which is an incorrect classification of a normal event as abnormal; (iii) true-negative, a correct classification of a normal event as normal; and (iv) false-negative, an incorrect classification of the abnormal event as normal. Here, true positive rates (TPR) are recorded as the algorithm is applied to classify fault and attack events. Note that the TPR positive rate for a given event is computed by calculating the true events divided by total events for line fault and cyberattacks. In this work, the two-area test system is considered as a case study, which is characterized by two distinct areas separated by a long transmission line corridor. Simulation parameters for this test system are provided in [26]. This configuration is particularly conducive to inter-area oscillations between two clusters of nodes. In this case, 1,200 randomized events are simulated, and the algorithm was found to perform satisfactorily across a wide range of operating conditions.

Figure 4.9 demonstrates a collection of 3D surface plots that show the TPR of the classification algorithm as a function of an important decision threshold parameter and length of the sliding window (number of rows in data matrix, X). For each element of the surface plot, the TPR is calculated with respect to threshold parameter and data window size for 600 fault and attack events. Finally, the surface plots of TPR are created for three different scenarios where the number of deployed PMUs is varied between 3, 7, and 11. For generating attack events, two of the deployed PMUs are compromised by the attacker. A purely data-driven approach is applied for training detection algorithm (GEP partitioning is not used), and features and signatures from three historical events are applied within the classification subroutine. Under these conditions, Figure 4.9 and Table 4.1 show that the performance of the classification is sensitive to the number of deployed PMUs. It shows a superior performance when 11 PMUs are deployed in the network, and its performance degrades with a decrease in the deployed PMUs. The TPR for cyberattack reduces to 0.28 when only 3 PMUs are deployed in the network.

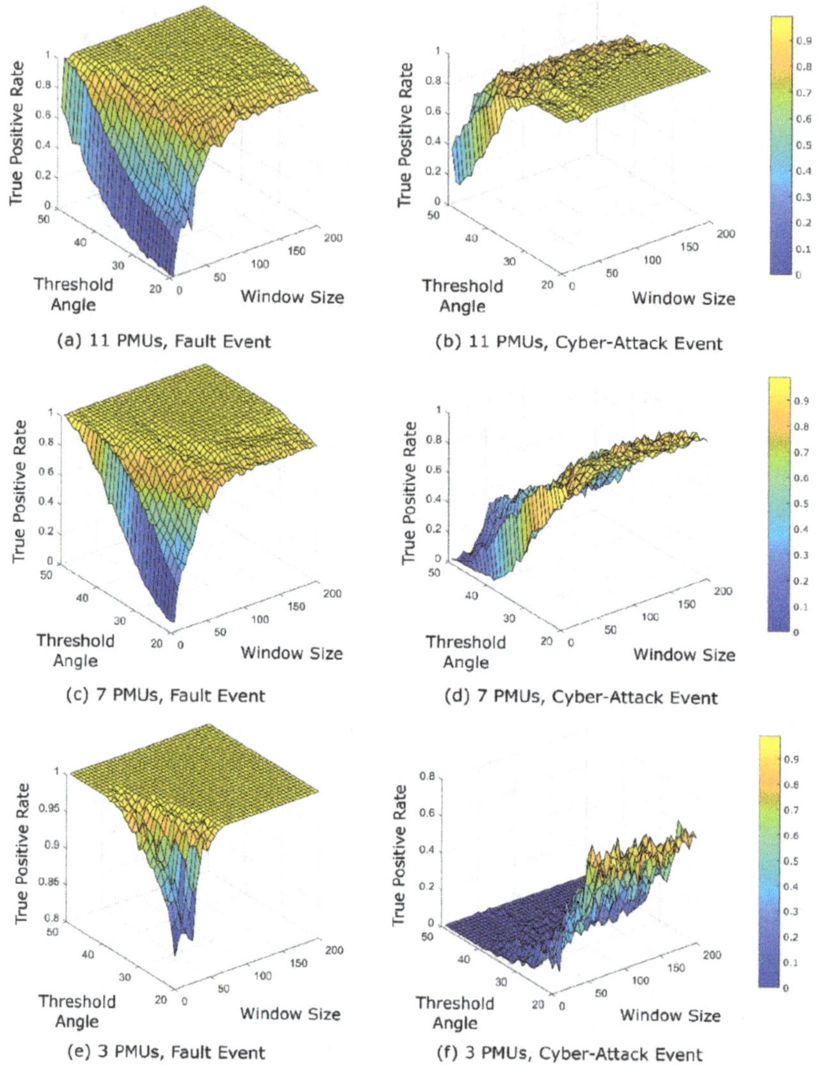

Figure 4.9 *TPRs for fault event and cyberattack event classification, across a range of threshold angle parameters, window sizes, and number of PMUs in the network*

4.4.3 Prototype demonstration using the commercial platform

A prototype version of the proposed detection algorithm for WAMS has been ported into a set of fully functioning Python modules, which enables a preliminary integration of the algorithm into a commercial WAMS monitoring software platform, courtesy of the GE. The integration of this algorithm is strictly permitted for demonstration purposes. It utilizes the Python interface within the platform to connect the

Table 4.1 *True-positive rate with the specified parameters (angle rotation threshold of 31 degrees and window size of 130 samples)*

	Fault TPR (TPR)	**Cyberattack TPR (TPR)**
11 PMUs	0.99	0.98
7 PMUs	1.0	0.8
3 PMUs	1.0	0.28

algorithm's back-end Python modules to the visualization and analysis features of GE-WAMS platform. This Python interface facilitates rapid prototyping and evaluation of new algorithms and data-driven features within the software framework.

To provide datasets for the demonstration purpose, synchrophasor-based synthetic datasets are pre-assembled. These simulated test and training datasets are converted into a JSIS format and integrated them into the platform. Using these synthetic datasets, the performance of the WAMS anomaly detection algorithm was demonstrated in this platform for replay attacks and physical line faults.

The WAMS anomaly detection algorithm is tested on two fault events that are randomly selected from a set of 20 fault events. In Figures 4.10 and 4.11 respectively, new fault events "T3" and "T6" are simulated after their datasets are converted to JSIS comma-separated values (CSV) format and loaded them into the GE-WAMS platform. It can be observed that the proposed algorithm successfully detects events T3 and T6 and correctly classifies them as actual faults, as highlighted through

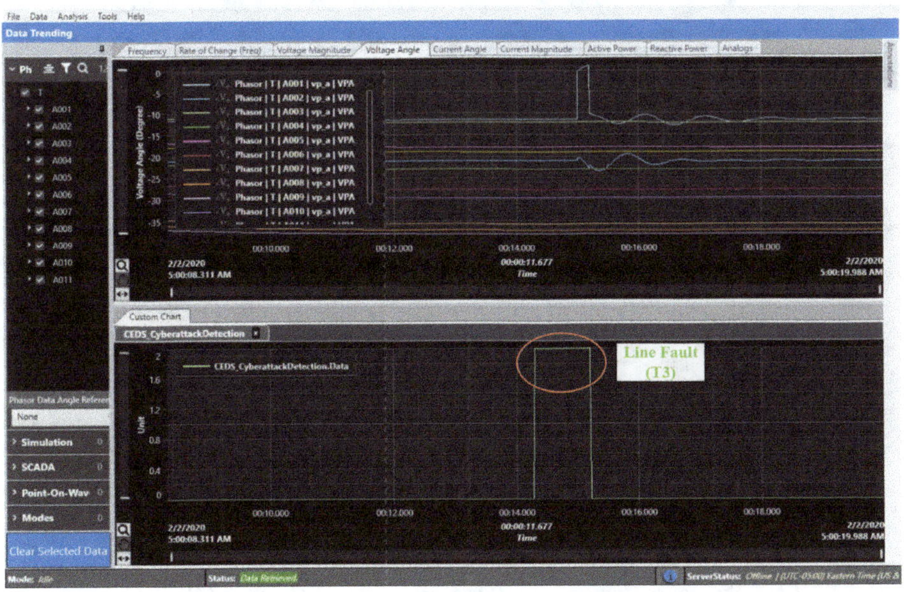

Figure 4.10 *Snapshot during a fault event (T3) in the GE-WAMS software*

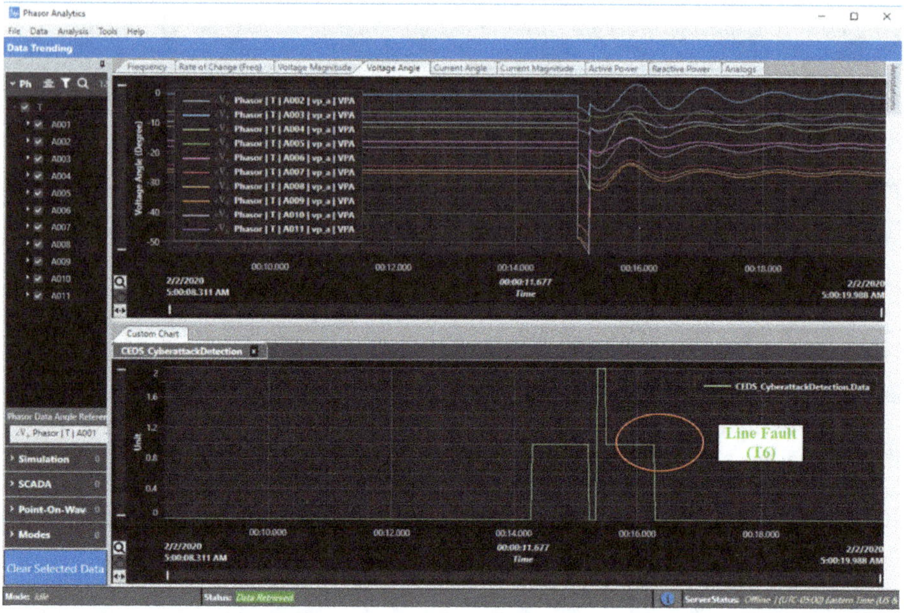

Figure 4.11 Snapshot during a fault event (T6) in the GE-WAMS software

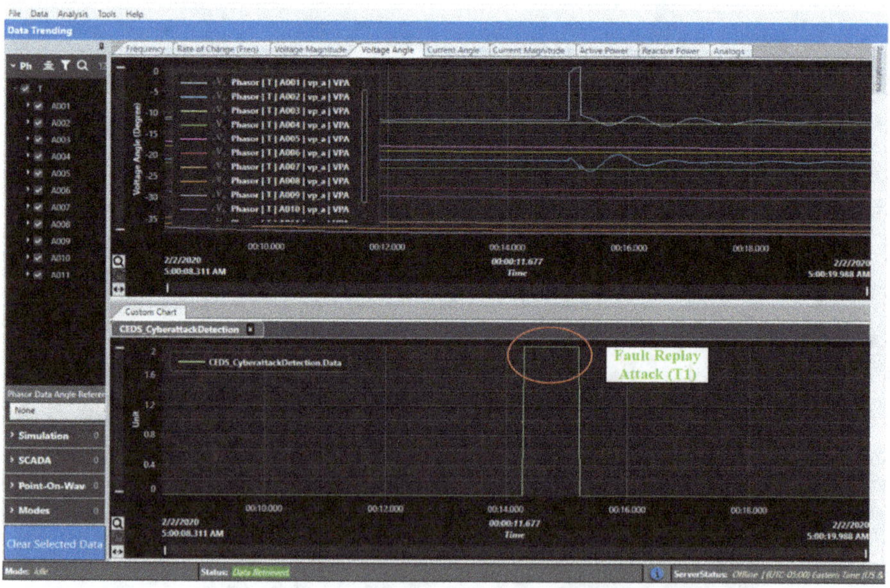

Figure 4.12 Snapshot during a fault replay attack (T1) in the GE-WAMS software

ovals. Additionally, a fault replay attack event "T1" is simulated after its dataset was converted to the JSIS CSV format and loaded into the GE-WAMS platform. In this case, Figure 4.12 shows how the algorithm accurately detects this event and correctly classifies it as a cyberattack, also highlighted through oval in the figure.

4.5 Cybersecurity training for situational awareness

Cybersecurity training is necessary and mandatory for system operators to learn about cyberspace in the smart grid and gain knowledge and expertise about interconnected IT and operational technology in the grid network. It provides hands-on learning experiences to grid operators, engineers, and analysts through simulated attack-defense experiments and grid security exercises, such as NERC GridEx [34] and DOE's CyberForce competition [35]. Furthermore, this training platform can be utilized for demonstrating state-of-the research works, orchestrating realistic cyberattacks, and experimentally testing, validating, and evaluating defense/detection measures. It can also be utilized as an educational platform for students and researchers to gain interdisciplinary knowledge and improve skill sets. Hence, cybersecurity training plays a crucial role in developing CSA for power system. In addition, the training can assist in performing forensic analysis and proper documentation of grid assets, patched vulnerabilities, and enrolling qualified individuals in organizations.

4.6 Conclusion

Developing CSA for the smart grid is a challenging task since it requires security-related big data and cutting-edge technologies to process, manage, and analyze heterogeneous and multi-dimensional datasets. This chapter presented a conceptual architecture of CSA by integrating anomaly detection algorithms, heterogeneous database system, IDS tools, and event visualization dashboard in the grid network. Initially, this chapter introduced the CSA, discussed its related work, and highlighted the existing challenges in developing a real-time CSA in the smart grid. We then discussed the significance of cybersecurity testbed for testing and validation, and briefly described several components of ISU PC testbed that emulate the real grid network. As a proof of concept, the synchrophasor-based anomaly detection algorithm is presented for WAMS to detect fault replay attacks. Experimental results illustrated that the proposed data-driven anomaly detection algorithm is efficient in detecting fault replay attacks and distinguishing them from physical line faults. Our case study also showed that the performance of this algorithm is directly affected by the number of deployed PMUs as the algorithm exhibits superior performance when more PMUs are deployed in the network. We also presented the prototype demonstration by integrating the proposed algorithm with the GE-WAMS using Python modules. A potential avenue for future work is to test and validate anomaly detection and mitigation algorithms using the industry-grade commercial software for several WAMPAC applications, including RAS, AGC, WAVC, etc.

References

[1] Terzija V., Valverde G., Cai D. 'Wide-area monitoring, protection, and control of future electric power networks'. *Proceedings of the IEEE.* 2011;99(1):80–93.

[2] Sridhar S., Hahn A., Govindarasu M. 'Cyber–physical system security for the electric power grid'. *Proceedings of the IEEE.* 2012;100(1):210–24.

[3] 'NERC reliability guideline: PMU placement and installation'. 2016.

[4] 'NERC critical infrastructure protection Committee (CIPC)'. *Cyber attack Task Force (CATF) Update, North American Electric Reliability Corporation (NERC).* 2011.

[5] Miller B., Rowe D.C. 'A survey of SCADA and critical infrastructure incidents'. *First Annual Conference on Research.* 2012:51–6.

[6] Kumar Singh V., Ozen A., Govindarasu M. 'Stealthy cyber attacks and impact analysis on wide-area protection of smart grid' in *2016 North American Power Symposium (NAPS).* Denver, CO; pp. 1–6.

[7] Falliere N*et al.* *"W32.stuxnet dossier," Technical Report, Symantec,.* 2011.

[8] SANS. 'Analysis of the cyber attack on the Ukrainian power grid—defense use case [online]'. 2016.

[9] The MITRE Corporation, Cyber security situational awareness. [online]. Available from https://www.mitre.org/capabilities/cybersecurity/situation-awareness.

[10] Singh V.K., Callupe S.P., Govindarasu M. Testbed-based Evaluation of SIEM Tool for Cyber Kill Chain Model in Power Grid SCADA System. *North American Power Symposium (NAPS)*; Wichita, KS, 13-15 Oct; 2019. pp. 1–6.

[11] Singh V.K., Ebrahem H., Govindarasu M. Security Evaluation of Two Intrusion Detection Systems in Smart Grid SCADA Environment. *2018 North American Power Symposium (NAPS)*; Fargo, ND, USA, 9-11 Sep; 2018. pp. 1–6.

[12] Quint R.D. 'NERC'. *Synchrophasors for Situational Awareness, Atlanta, GA: North American Synchrophasor Initiative.* 2015.

[13] 'National Institute of Standards and Technology (US.)'. *Situational Awareness a New Way to Attack Cybersecurity Issues Rather Than Using a System Defense Approach.* 2013.

[14] Endsley M.R. 'Design and evaluation for situation awareness enhancement'. *Proceedings of the Human Factors Society Annual Meeting.* 1988;32(2):97–101.

[15] Bass T. 'Multisensor data fusion for next generation distributed intrusion detection systems'. 1999.

[16] Catherine A. 'Splunk-industry perspective'. *Gaining the Situational Awareness needed to Mitigate Cyberthreats.* Washington, DC; 2015.

[17] Wu J., Ota K., Dong M., Li J., Wang H. 'Big data analysis-based security situational awareness for smart grid'. *IEEE Transactions on Big Data.* 2018;4(3):408–17.

[18] Singh V.K., Ozen A., Govindarasu M. A Hierarchical Multi-Agent Based Anomaly Detection for Wide-Area Protection in Smart Grid. Resilience Week (RWS); Denver, CO; 2018. pp. 63–9.

[19] Chen H., Zhang L., Mo J., Martin K.E. 'Synchrophasor-based real-time state estimation and situational awareness system for power system operation'. *Journal of Modern Power Systems and Clean Energy*. 2016;4(3):370–82.

[20] Basu C. 'A knowledge-based approach to situational awareness for the power grid using synchrophasors'. *IEEE PES, Green Mountain Chapter, Phasor Measurement Applications*. October 12, 2016.

[21] Singh V.K., Govindarasu M. 'Decision tree based anomaly detection for remedial action scheme in smart grid using PMU data'. *2018 IEEE Power & Energy Society General Meeting (PESGM), Portland*. 2018:1–5.

[22] Singh V.K., Vaughan E., Rivera J., Hasandka A. HIDES: Hybrid Intrusion Detector for Energy Systems. IEEE Texas Power and Energy Conference (TPEC); College Station, TX; 2020. pp. 1–6.

[23] Singh V.K., Govindarasu M. 'Cyber Kill Chain-Based Hybrid Intrusion Detection System for Smart Grid' in Haes Alhelou H., Abdelaziz A.Y., Siano P. (eds.). *Wide Area Power Systems Stability, Protection, and Security. Power Systems. Cham: Springer*; 2021.

[24] Singh V.K., Vaughan E., Rivera J. 'SHARP-Net: platform for self-healing and attack resilient PMU networks'. *IEEE Power & Energy Society Innovative Smart Grid Technologies Conference (ISGT)*; Washington, DC; 2020. pp. 1–5.

[25] Singh V.K., Govindarasu M. A Novel Architecture for Attack-Resilient Wide-Area Protection and Control System in Smart Grid. *Resilience Week (RWS)*; Salt Lake City, ID; 2020. pp. 41–7.

[26] Kundur P., Balu N.J., Lauby M.G. *Power System Stability and Control*. New York: McGraw-Hill; 1994.

[27] DeMarco C., Wassner J. A generalized eigenvalue perturbation approach to coherency. Proceedings of the 4th IEEE Conference on Control Applications,; Sep. 1995. pp. 611–7.

[28] Chow J. 'Time-Scale modeling of dynamic networks with applications to power systems'. *Berlin and Heidelberg GmbH: Springer Verlag;*. 1982.

[29] Podmore R. 'Identification of coherent generators for dynamic equivalents'. *IEEE Transactions on Power Apparatus and Systems*. 1978;PAS-97(4):1344–54.

[30] Anaparthi K.K., Chaudhuri B., Thornhill N.F., Pal B.C. 'Coherency identification in power systems through principal component analysis'. *IEEE Transactions on Power Systems*. 2005;20(3):1658–60.

[31] Hart P.J., Lasseter R.H., Jahns T.M. 'Coherency identification and aggregation in grid-forming droop-controlled inverter networks'. *IEEE Transactions on Industry Applications*. 2019;55(3):2219–31.

[32] Hart P.J., Acharya S., Wang H. 'Coherency-based detection algorithm for synchrophasor cyberattacks'. *North American Power Symposium (NAPS)*; Wichita, KS; 2019. pp. 1–6.

[33] Overholt P., Ortiz D., Silverstein A. 'Synchrophasor technology and the DOE: exciting opportunities lie ahead in development and deployment'. *IEEE Power and Energy Magazine*. 2015;13(5):14–17.
[34] NERC. *Grid Security Exercise (GridEx)*. [online]. Available from https://www.nerc.com/pa/CI/CIPOutreach/Pages/GridEx.aspx.
[35] Department of Energy (DoE). *CyberForce Competition* [online]. Available from https://cyberforcecompetition.com/.

Chapter 5

Artificial intelligence-aided detection of data manipulation attacks in smart grid

Kush Khanna[1], Bijaya Ketan Panigrahi[2], and Anupam Joshi[3]

The electrical grid is rapidly changing. Microgrids, distributed energy resources, and increased automation in decision-making enabled by smart sensors have bridged the gap between utilities and end users by increasing transparency and allowing consumers to monitor and control their electricity usage efficiently via advanced metering infrastructure. The intelligent sensors have enabled system operators to continually monitor the power system in real time and take appropriate measures to prevent system breakdowns. The power grid's cyber-physical security has gained more attention in the research community in the recent past. Smart devices and communication infrastructure are vulnerable to cyber threats, and if their security is compromised, then the consequence can be devastating. This chapter presents the basics of data manipulation attacks on the smart grid targeting power system state estimation (PSSE), a core function of the energy management system (EMS) on which power system operations and security analysis heavily depend; and artificial intelligence (AI)-aided methods to detect the data manipulation in real time.

5.1 Introduction

5.1.1 Power grid as a cyber-physical system

Electric power must be delivered to customers efficiently and reliably. More sophisticated communication technologies are integrated to enable quick decision-making and make conventional power systems scalable, smart, and robust in the quest for reliability. Generation, transmission, and distribution are the three core domains of the conventional power system. The generator units (power plants) are usually

[1]Hitachi ABB Power Grids, Bangalore, India
[2]Department of Electrical Engineering, Indian Institute of Technology, Delhi, India
[3]Department of Computer Science and Electrical Engineering, University of Maryland, Baltimore County, MD, USA

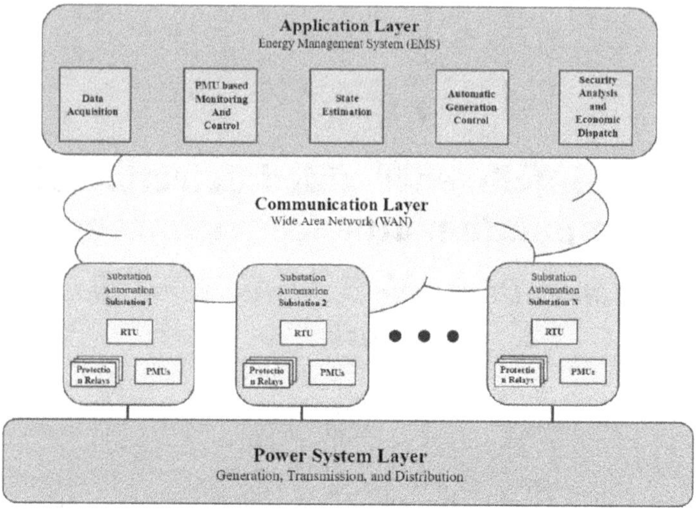

Figure 5.1 Conceptual three-layer architecture of smart grid

situated in a remote location, away from the end users. To efficiently transfer electrical power, the voltage level at the generation site is stepped up using power transformers. The transmission system transports high-voltage electrical energy to the distribution system, which then distributes it to customers (industries, commercial buildings, hospitals, and households).

Smart grids can be conceptualized as a three-layered architecture, as shown in Figure 5.1. The bottom layer is the *power system layer*, which is the interconnection of various power system components (generators, transmission lines, transformers, power electronics converters/devices, etc.), all working in synchronization to deliver electrical power to the end users. The application layer is the top layer, which provides a platform for running advanced algorithms to ensure reliable and efficient operation and control of the power electronic-dominated grid. The communication layer facilitates the flow of measurements and control signals between the physical system and the application layer. The substations act as a bridge between the physical power system layer and the communication layer by transmitting the measured voltages and currents phasors to the control center to facilitate wide-area monitoring and control.

Higher dependability, efficiency, improved services, and ease of use for consumers and utilities in every area have emerged from the merger of information technology (IT) and operational Technology (OT), allowing energy systems to enter the cyber-physical systems (CPS) arena. Because of the integration of cyber and physical technologies, the electrical energy sector has recently undergone substantial modernization [1]. In the quest of expanding and advancing the cyber-physical infrastructure, power and energy systems are becoming more exposed and vulnerable to cyber-induced power failures and outages. Cybersecurity has become the primary concern for power engineers, and cyber threats are expected to grow with the increase in the attack surface.

5.1.2 Cybersecurity concerns and related work

Securing the cyber layer entails securing all communication routes between physical power system components and substation remote terminal units (RTUs), as well as RTUs and control centers. Because of the interwoven cyber-physical infrastructure, any flaw in the power system's cybersecurity architecture can jeopardize the power system's physical security [2–4]. Smart grids inherently have many vulnerabilities that can be exploited by the adversaries to carry out system-wide blackouts [5, 6]. Recent attacks on the Ukrainian power grid are the first publicly acknowledged incidents to result in power outages, which affected 225,000 customers in 2015. The control centers are generally highly secured; however, a highly motivated adversary can breach the security and pose a more significant threat to the operation and control of the interconnected power system.

The various attack vectors in a smart grid environment are shown in Figure 5.2. An adversary can get into the OT environment even if the OT network is not connected to the IT network. The adversary can also target third-party vendors that supply software to substations and control centers; this is referred to as a 'supply chain attack'. The SolarWinds Orion supply chain compromise is a recent example of a highly sophisticated cyberattack. The requirement of periodic system/security updates of the OT network if installed remotely by the manufacturer, or if the

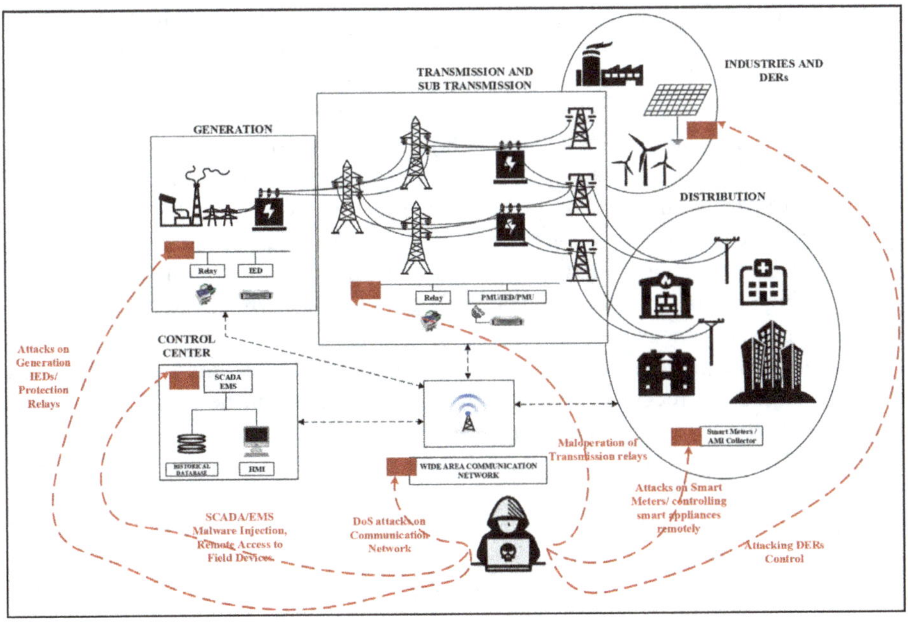

Figure 5.2 Cyber threats in a smart grid environment

control center/substation applications are accessed remotely by the system operator, makes the OT network vulnerable to cyberattacks. The adversary can launch a series of attacks on the power network, as shown in Figure 5.2, thereby affecting the power grid infrastructure.

The research in cybersecurity for power systems is mainly focused on analyzing the possible impacts of cyber intrusion on the operation and control of power systems. Considering the core components of the information security triad, confidentiality, integrity, and availability, the cyberattack vector can be generalized to manipulate any information security component or its combination. The real-time monitoring of the events in the power system is very crucial for its secure operation. The core EMS applications such as automatic generation control (AGC), contingency analysis, economic dispatch, and voltage stability assessment are dependent on accurate estimation of the operating snapshot of the power system. EMS uses state estimation (SE) to estimate complex voltage ($V\angle\theta$) [7]. Since the SE uses field sensor data collected via SCADA (supervisory control and data acquisition), the SE application can be compromised by random or precise manipulation of the measurements sent by the substation RTUs or once received at the control center. These data/measurement manipulations are also referred to as false data injection attacks (FDIAs). FDIAs can be launched by injecting small errors in the measurements within the tolerance band of the bad data detection, used by SE to filter out sensor noise or telemetry errors [8]. Vulnerability assessment of the SE against the FDIAs provides an insight into the measurements, which can cause significant errors in the state estimates ($V\angle\theta$) and needs protection [9–13].

FDIAs can be formulated for precisely targeting the said EMS applications by measurement data manipulations. Load redistribution attacks, a class of FDIAs, can be launched by manipulating the power demand at different locations seen by the control center. The modified power demand measurements can cause incorrect generation dispatch, resulting in an uneconomic and insecure power system operation, load shedding, and in the worst case, system-wide disruptions or power system blackouts [14]. Instead of modifying analog measurements, the adversary can also launch FDIAs by modifying the power network's status measurements to deceive system operators with falsified network topology [15]. The coordinated analog and status measurement manipulation can lead to undetectable transmission line outages, making system operators unaware of contingencies, which can further lead to cascading power outages [16–18].

To secure the power grid from possible cyber threats, researchers have proposed various methods in recent years. The possible solutions can be broadly classified into two categories: *protection of the high-risk assets* and *real-time detection of the cyberattacks*. In the former approach, a detailed risk assessment is necessary to protect the high-risk assets. For assessing the risks, the impact of each vulnerable sensor or target's failure needs to be quantified. The high risk assets are thus ranked, and a protection strategy is proposed to address the cybersecurity concerns. The impact analysis depends not only on the vulnerable sensor but also on the applications that the adversary can target by corrupting a set of vulnerable measurement sensors and control set-points. After assessing the risk, the cybersecurity practices, such as watermarking [19], deployment of secured phasor measurement units (PMUs) [20], encryption of communication channels, and so on, are reinforced to manage the cyber-physical vulnerabilities. Machine learning (ML) and the statistics of the power system measurements can detect certain FDIAs if the injected bias in the measurements is significant [21–25]. Anomalies can be separated out from the measurements using the Bayesian framework and sparse optimization. Advanced ML models can be used for short-term forecasting of the power system states and measurements to improve detection efficiency.

In the following sections, a case study to quantify the impact on the operation and control of power systems will be presented in detail with the data manipulation attacks on SE. Subsequently, a thorough discussion will follow on protection-based and detection-based techniques.

5.2 Data manipulation attacks on PSSE

5.2.1 Brief overview of state estimation

Metered data is received by the SCADA system, which is used to monitor the operation of the power system. State estimation (SE) gives an accurate state of the power system in real time based on measurements collected from the SCADA system. The power system's stability and security studies are heavily reliant on these estimated states. The SE generally uses redundant measurements to avoid inaccuracies related to sensor noise and telemetry errors in order to assure the correctness of the estimated states [7]. Estimation of the voltages on each bus (node) of the power grid is the key goal of the SE. Measurements needed for the accurate estimation of $V\angle\theta$ (complex voltage) at each bus are real and reactive power injections; actual and reactive power flows in the transmission lines; and voltage measurements at the generator buses.

For N bus power system, the total number of the states to be estimated is $2N-1$; i.e., $N-1$ angles (one angle is considered as reference angle) and N voltage magnitudes. The state vector is given as

$$\boldsymbol{x} = [\theta_2, \theta_3, \ldots, \theta_N, V_1, V_2, \ldots, V_N] \tag{5.1}$$

After the estimation of power system states, a security and contingency analysis for the power system is performed. The EMS is responsible for unit commitment, economic load dispatch, AGC, and many other critical functions of power grid.

The conventional weighted least square method is the predominant approach for estimating the power system states [26]. The measurement z_i can be calculated using the measurement function $h_i(x)$ and an error e_i as given in the following equation:

$$z_i = h_i(x) + e_i \tag{5.2}$$

where

x: state vector to be estimated (size $n = 2N-1$);
z_i: i^{th} measurement;
h_i: i^{th} measurement function;
e_i: error in the ith measurement.

The measurement function $h(x)$ for AC SE consists of both real/reactive power injection and power flow measurements. The measurement function for real and reactive power injection measurements is defined as

$$P_i = V_i \sum_{j=1}^{N} V_j(G_{ij}\cos\theta_{ij} + B_{ij}\sin\theta_{ij}) \quad i = 1, 2, \ldots, N \tag{5.3}$$

$$Q_i = V_i \sum_{j=1}^{N} V_j(G_{ij}\sin\theta_{ij} - B_{ij}\cos\theta_{ij}) \quad i = 1, 2, \ldots, N \tag{5.4}$$

Here G_{ij}, B_{ij} are the real and imaginary components of $(i,j)^{th}$ element of bus admittance matrix, respectively. P_i and Q_i are the real and reactive power injections at bus i. Similarly, for the line connected between bus i and bus j, the real and reactive power flow measurements are defined as

$$P_{ij} = V_i^2(g_{si} + g_{ij}) - V_iV_j(g_{ij}\cos\theta_{ij} + b_{ij}\sin\theta_{ij}) \tag{5.5}$$

$$Q_{ij} = -V_i^2(b_{si} + b_{ij}) - V_iV_j(g_{ij}\sin\theta_{ij} - b_{ij}\cos\theta_{ij}) \tag{5.6}$$

Here P_{ij}, Q_{ij} are the real and reactive power flow in the line $i-j$. g_{si}, b_{si} are the shunt conductance and shunt susceptance at bus i. g_{ij}, b_{ij} are the conductance and susceptance of line $i-j$.

Measurement errors are generally assumed to follow normal distribution with zero mean; therefore, a co-variance matrix R can be formulated as

$$R_{m,m} = \begin{bmatrix} \sigma_1^2 & 0 & \cdots & 0 \\ 0 & \sigma_2^2 & \cdots & 0 \\ \vdots & \vdots & \ddots & \vdots \\ 0 & 0 & \cdots & \sigma_m^2 \end{bmatrix} \tag{5.7}$$

where,

m: number of measurements;

σ: standard deviation of the error for each measurement.

For estimating the power system states, the weighted error residue $J(x)$ is minimized, given as

$$J(x) = \sum_{i=1}^{m} W_i * r_i^2 \tag{5.8}$$

where $W_i = R^{-1}$ known as weighting matrix and r_i is the measurement residual calculated as $r_i = z_i - h_i(x)$.

Bad data detection, a core function of the SE, detects telemetry/sensor errors and labels the measurements as *bad* if the normalized error residue is higher than a predefined threshold. The adversary can inject the calculated errors in the measurements to bypass the bad data detection stage, and this will be explained in the following sections.

5.2.2 Attack formulation

In order to deceive and force the system operator to take inappropriate decisions in real time, the attacker must inject the errors in the measurements without changing the measurement error residual or largest normalized residual (LNR).

The measurement vector after malicious data injection can be represented by $z_{bad} = z + a$, where a is the error magnitude vector or 'attack vector' injected into the actual measurement z. The general FDIA can be launched by finding a suitable attack vector as long as the normalized residue is less than the threshold for bad data detection, as shown below:

$$J(\hat{x}_{bad}) = \sum_{i=1}^{m}(z_i - h(\hat{x}) + a_i)^2/\sigma_i^2 \leq \tau \tag{5.9}$$

where a_i is the injected error in the measurement z_i. This problem can be solved for the set of meters accessible by the adversary. The feasibility of launching the attack with the specified set of accessible meters are not guaranteed; however, possible combination of meters can be found out to form a suitable attack vector with small errors, which are tolerated by SE algorithm.

Stealthy FDIAs

FDI attacks bypasses the bad data detection of the SE if the LNR considering malicious measurements is same as that for actual measurements [8, 27]. In other words, if LNR_{bad} is denoted as LNR considering the corrupted measurements and $\hat{x}_{bad}$ is the incorrect estimated system state, then to bypass bad data detection, $a = Hc$, as shown below:

$$\begin{aligned} LNR_{bad} &= \|z_{bad} - H\hat{x}_{bad}\| \\ &= \|z + a - H(\hat{x} + (H^T W H)^{-1} H^T W a)\| \\ &= \|z - H\hat{x} + a - H(H^T W H)^{-1} H^T W a)\| \\ &= \|z - H\hat{x} + (a - Hc)\| \end{aligned} \tag{5.10}$$

Here, H is the measurement Jacobian, a is the attack vector (error injected in the measurement z_i), and c is the deviation in the state vector due to attack vector a.

It is also worth noting that even if $a \neq Hc$, the adversary can still inject small errors within the tolerance band, i.e., keeping LNR as low as possible. Based on the $a = Hc$ criteria, attacks can also be launched with limited information [28, 29].

5.2.3 Impact analysis

The falsified power system states, $\hat{x}_{bad}$, can allow an adversary to modify the power flows without getting noticed by the system operator. These modifications can cause discrepancies in electricity markets and an attack can leverage incorrect real-time locational marginal prices (LMPs) for financial misconducts. Furthermore, certain part of the power grid can be overloaded, and if masked by the FDIAs can lead to power system outages. The power system is generally operating in compliance with the $N-1$ reliability criteria, i.e., a loss of one component, generator, or transmission line does not cause cascading outages in the power grid. However, if an attacker injects an intelligently crafted attack vector forcing system operator to re-dispatch generators incorrectly, power system may no longer be $N-1$ compliant, thus resulting in system-wide cascading failures.

To launch such a stealthy attack, adversary requires full system information, including network topology, generation/demand profiles, and measurements. To have all this information, adversary must get a remote access to control center, and a highly motivated actor will have all the resources to exploit the vulnerabilities of the control center to study the system behavior before launching an attack. Although the probability of finding and exploiting the vulnerability in a highly secured control center infrastructure is very low, the huge impact of such an attack on power systems operation and control can make it highly significant.

The attack to make power grid $N-1$ in-compliant is launched by formulating an attack vector by solving a bi-level optimization problem, as shown in Figure 5.3. The inner level finds the economic and $N-1$ reliable generator dispatches for the load measurements received from the outer layer (SCADA measurements). The outer layer modifies the load measurements such that the inner level dispatch is not $N-1$ compliant for actual load measurements.

In the outer level (L1), the minimum number of measurements, which requires modification to launch the attack, is calculated by solving the optimization problem (5.11) with ΔP_D as the decision variable:

$$\textbf{(L1)} \quad \min_{\Delta P_D}\{\sum_{\forall d \in N_D} L_d + 2\sum_{\forall l \in N_L} FL_l\} \tag{5.11}$$

where, N_D is the set of all the buses with load, N_L is the set of all the branches, L_d and FL_l are the binary variables where '1' indicates whether the change in load (bus)

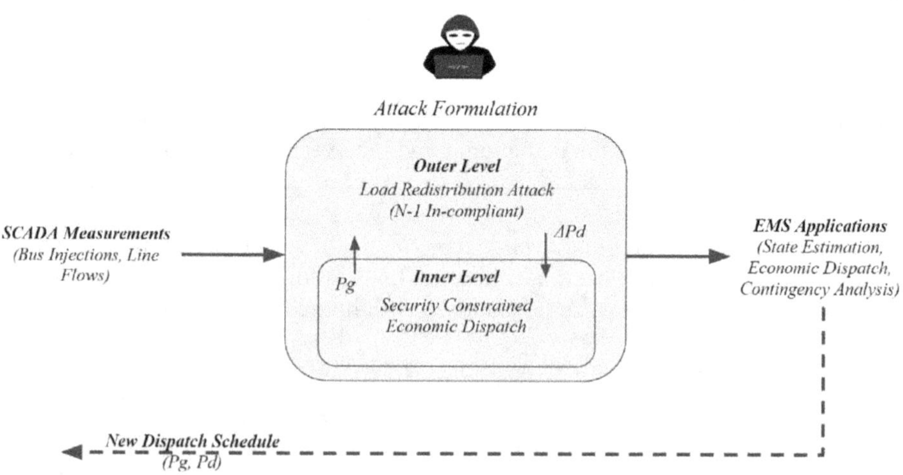

Figure 5.3 Attack formulation as a bi-level optimization problem

and change in flow (branch), respectively, is not equal to zero. The optimization problem in (5.11) is subjected to the constraints (5.12)–(5.21).

$$L_d = \begin{cases} 1 & \Delta P_{Dd} \neq 0, \\ 0 & \Delta P_{Dd} = 0 \end{cases} \quad \forall d \in N_D \tag{5.12}$$

$$FL_l = \begin{cases} 1 & \Delta F_l \neq 0, \\ 0 & \Delta F_l = 0 \end{cases} \quad \forall l \in N_L \tag{5.13}$$

$$-\epsilon P_{Dd} \leq \Delta P_{Dd} \leq \epsilon P_{Dd} \quad \forall d \in N_D \tag{5.14}$$

$$\sum_{\forall d \in N_D} \Delta P_{Dd} = 0 \tag{5.15}$$

where, P_D is the real power demand vector, ϵ is the fraction of allowable load change as decided by the attacker, ΔP_{Dd} is the change in load at d^{th} bus, and ΔF_l is change in active power flow on l^{th} caused by the attack.

$$\Delta F = S \cdot D \cdot \Delta P_D \tag{5.16}$$

$$F^N = S \cdot (G \cdot P'_g - D \cdot P_D) \tag{5.17}$$

$$\underline{F_l} \leq F_l^N \leq \overline{F_l} \quad \forall l \in N_L \tag{5.18}$$

$$F^{N-1(cl)} = S^{cl} \cdot (G \cdot P'_g - \cdot D \cdot P_D) \quad \forall cl \in N_L \tag{5.19}$$

$$B_{line} = \sum_{\forall l, cl \in N_L} \begin{cases} B^l_{cl} = 1 & \overline{F_l} - |F_l^{N-1(cl)}| < 0 \\ B^l_{cl} = 0 & \overline{F_l} - |F_l^{N-1(cl)}| \geq 0 \end{cases} \tag{5.20}$$

$$B_{line} > 0 \tag{5.21}$$

The number of bus injection measurements and line flow measurements to be modified based on change in the load measurement vector ΔP_D is formulated in (5.12) and (5.13), respectively. The constraint (5.14) ensures the change in the load confined within the limits $\pm\epsilon P_{Dd}$. Constraint (5.15) guarantees that the net change in system load is zero, i.e., load is redistributed, because a non-zero value implies a rise or reduction in load, which should activate frequency control (for genuine load change), revealing the attack to the system operator.

The changes in the line flow measurements ΔF for the change in ΔP_D are calculated from (5.16). For system without line contingency, the line flows with perturbed load measurements are obtained using (5.17). The constraint (5.18) checks the line flow limits for altered load measurements without taking into account the line contingencies. Shift factor matrix, bus-generator incidence matrix, and bus-load incidence matrix are denoted by S, G, and D, respectively. The constraint (5.18) ensures that even after the attack, the line flows are still within the limits for the system without contingency. For each line contingency cl, the line flows for perturbed load measurements are calculated from (5.19). To guarantee that the new dispatch does not violate $N-1$, in each line outage, (5.20) counts the number of lines where the power flow is above the limit. Constraint (5.21) requires that this count be larger than zero before the attack can be launched.

The inner level (L2) optimization problem (5.22) calculates the new generator dispatch schedules taking the deviation in the load demand vector, ΔP_D, into account. The problem is subjected to constraints (5.23)–(5.28).

$$\textbf{(L2)} \quad \{P'_g\} = arg\{\min_{P_g} \sum_{\forall i \in N_G} C_{gi}(P_{gi})\} \tag{5.22}$$

subject to:

$$\underline{P_{gi}} \leq P_{gi} \leq \overline{P_{gi}} \tag{5.23}$$

$$F = S \cdot [G \cdot P_g - D \cdot (P_D + \Delta P_D)] \tag{5.24}$$

$$F^{cl} = S^{cl} \cdot [G \cdot P_g - D \cdot (P_D + \Delta P_D)] \quad \forall cl \in N_L \tag{5.25}$$

$$\underline{F_l} \leq F_l \leq \overline{F_l} \quad \forall l \in N_L \tag{5.26}$$

$$\underline{F_l} \leq F_l^{cl} \leq \overline{F_l} \quad \forall l, cl \in N_L \tag{5.27}$$

$$\sum_{\forall i \in N_G} P_{gi} - \sum_{\forall d \in N_D} (P_{Dd} + \Delta P_{Dd}) = 0 \tag{5.28}$$

The security constrained optimal power flow is formulated in the inner loop considering the line contingencies only. The constraint (5.23) limits generation between $\underline{P_{gi}}$ and $\overline{P_{gi}}$. The power flows without considering line flows are obtained using (5.24). Similarly, for contingency *cl*, the line flows are obtained using (5.25). The constraints (5.26) and (5.27) limit the line flows as per the line limits, while (5.28) maintains generation and load balance.

After solving (5.11), we get an attack vector with the required change in the load measurements ΔP_D and line flows ΔF, along with minimum number of measurements required to successfully launch this attack (as the attack vector here is function of measurement function, since ΔF is calculated by using S, D, and ΔP_D, the attack bypasses the bad data detection and hence can be termed as a 'successful attack').

Three-bus system comprises two generators and three transmission lines, as shown in Figure 5.4. The details of the system (network parameters, generation cost, etc.) are given in [14]. The limits for the lines, 1–2, 2–3, and 1–3 are 500, 300, and 300 MW, respectively. Before the attack, the generator dispatch and line flows for the system are shown in Figure 5.4. Post contingency, i.e., when the transmission line 1–2 is out, the line flows are still under the limits. Now considering an attack scenario, where the attacker modifies the line flows and bus injection measurements, the modified measurements are shown in Figure 5.4 in red font. For revised load measurements, the new optimal generator dispatch for generators 1 and 2 is 465 and 385 MW, respectively. The system is within limitations under typical working conditions (without contingencies), as indicated in Table 5.1. However, if a coordinated attack is launched and line 1–2 is tripped, these new dispatches cause line 1–3 to be overloaded as illustrated in Figure 5.4, resulting in cascaded line tripping.

It is shown how adversaries can change the load measurements, which trigger the system operator to incorrectly re-dispatch the generators and thus make the $N-1$ power system insecure. Through modeling the attack as a two-level optimization problem, the minimum number of measurements needed by the attacker to initiate the attack is achieved. In addition, knowledge of measurements can be of

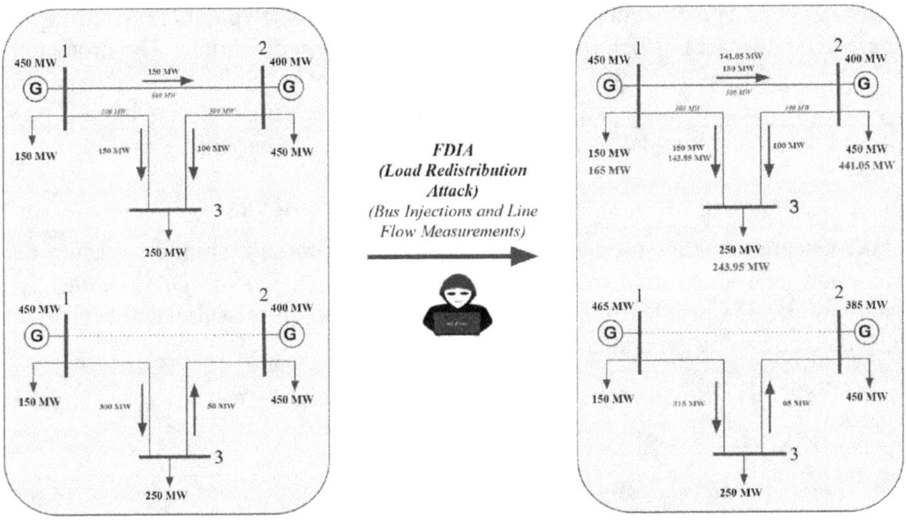

Figure 5.4 Cascading outage on a three-bus system

assistance to the system operator in upgrading cybersecurity practices to minimize the risk of an attack. The generalized defense strategies against FDIAs for protecting and preserving the credibility of power system operations will be addressed in the rest of the pages.

5.3 Defense strategies against FDIAs

By making sure that the attack vector *a* satisfies (5.2.2), an adversary may execute an attack on any state variable or multiple states. The FDIAs can be launched as seen in [28], even with limited network knowledge or restricted access to SCADA

Table 5.1 Generator dispatch, load, and line flows before and after attack for a three-bus system

		Pre-attack	**Attack**	**SCED (Post-attack)**	**N−1**
P_g	G1	450 MW	–	465 MW	–
	G2	400 MW	–	385 MW	–
P_D	P_{D1}	150 MW	165 MW	150 MW	–
	P_{D2}	450 MW	441.05 MW	450 MW	–
	P_{D3}	250 MW	243.95 MW	250 MW	–
F	F_{1-2}	150 MW	141.05 MW	160.38 MW	Out
	F_{2-3}	100 MW	100 MW	95.38 MW	65 MW
	F_{1-3}	150 MW	143.95 MW	154.62 MW	315 MW

measurements. The immediate change detected from the viewpoint of the intruder is altered system states, which further affects power system operations. The protection must be capable of defending all states against the FDIAs, so it is critical to ensure that the current defense plan defines and secures the measurements so that it is possible to detect and prevent an attack on even a single state variable.

5.3.1 Securing measurements: merits and demerits

Many researchers have proposed various techniques in the literature to secure the minimum set of measurements or sensors to minimize the risk of FDIAs against SE. The optimization problem for minimization of number of measurement sensors to be protected can be generalized as

$$\min_{\psi \in \{0,1\}} \sum_{\forall i \in \mathcal{M}} \psi_i \tag{5.29}$$

subjected to

$$[1]_{|\mathcal{N}|\times 1} - H^{*T}\psi \leq 0 \tag{5.30}$$

where, ψ is a binary variable to select the sensor/measurement for protection, i.e., if $\psi_i = 1$, then the ith sensor is protected, and if $\psi_i = 0$, then for i^{th} sensor, sensor is unprotected. The constraint (5.30) makes sure that for each state, at least one measurement is protected. The H^* is a binary matrix for the measurement Jacobian matrix H with all the non-zero entries considered as ones.

However, assuming a binary Jacobian matrix (H^*), there are several solutions to the optimization problem. In addition, an attack can also be initiated in some situations by inserting a measured error in unprotected measurements [30]. The measurement Jacobian gives the sensitivity of each state on the corresponding measurement. The higher the $H(i,j)$ value, more is the sensitivity of j state on the i measurement. Taking this into account, a defense strategy by securing a minimum set of both SCADA and PMU measurements can be formulated [30].

In order to pick and secure the most important measurement, a normalized Jacobian H_s^* matrix (5.31) can be used instead of binary matrix for measurement selection:

$$H_s^*(i,j) = |H(i,j)|/\sum_i |H(i,j)| \quad \forall i \in \mathcal{M}, j \in \mathcal{N} \tag{5.31}$$

H_s^* is defined in (5.32) for the PJM five-bus system (MATPOWER [31] 'case5.m'). The jth column entries of H_s^* reflect the effect of each of the 'i' measurements on the 'j' state. The higher the value of the $H(i,j)$, the greater the effect of the 'i' calculation on the 'j' condition.

$$
H_s^* = \begin{array}{c} \\ P_1 \\ P_2 \\ P_3 \\ P_4 \\ P_5 \\ P_{12} \\ P_{14} \\ P_{15} \\ P_{23} \\ P_{34} \\ P_{45} \end{array}
\begin{array}{c} \begin{array}{cccc} \theta_2 & \theta_3 & \theta_4 & \theta_5 \end{array} \\
\begin{pmatrix}
0.0922 & 0 & 0.1090 & 0.2733 \\
0.3322 & 0.2436 & 0 & 0 \\
0.2399 & 0.3322 & 0.1116 & 0 \\
0 & 0.0885 & 0.3322 & 0.0589 \\
0 & 0 & 0.1116 & 0.3322 \\
0.0931 & 0 & 0 & 0 \\
0 & 0 & 0.1101 & 0 \\
0 & 0 & 0 & 0.2760 \\
0.2423 & 0.2460 & 0 & 0 \\
0 & 0.0849 & 0.1127 & 0 \\
0 & 0 & 0.1127 & 0.0594
\end{pmatrix} \end{array} \tag{5.32}
$$

It is possible to obtain the minimum number of sensors to be protected to cover all states by solving the optimization problem given in (5.33).

$$
\min_{\psi \in \{0,1\}, l \geq 0} \sum_{\forall i \in \mathcal{M}} l_i \tag{5.33}
$$

subject to,

$$
\sum_{\forall i \in \mathcal{M}} \psi_i - M \leq 0 \tag{5.34}
$$

$$
[1]_{|\mathcal{N}| \times 1} - H_s^{*T}(\psi \cdot l) \leq 0 \tag{5.35}
$$

The (5.33)–(5.35) problem minimizes the number of sensors that are to be secured. For the ith measurements, a positive variable 'l_i' is used as a coupling variable to select the measurements. The higher the value of $H_s^*(i,j)$, the lower the value of l_i, and the most important sensors are therefore chosen by the problem. The (5.34) limitation restricts the problem solution to a minimum number of M measurements, while (5.35) guarantees security for each state variable.

The set of measurements once identified must be secured both at the substation and at the control center side to mitigate the risk associated with malicious data injection. Although the aforementioned algorithms generally identify one measurement for protection at important substations, but from a security standpoint, there is no way to secure just one measurement, as all the measurements are sent as one DNP3 payload to the control center. Furthermore, it is unrealistic to upgrade the communication infrastructure identified substations to send the measurements and control signals via secured DNP3 (enabled with encryption and other security measures), as it will have its complications at the control center where these measurements will be received and decryption is needed. Therefore, although security upgradation will happen in the future, it becomes simultaneously necessary to devise new real-time detection techniques against FDIAs.

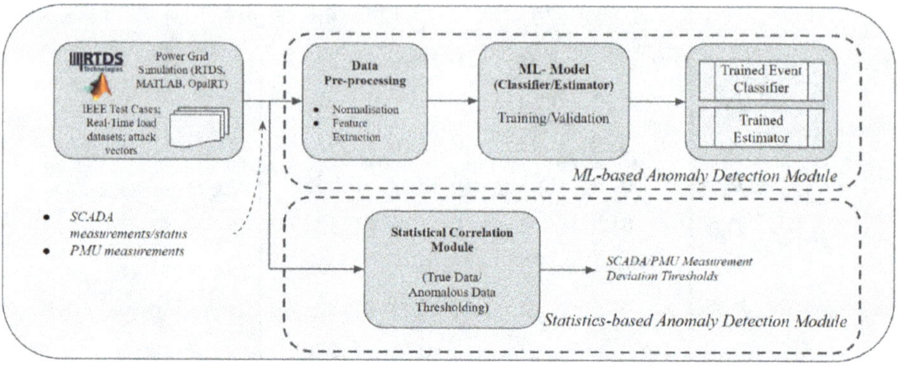

Figure 5.5 ML- and statistics-based anomaly detection

5.3.2 Hybrid statistical-AI-based detection

The literature shows that by tracking the dynamics of measurement variations, FDIAs can be effectively detected [32]. Rapid advancement in the methodology of detection has improved the accuracy of detection significantly. The change in the targeted states is considerably small for the attacks where the adversary's motive is to gain momentary economic benefits ($\approx \pm 1\%$ of true value) [14]. Therefore, to detect such FDIAs, a statistical technique based on measurement transformation is used [33]. For mitigation, ML techniques are used to maintain the integrity of the estimation of the power system states for short-term load and state forecasting [34].

Different statistical distance/divergence metrics can track the dynamics of the measurement variance [33]. As shown in Figure 5.5, to reliably distinguish regular and anomalous measurement samples, the statistical correlation module takes real-time measurements as well as historical measurements from SCADA to generate thresholds. In the field of image processing, log transformation is used for intensity transformation by expanding the value of dark pixels in an image thus compressing higher intensity values. Log transformation maps small ranges of low input measurement values into a broader range of output values and can be implemented for smart grid-related applications. The log transformations can be defined as

$$s = c \cdot log(1 + r) \tag{5.36}$$

Here, r and s are measurement values before and after transformation, respectively. The Kullback-Leibler distance (KLD) is used to measure the deviation of the measurement variation of the real-time sample from the historical measurement variation [33].

The KLD value is compared to a predefined threshold value that is measured without a false data injection attack being considered. If the real-time KLD is greater than the threshold value, for that specific time step, the adversary has launched an FDI attack. The threshold value is a significant parameter and is determined by comparing the KLD values for actual measurement samples to those for historical measurement samples. If a high threshold value is selected, some of the attacks may

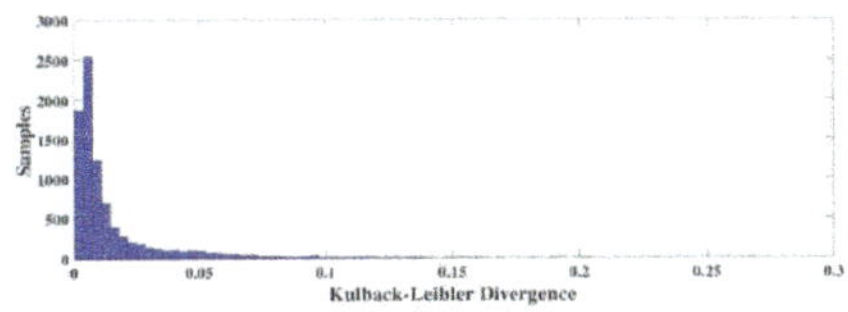

(a) Histogram of KLDs: July (No-attack).

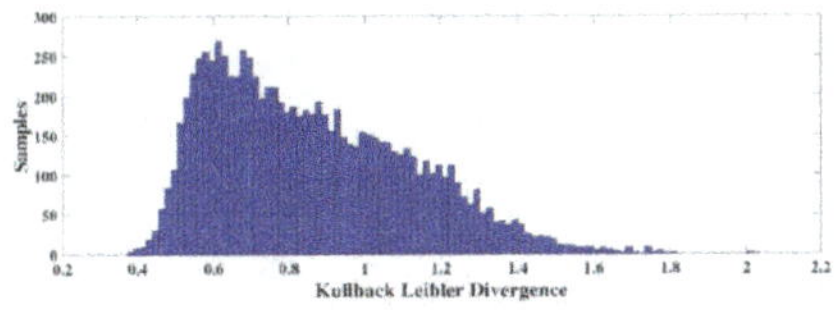

(b) Histogram of KLDs: August (Attack on θ_2 with $c = -10\%$).

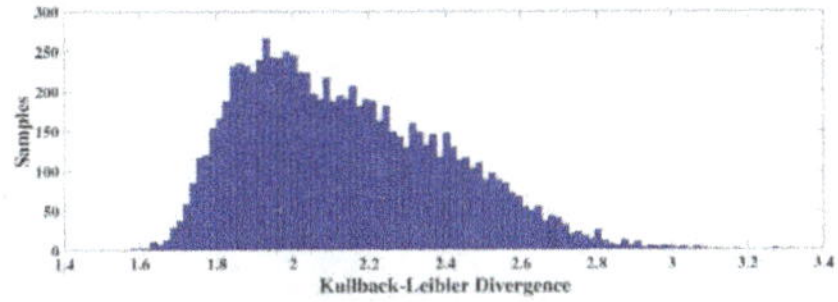

(c) Histogram of KLDs: August (Attack on V_2 with $c = -10\%$).

Figure 5.6 Histogram of KLDs considering no-attack and attack scenarios

not be detected, and similarly, if a low threshold value is chosen, actual measurements may be categorized as compromised measurements and will lead to high false positives.

After mapping all the real-time loads for 11 load zones in New York Independent System Operator (NYISO) to IEEE 14 bus system [14], the KLD for each measurement sample in July is calculated and plotted as a histogram shown in Figure 5.6(a) to decide the threshold. The threshold is the value of KLD in the histogram higher than 99% of the values, i.e., equal to 0.1382. The attack is simulated for each five-minute time interval for the complete month of August. The measurement variation for the month of August considering attack is calculated as $z^k_{bad} - z^{k-1}_{true}$, where z^k_{bad} is attacked measurement sample at time sample k, and z^{k-1}_{true} is the true measurement sample at the previous time sample $k - 1$.

The histogram of run-time KLD for each sample k in August for attack in θ_2 and V_2 with percentage change in targeted state variable -10% ($c = -10\%$) is shown in Figure 5.6(b) and (c), respectively. It is clearly observed that the run-time KLD (KLD_k) for all the measurement samples for the month of August, considering an attack on θ_2 and V_2 is greater than the threshold 0.1382. It is clearly observed that the method detects the FDI attack for each time sample in the month of August. However, the method fails to identify the measurements corrupted by the adversary. The AI-based ML model [34] is used to identify the anomalous measurements, as shown in Figure 5.7. The SCADA measurements for the previous sample ($k - 1$) is used to predict the measurements for the kth sample. The deviation of the SCADA measurements from the predicted measurements is compared with a pre-defined threshold to identify the anomalous measurements. These anomalous measurements are then replaced with the predicted measurements to estimate the power system state. The method accurately identifies the incorrect measurements with high detection accuracy and significantly lower false positives.

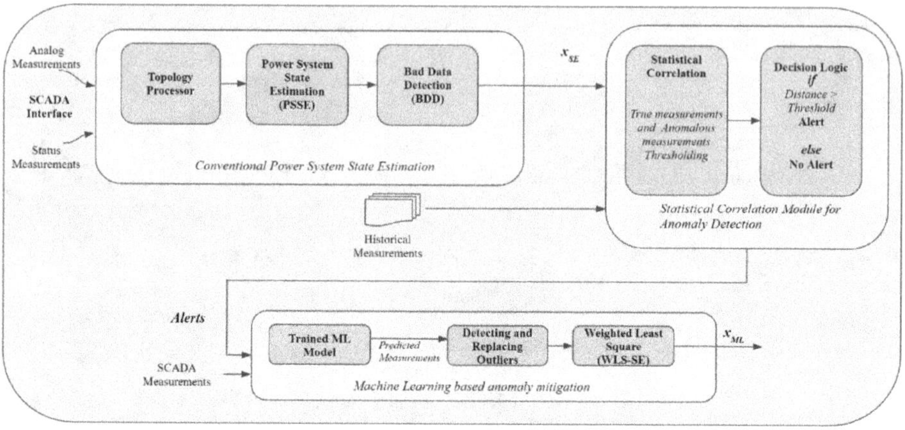

Figure 5.7 Anomaly detection and mitigation system

If the total injected error in the state variable is greater than ±1%, the hybrid ML and statistics-based approach will classify the attacks. However, this attack might not be detected if the attacker injects the errors within the limits of the predicted load patterns. In addition, it should be noted that injecting errors within the limits of predicted load patterns may not result in a substantial shift in the state of the system. One might argue, however, that attack vector cumulatively injected over time (if the measurements are continually attacked) may have a major effect, but the feasibility of such attack depends on the condition that loads, transformer taps, network topology, and generator schedules remain for the entire duration of the attack as per the adversary's forecast, but such assumption is rather idealistic.

5.4 Summary

The resilient power grid is the cornerstone for the entire nation's social and economic growth. Cyber-physical defense is a daunting challenge owing to the complex network of transmission lines and communication networks of the far-reaching power grid. In the event of data manipulation attacks, this chapter poses an immediate need to examine the threats involved with cyber-physical power systems and to devise algorithms that would be scalable to the actual power system. While in all the present literature the potential attacking scenarios have been extensively studied, there are still possibilities to examine scenarios that are not included yet. The content of the chapter is limited to the consequences on the power transmission systems of false data injection attacks. Similar concepts can be used to quantify the impacts of coordinated false data injection attacks and denial of service attacks (DoS) on the distribution grid and microgrid environment to understand, mitigate, and alleviate the threats.

References

[1] Meyers J. 'How the convergence of IT and OT enables smart grid development'. *Schneider Electric White Paper Revision*. 2013:1–7.

[2] Sridhar S., Hahn A., Govindarasu M. 'Cyber–physical system security for the electric power grid'. *Proceedings of the IEEE*. 2012;100(1):210–24.

[3] Liu C.-C., Stefanov A., Hong J., Panciatici P. 'Intruders in the grid'. *IEEE Power and Energy Magazine*. 2012;10(1):58–66.

[4] Ashok A., Hahn A., Govindarasu M. 'Cyber-physical security of wide-area monitoring, protection and control in a smart grid environment'. *Journal of Advanced Research*. 2014;5(4):481–9.

[5] Morison K., Wang L., Kundur P. 'Power system security assessment'. *IEEE Power and Energy Magazine*. 2004;2(5):30–9.

[6] Moslehi K., Kumar R. 'A reliability perspective of the smart grid'. *IEEE Transactions on Smart Grid*. 2010;1(1):57–64.

[7] Abur A., Exposito A.G. *Power System State Estimation: Theory and Implementation*. Boca Raton, FL: CRC Press; 2004.

[8] Liu Y., Ning P., Reiter M.K. 'False data injection attacks against state estimation in electric power grids'. *ACM Transactions on Information and System Security*. 2011;14(1):1–33.

[9] Hug G., Giampapa J.A. 'IEEE transactions on smart grid'. *Vulnerability Assessment of AC State Estimation With Respect to False Data Injection Cyberattacks*. 2012;3(3):1362–70.

[10] Göl M., Abur A. 'Identifying vulnerabilities of state estimators against cyberattacks'. *2013 IEEE Grenoble Conference IEEE*; 2013. pp. 1–4.

[11] Liang J., Sankar L., Kosut O. 'Vulnerability analysis and consequences of false data injection attack on power system state estimation'. *IEEE Power Energy Society General Meeting*; IEEE. 2017. pp. 1–1.

[12] Pan K., Teixeira A., Cvetkovic M., Palensky P. 'Cyber risk analysis of combined data attacks against power system state estimation'. *IEEE Transactions on Smart Grid*. 2019;10(3):3044–56.

[13] Pan K., Teixeira A., Cvetkovic M., *et al:* 'Data attacks on power system state estimation: Limited adversarial knowledge vs. limited attack resources'. *IECON 2017 - 43rd Annual Conference of the IEEE Industrial Electronics Society*; 2017. pp. 4313–18.

[14] Khanna K., Panigrahi B.K., Joshi A. 'Bi-level modelling of false data injection attacks on security constrained optimal power flow'. *IET Generation, Transmission & Distribution*. 2017;11(14):3586–93.

[15] Kim J., Tong L. 'On topology attack of a smart grid: undetectable attacks and countermeasures'. *IEEE Journal on Selected Areas in Communications*. 2013;31(7):1294–305.

[16] Li Z., Shahidehpour M., Alabdulwahab A., Abusorrah A. 'Bilevel model for analyzing coordinated cyber-physical attacks on power systems'. *IEEE Transactions on Smart Grid*. 2016;7(5):2260–72.

[17] Zhang J., Sankar L. 'Physical system consequences of unobservable state-and-topology cyber-physical attacks'. *IEEE Transactions on Smart Grid.* 2016;7(4):1–10.

[18] Liu X., Li Z., Liu X., Li Z. 'Masking transmission line outages via false data injection attacks'. *IEEE Transactions on Information Forensics and Security.* 2016;11(7):1592–602.

[19] Mo Y., Weerakkody S., Sinopoli B. 'Physical authentication of control systems: designing watermarked control inputs to detect counterfeit sensor outputs'. *IEEE Control Systems Magazine.* 2015;35(1):93–109.

[20] Kim T.T., Poor H.V. 'Strategic protection against data injection attacks on power grids'. *IEEE Transactions on Smart Grid.* 2011;2(2):326–33.

[21] Li S., Yılmaz Y., Wang X. 'Quickest detection of false data injection attack in wide-area smart grids'. *IEEE Transactions on Smart Grid.* 2015;6(6):2725–35.

[22] Liu X., Zhu P., Zhang Y., Chen K. 'A collaborative intrusion detection mechanism against false data injection attack in advanced Metering infrastructure'. *IEEE Transactions on Smart Grid.* 2015;6(5):2435–43.

[23] Liu L., Esmalifalak M., Ding Q., Emesih V.A., Han Z. 'Detecting false data injection attacks on power grid by sparse optimization'. *IEEE Transactions on Smart Grid.* 2014;5(2):612–21.

[24] Huang Y., Tang J., Cheng Y., Li H., Campbell K.A., Han Z. 'Real-Time detection of false data injection in smart grid networks: an adaptive CUSUM method and analysis'. *IEEE Systems Journal.* 2016;10(2):532–43.

[25] Cui S., Han Z., Kar S., *et al.* 'Coordinated data-injection attack and detection in the smart grid: a detailed look at enriching detection solutions'. *IEEE Signal Processing Magazine.* 2012;29(5):106–15.

[26] Schweppe F. 'Power system state estimation, parts I, II and III'. *IEEE Transactions on Power Apparatus and Systems.* 1970;89:120–35.

[27] Liang G., Zhao J., Luo F., *et al.* 'IEEE transactions on smart grid'. *A Review of False Data Injection Attacks Against Modern Power Systems.* 2016;PP(99):1–1.

[28] Liu X., Li Z. 'Local load redistribution attacks in power systems with incomplete network information'. *IEEE Transactions on Smart Grid.* 2014;5(4):1665–76.

[29] Srivastava A., Morris T., Ernster T., Vellaithurai C., Pan S., Adhikari U. 'Modeling cyber-physical vulnerability of the smart grid with incomplete information'. *IEEE Transactions on Smart Grid.* 2013;4(1):235–44.

[30] Khanna K., Panigrahi B.K., Joshi A. 'Priority-based protection against the Malicious data injection attacks on state estimation'. *IEEE Systems Journal.* 2019.

[31] Zimmerman R.D., Murillo-Sanchez C.E., Thomas R.J. 'MATPOWER: steady-state operations, planning, and analysis tools for power systems research and education'. *IEEE Transactions on Power Systems.* 2011;26(1):12–19.

[32] Chaojun G., Jirutitijaroen P., Motani M. 'Detecting false data injection attacks in AC state estimation'. *IEEE Transactions on Smart Grid.* 2015;6(5):2476–83.

[33] Singh S.K., Khanna K., Bose R., Panigrahi B.K., Joshi A. 'Joint-transformation-based detection of false data injection attacks in smart grid'. *IEEE Transactions on Industrial Informatics*. 2018;14(1):89–97.
[34] Khanna K., Panigrahi B.K., Joshi A. 'AI-based approach to identify compromised meters in data integrity attacks on smart grid'. *IET Generation, Transmission & Distribution*. 2018;12(5):1052–66.

Chapter 6
Cyber security threats in multi-agent microgrids

Deepak Pullaguram[1] and Subham Sahoo[2]

The conventional power systems were designed to have centralized power generation with more of a passive distribution system with a unidirectional power flow. Significant modifications have since taken place in the power system for the past few decades with the development of distributed energy resources (DERs), energy storage, active power management, microgrids (MGs), and smart-grid control using the information and communication technologies (ICTs). In modern power systems, the large penetration of DERs remains a challenge due to the voltage stability and power flow limitations. However, these large penetrations may also favor flexible, reliable, and efficient operation of the grid if coordinated properly with conventional generation. Normally, DERs are connected to distribution networks close to consumer premises, thus reducing transmission losses and improving reliability. Furthermore, with adequately designed control strategies for DERs in combination with storage systems, a broad scope of ancillary services may be provided along with the improvement in the continuity of supply by forming active islands with multiple local DERs, namely MGs [1, 2].

The MG concept has been developed as part of a realistic approach to handle the penetration of DERs so that the final customers can manage part of the energy to be consumed by generating and storing the energy locally. The DERs along with local loads are collectively connected through a network to form AC, DC, or AC-DC MGs. For robust and reliable operation, MG systems need a well-structured hierarchical control scheme realized through ICTs [3, 4]. These MGs can be viewed as cyber-physical systems (CPSs) when integrated with centralized or distributed communication networks for better system coordination.

Various researchers have explored the application of ICTs in the MG system. MG controls include hierarchical centralized control, where a central controller dispatches new setpoints to several units based on various measurements and distributed control in which each DER interacts with its neighbor and there is no single point of contact. Owing to the single point of failure and high communication band width requirements of centralized controller [4–6], a distributed network controller

[1]Department of Electrical Engineering, National Institute of Technology Warangal, India
[2]Department of Energy, Aalborg University, Denmark

is developed in which each DER was considered as an independent agent and shares its own voltage and power measurements with its communication neighbors [7]. A synchronized controller is developed in Reference 8 to achieve a proportional power-sharing and frequency restoration, by considering Kuramoto equivalent modeling for power–frequency droop controllers. This method might not be able to ensure voltage restoration as it is not considered while designing the controller. A multi-agent distributed cooperative control scheme to restore both frequency and voltage of the MG is developed in Reference 9. But effective reactive power-sharing is challenging using these methods. Similar work is extended in References 10–12 by replacing the conventional single integral distributed cooperative control with a finite-time distributed secondary control. The finite-time distributed control can accelerate the convergence and also ensure improved dynamic performance. However, these methods were not designed to achieve optimal active power dispatch and proportional reactive power-sharing. To achieve optimal dispatch in the MG, a tertiary controller that focuses on the minimization of individual generation cost is conventionally used. To have a dynamic economic dispatch, distributed economic load dispatch is developed in Reference 13, where optimal generation is achieved by using each DER local information and its respective neighbor's incremental cost information. A fully distributed hierarchical control scheme is developed in Reference 14, which focuses on both frequency and average voltage restoration while considering the optimum economic dispatch and simultaneous reactive power-sharing. The MG system utilizes ICTs to successfully achieve the desired performance and operating conditions.

Although integration of ICTs improves the system performance, at the same time, it makes the system highly susceptible to malicious intrusions and cyberattacks [15]. A typical multi-agent MG system with ICTs is modeled as a network graph with active DER sources (agents) represented by nodes of the graph and communication links as edges connecting nodes [9] forming a CPS. Using this network graph theory, the cooperative distributed controllers are designed to achieve the desired performance in a physical system. Although the distributed control in MGs is eminent in offering scalability and reliability, the ubiquitous nature of communication links makes them vulnerable to cyberattacks. Thus, in MG CPS system, it is very important to understand and analyze all possible attacks and thereby can use this information to achieve attack-resilient operation, data integrity, and confidentiality. This chapter provides a brief insight into the cyber-physical MG system and the designing of the distributed control schemes and provides the details of possible vulnerabilities in the cyber-physical MG system.

6.1 Cyber-physical MG system architecture

Cyber-physical MGs consists of three layers: (i) physical layer, (ii) control layer, and (iii) communication (cyber) layer, as shown in Figure 6.1. The physical layer includes the physical electrical distribution system with multiple DERs interconnected through a distribution network to meet the load demand connected to the

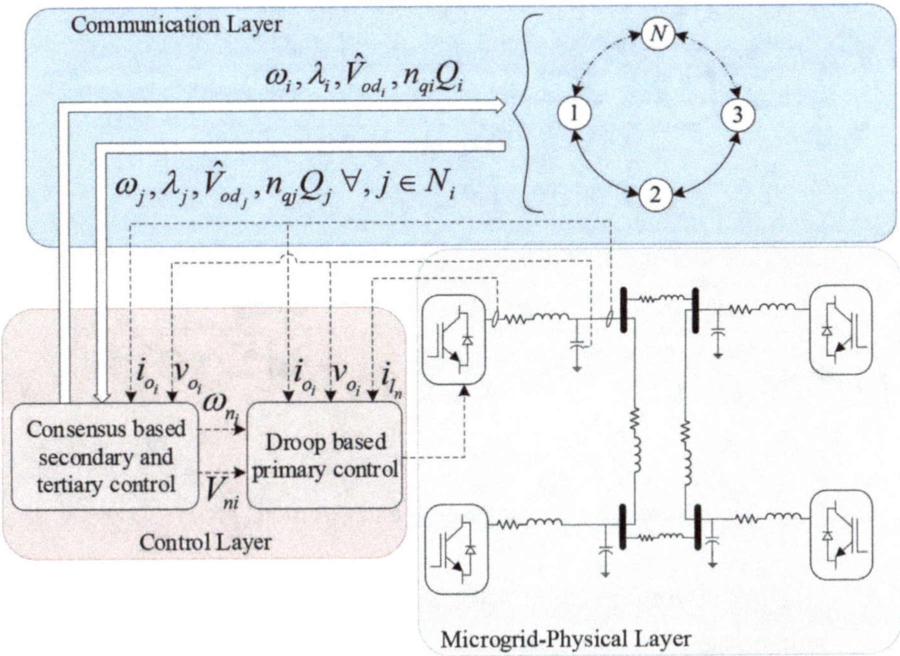

Figure 6.1 Cyber-physical schematic of an exemplary AC MG

same network. The control layer consists of the sensing/measuring units and controllers in which the controller reads the physical layer operating status from sensing units and provides the necessary control signals to the physical system to achieve the desired system performance. Lastly, the cyber layer includes the communication network (distributed/centralized) and is used for exchanging information among multiple DER controllers (or a centralized control unit) using a sparse (or a full) communication network to ensure coordinated operation of the MG.

6.1.1 Microgrid physical layer architecture

The complete MG electrical distribution system with passive (distribution network and loads) and active elements (DERs) is considered a physical layer. In an inverter dominant MG system, the majority of the DER sources are voltage source inverter (VSI)-driven sources. Each VSI is connected to the MG network through a series of inductive filter and shunt capacitive filter circuits, as shown in Figure 6.2. The R_f and L_f represent the per phase resistance and inductance of the series filter, and C_f indicates the capacitance of the filter circuit. Here, v_{od} and v_{oq} represent the *dq* transformation of the voltage across the filter capacitor C_f, and i_{od} and i_{oq} indicate the *dq* transformation of the output current injected into the MG measured after the shunt capacitor C_f filter.

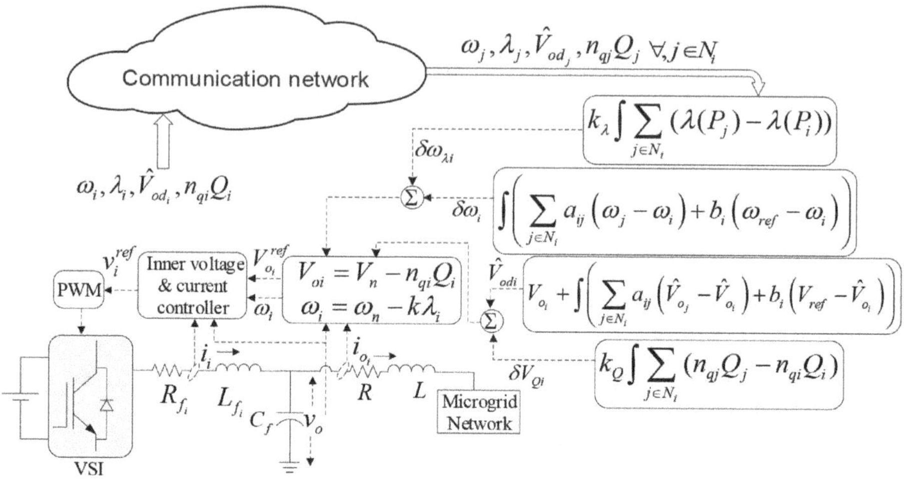

Figure 6.2 Full control of ith DER in AC MG

6.1.2 Cyber-communication layer

In a cyber-physical MG system, an undirected cyber-communication network configuration is considered. This communication network is mathematically represented by using graph theory as $G = (\mathcal{V}, \mathcal{E})$. Here, $\mathcal{V} = (1, 2...n)$ represents the set of vertices corresponding to the DER nodes of MG. $\mathcal{E} = \{(i,j) \in n \times n\}$ represents the edge set where i,j are called as adjacents and j is considered as one of the neighbors to i that represents the complete neighbor set of ith DER. An adjacency matrix $\mathcal{A}$ is defined as $\mathcal{A} = a_{ij} \in \mathbb{R}^{n\times n}$, the a_{ij} will be 1 if and only if i,j are adjacents else $a_{i,j}$ is given as 0 . Based on the of cardinality of neighbor set, a degree matrix $\mathcal{D}$ of the cyber-graph G is defined as $\mathcal{D} = \text{diag}\{d_1, d_2, ..., d_n\}$, where d_i indicates cardinality of $\mathcal{N}_i$ of ith DER. Using the adjacency matrix $\mathcal{A}$ and degree matrix $\mathcal{D}$, a Laplacian matrix $\mathcal{L}$ is defined as $\mathcal{L} = \mathcal{D} - \mathcal{A}$. The Laplacian matrix $\mathcal{L}$ is symmetric and positive semi-definite for a undirected communication graph forming a spanning tree. The eigenvalues of the $\mathcal{L}$ will be $0 = \bar{\lambda}_1 \leq \bar{\lambda}_2 \leq \,..\, \leq \bar{\lambda}_n = \bar{\lambda}_{max}$. Further to have synchronization and frequency restoration in MG system, their corresponding references values are fed to few DERs using pinning strategy through a pinning gain b_i. The b_i for ith DER will be 1 if the DER is provided with the reference value else it is assigned to 0. The pinning gain matrix is defined as $\mathcal{B}$, which is also a positive semi-definite matrix.

6.1.3 Control layer

The control layer consists of a hierarchical control strategy with a three-level hierarchy as follows:

- **Primary droop control** to achieve active and reactive power sharing using a droop strategy that provides the operating reference to device level voltage control [16].

- **Secondary control** to restore the voltage and frequency deviations caused by the primary control to zero and also used as synchronizing unit to connect MG with grid or other MGs.
- **Tertiary control** to optimal economical active power dispatch and also have proportional reactive power dispatch.

In a typical MG system, the dynamics response time of the device level voltage controller and filter dynamics of a DER is significantly faster compared to the primary droop controller. Thus, only conventional droop controller dynamics are considered for designing the secondary control. The primary droop-based power-sharing control law can be written as

$$\omega_i = \omega_n - m_{pi}P_i \tag{6.1}$$

$$V_{oi} = V_n - n_{qi}Q_i \tag{6.2}$$

where ω_i and V_{oi} are the frequency and voltage magnitude of an *i*th DER in MG, respectively. Here, ω_n and V_n are nominal frequency and voltage values, respectively. m_{pi} is active power and frequency $P-f$ droop constant, and n_{qi} is reactive power and voltage $Q-V$ droop constant for the *i*th DER, respectively. The P_i and Q_i are the filtered active and reactive power supplied by the *i*th DER, respectively. The dynamics of the filter active power and reactive power are given by

$$\dot{P}_i = -\omega_c P_i + \omega_c(v_{od}i_{od} + v_{oq}i_{oq}) \tag{6.3}$$

$$\dot{Q}_i = -\omega_c Q_i + \omega_c(v_{oq}i_{od} - v_{od}i_{oq}) \tag{6.4}$$

$$\dot{\theta}_i = \omega_i - \omega_{com} \tag{6.5}$$

Here, ω_c represents the low pass filter cutoff frequency in *rad/s*, which is the power angle *i*th DER obtained with respect to a common reference having a frequency ω_{com}. Usually, the first DER frequency is considered as the common reference frequency (i.e., $\omega_{com} = \omega_1$). The $P-f$ droop constant defined in (6.1) allows individual DERs to share power in proportion to their droop constant and might not be optimal economical dispatch.

In the conventional power system, tertiary controllers are used for optimal economic dispatch. To apply a similar dispatch strategy for the MG, an appropriate valid cost function is to be defined for all dispatchable DER sources, such as microturbines and fuel cells. Typically, the generation cost depends on the fuel consumed, maintenance, emission penalties, and incentives for generating a certain amount of power from the generator. This generation cost can be approximated by the quadratic cost function [17] given by

$$C_i(P_i) = a_i P_i^2 + b_i P_i + c_i \tag{6.6}$$

where a_i, b_i, and c_i are the quadratic cost coefficients that are empirically defined by generation companies.

While solving the economic dispatch problem, it is mandatory to consider the generation and demand power balance along with the generation limits of each DER. This can be considered by using equality and inequality constraints given by

$$\sum_{i=1}^{N} P_i = \sum_{j=1}^{m} P_{L_j} - P_{R_j} = P_D - P_{Ren}$$
$$P_{i_{\min}} \leq P_i \leq P_{i_{\max}} \tag{6.7}$$

where P_{L_j} and P_{R_j} are the active power demand and power generated by renewables at jth bus. P_D indicates the total active power demand and P_{Ren} is the total renewable generation in the MG. The minimum and maximum generation capacities of dispatchable ith DER unit are represented as $P_{i_{\min}}$ and $P_{i_{\max}}$, respectively. By using equality constraints, a Lagrangian function is defined as

$$\mathbb{L} = \sum_{i=1}^{N} C_i(P_i) + \lambda \left(\sum_{j=1}^{m} P_{L_j} - \sum_{i=1}^{N} P_i \right) \tag{6.8}$$

From Karush–Kuhn–Tucker conditions [18], the optimal solution can be obtained using

$$\frac{\partial C_i}{\partial P_i} = \lambda_i \ \forall \ i = 1, 2, ..N \tag{6.9}$$

Here, λ_i is called as a Lagrange multiplier. For a system with quadratic generation cost functions, the λ_i can be calculated using the linear function of P_i.

For varying load and generation, the lambda λ_i may be different for different generators. The optimal dispatch is said to be achieved only if the Lagrange multiplier of all DERs converges to the same value (i.e., $\lambda_1 = \lambda_2 = ... = \lambda_N$) while each generated power lies within its minimum and maximum limits.

Owing to this condition, an incremental cost function-based droop control (6.10) is developed to operate the MG at optimal power dispatch [19].

$$\omega_i = \omega_{ni} - k\lambda_i = \omega_{ni} - k(2a_iP_i + b_i) \tag{6.10}$$

Here, λ_i is the incremental cost function and k is the droop constant.

Using the economic droop control, a fully distributed hierarchical control is developed with following objectives,

$$\begin{aligned}
&\lim_{t\to\infty} \left|\omega_i(t) - \omega_{ref}\right| = 0, \ \forall \ i \in V; \\
&\lim_{t\to\infty} \left|\lambda_i(t)) - \lambda_j(t))\right| = 0, \ \forall \ j \in N_i, \ i \in V; \\
&\lim_{t\to\infty} \left|\hat{V}_i(t) - V_{ref}\right| = 0, \ \forall \ i \in V; \\
&\lim_{t\to\infty} \left|Q_i(t) - Q_j(t)\right| = 0, \ \forall \ j \in N_i, \ i \in V;
\end{aligned} \tag{6.11}$$

where $P_i(t)$ and $Q_i(t)$ are active and reactive power, and $\hat{V}_i$ is the network average voltage estimation at ith DER.

6.2 Distributed control for AC MG

While designing the secondary control in an AC MG, it is to be noted that implementing only frequency- and voltage-based secondary control without the power control or economic dispatch control may lead to inaccurate and non-optimal power sharing. Thus, a combined secondary and tertiary control is defined to achieve the objectives in (6.11),

$$\dot{\omega}_{ni} = \delta\dot{\omega}_i + k\delta\dot{\omega}_{\lambda_i} = u_{\omega_i} + ku_{\lambda_i} \tag{6.12}$$

$$\dot{V}_{ni} = \delta\dot{\hat{V}}_i + n_{qi}\delta\dot{V}_{Q_i} = u_{\hat{V}_i} + u_{Qi} \tag{6.13}$$

where u_{ω_i} and $u_{\hat{V}_i}$ are the control inputs from the secondary to primary control that used to achieve frequency and voltage restoration. u_{λ_i} and u_{Q_i} are the control inputs from tertiary control to primary achieving economic active power dispatch and reactive power sharing, respectively.

6.2.1 Optimal dispatch control along with frequency restoration

The economic active power control input u_{λ_i} is designed using the incremental cost function of its own DER and its neighboring DERs, and it is given by

$$\begin{aligned} u_{\lambda_i} = \delta\dot{\omega}_{\lambda_i} &= \textstyle\sum_{j\in N_i} a_{ij}\left(\lambda_j - \lambda_i\right) \\ \delta\dot{\omega}_{\lambda} &= -\mathcal{L}\lambda \end{aligned} \tag{6.14}$$

Similarly, the frequency restoration control input u_{ω_i} for *i*th is obtained using its own frequency and its neighboring DER frequencies. Moreover, it also depends on the pinning information if the *i*th is provided with ω_{ref} . The frequency restoration control input is given by

$$\begin{aligned} u_{\omega_i} &= \delta\omega_i = \sum_{j\in N_i}(\omega_j - \omega_i) + (\omega_{ref} - \omega_i) \\ \delta\dot{\omega} &= \mathscr{L}\omega + \mathscr{B}(\omega_{ref} - \omega) \end{aligned} \tag{6.15}$$

The convergence of the above cooperative controller can be proved using Lyapunov stability criteria [20]. For this, a suitable Lyapunov function is needed to be defined and should ensure that the defined Lyapunov function decreases along the trajectories resulting that the system converges to its equilibrium. Additionally, the control can also be designed at a faster rate of convergence by adjusting the controller input gains thereby forcing the Lyapunov function to decrease to zero faster.

Consider two error variables defined as

$$\mathbf{e}_{\omega} = \left(\omega_1 - \omega_{ref}, \omega_2 - \omega_{ref}, ..., \omega_N - \omega_{ref}\right) \tag{6.16}$$

$$\mathbf{e}_{\lambda} = \left(\lambda_1 - \lambda_{ss}, \lambda_2 - \lambda_{ss}, ..., \lambda_N - \lambda_{ss}\right) \tag{6.17}$$

where

$$\lambda_{ss} = \lim_{t\to\infty} \tfrac{1}{N}\sum_{i=1}^{N}\lambda_i(t)$$

Using the error variables, a Lyapunov energy function $\mathbb{V}$ is defined as

$$\mathbb{V} = \tfrac{1}{2}\sum_{i=1}^{n} e_{\omega_i}{}^2 + \tfrac{1}{2}\sum_{i=1}^{n} e_{\lambda_i}{}^2 \geq 0, \tag{6.18}$$

differentiating $\mathbb{V}$ in (6.18) with respect to time one can obtain,

$$\dot{\mathbb{V}} = \sum_{i=1}^{n} e_{\omega_i}\dot{e}_{\omega_i} + \sum_{i=1}^{n} e_{\lambda_i}\dot{e}_{\lambda_i} \tag{6.19}$$

Using error vector 'e_λ' and 'e_ω' from (6.16) and (6.17) in (6.14) and (6.15), respectively, the system dynamics can be reformulated as

$$\begin{aligned}\dot{e}_{\lambda_i} &= \sum_{j\in N_i} a_{ij}\left(e_{\lambda_j} - e_{\lambda_i}\right)\\ \dot{e}_{\omega_i} &= \sum_{j\in N_i} a_{ij}\left(e_{\omega_j} - e_{\omega_i}\right) - b_i e_{\omega_i}\end{aligned} \tag{6.20}$$

Substituting (6.20) in (6.19)

$$\begin{aligned}\dot{\mathbb{V}} &= \sum_{i=1}^{n} e_{\omega_i} \sum_{j\in N_i} a_{ij}(e_{\omega_j} - e_{\omega_i}) - \sum_{i=1}^{n} b_i e_{\omega_i}^2\\ &\quad + \sum_{i=1}^{n} e_{\lambda_i} \sum_{j\in N_i} a_{ij}(e_{\lambda_j} - e_{\lambda_i})\\ &\le \sum_{i=1}^{n} |e_{\omega_i}| \sum_{j\in N_i} a_{ij}(|e_{\omega_j}| - |e_{\omega_i}|) - \sum_{i=1}^{n} b_i |e_{\omega_i}|^2\\ &\quad + \sum_{i=1}^{n} |e_{\lambda_i}| \sum_{j\in N_i} a_{ij}(|e_{\lambda_j}| - |e_{\lambda_i}|)\\ &\le -|\mathbf{e}_\omega|^T \mathcal{L}|\mathbf{e}_\omega| - |\mathbf{e}_\omega|^T \mathcal{B}|\mathbf{e}_\omega| - |\mathbf{e}_\lambda|^T \mathcal{L}|\mathbf{e}_\lambda|\\ &\le -|\mathbf{e}_\omega|^T (\mathcal{L}+\mathcal{B})|\mathbf{e}_\omega| - |\mathbf{e}_\lambda|^T \mathcal{L}|\mathbf{e}_\lambda|\end{aligned} \tag{6.21}$$

The Laplacian matrix of an undirected communication network that forms a basic spanning tree, $\mathcal{L}$ and $\mathcal{L}+\mathcal{B}$, is always positive semi-definite. Thus, making $\dot{\mathbb{V}} \le 0$ implies that error variables considered converge to zero, proving that the consensus among the DERs in achieving the optimal power and frequency restoration.

6.2.2 Reactive power sharing along with network voltage restoration

In an MG system, due to the presence of inverter-based DER systems, the number of voltage compensation units to provide reactive power support may be limited. Thus, all DERs connected to the network are expected to control their reactive power to ensure the proper voltage regulation in the distribution network, while avoiding inaccurate reactive power-sharing and overloading of individual DER. Due to the coupled relationship between the terminal voltage and unit reactive power output [18], it may not be possible to achieve simultaneous reactive power sharing and terminal voltage regulation. Therefore, to achieve both network voltage restoration and reactive power sharing, the hard equality voltage constraints $V_{o_i} = V_{ref}$ on individual DER can be converted to hard inequality constraints, i.e., $V_{min} < V_{oi} < V_{max}$ such that the complete network average voltage maintained at reference value, i.e., $\left(\hat{V}_{o_1} = \hat{V}_{o_2} = ... = \hat{V}_{o_N} = V_{ref}\right)$. The control input is accordingly, defined as

$$\begin{aligned}u_{\dot{\hat{V}}_{oi}} &= \delta\dot{\hat{V}}_{o_i} = \dot{V}_{o_i} + a_{ij}\sum\nolimits_{i\in N_i}\left(\hat{V}_{oj} - \hat{V}_{o_i}\right) + b_i\left(V_{ref} - \hat{V}_{o_i}\right)\\ \delta\dot{\hat{V}}_o &= \dot{V}_o - \mathcal{L}\hat{V}_o + \mathcal{B}\left(V_{ref} - \hat{V}_o\right)\end{aligned} \tag{6.22}$$

For reactive power sharing, the control input is

$$\begin{aligned} u_{n_{qi}Q_i} &= \sum_{j\in N_i} a_{ij}(n_{qi}Q_j - n_{qi}Q_i) \\ \delta\dot{V}_Q &= -\mathcal{L}(n_q Q) \end{aligned} \tag{6.23}$$

Two error terms are defined as

$$\mathbf{e}_{\hat{V}} = \left(\hat{V}_{o_1} - V_{ref}, \hat{V}_{o_2} - V_{ref}, ..., \hat{V}_{o_N} - V_{ref}\right) \tag{6.24}$$

$$\mathbf{e}_Q = \left(n_{q1}Q_1 - Q_{ss}, n_{q2}Q_2 - Q_{ss}, ..., n_{qN}Q_N - Q_{ss}\right) \tag{6.25}$$

where $Q_{ss} = \lim_{t\to\infty} \frac{1}{N}\sum_{i=1}^{N} n_{qi}Q_i$.

Using (6.24) in (6.22), the system dynamics are modified as,

$$\dot{\mathrm{e}}_{\hat{V}} = \dot{V}_o - \mathcal{L}\dot{\mathrm{e}}_{\hat{V}} + \mathcal{B}\dot{\mathrm{e}}_{\hat{V}} \tag{6.26}$$

The solution of (6.26) with $(L + G)$ being positive semi-definite can be obtained as [21],

$$\begin{aligned} \dot{\mathrm{e}}_{\hat{V}}(t) &= \exp(-(\mathcal{L}+\mathcal{B})(t-t_0))\dot{\mathrm{e}}_{\hat{V}}(t_0) \\ &\quad + \int_{t_0}^{t} \exp(-(\mathcal{L}+\mathcal{B})(t-\tau))\dot{V}_o(\tau)d\tau \end{aligned} \tag{6.27}$$

using the boundary condition

$$\|\exp -(\mathcal{L}+\mathcal{B})(t-t_0)\| \leq k_1 \exp(-\gamma(t-t_0)), \tag{6.28}$$

(6.27) is modified as,

$$\begin{aligned} \|\dot{\mathrm{e}}_{\hat{V}}(t)\| &\leq \exp(-\gamma(t-t_0))\|\dot{\mathrm{e}}_{\hat{V}}(t_0)\| \\ &\quad + \int_{t_0}^{t} \exp(-\gamma(t-\tau))\|\dot{V}_o(\tau)\|d\tau \end{aligned} \tag{6.29}$$

where γ indicates the eigenvalue of the matrix $(\mathcal{L}+\mathcal{B})$. At steady state, i.e., $t \to \infty$,

$$\|\dot{\mathrm{e}}_{\hat{V}}(t)\| \leq \exp(-\gamma(t-t_0))\|\dot{\mathrm{e}}_{\hat{V}}(t_0)\| + \frac{1}{\gamma}\|\dot{V}_o\| \tag{6.30}$$

As $(\mathcal{L}+\mathcal{B})$ is positive semi-definite, the eigenvalues 'γ' corresponding to this matrix consists of one zero and the rest are positive values. Thus, (6.30) will be stable and converge to a bounded solution if $\|\dot{V}_o\|$ bounded.

Furthermore, the reactive power control input $u_{n_{qi}}Q_i$ is similar to that of the optimal active power control input. Thus, a Lyapunov energy function $\mathbb{V}_Q = 1/2\sum_{i=1}^{N} {e_{Q_i}}^2$ is considered, whose derivative is

$$\dot{\mathbb{V}}_Q \leq -|\mathbf{e}_Q|^T L |\mathbf{e}_Q| \leq 0 \tag{6.31}$$

6.3 Cyber threats in multi-agent–based MGs

In a cyber-physical MG system, there are multiple factors at different layers that render the operation, as detailed in Table 6.1. However, with the increase in the number of DERs, the requirement for communication resources rises, and therewith concern about cyberattacks, especially, when there is no central unit that can track the activities of DERs. Cyber intrusions can happen at the control layer, especially at sensors and actuators, or at the communication layer, as shown in Figure 6.3. The intrusions

Table 6.1 Risk associated in MG operation and solutions [22]

Microgrid layers	Vulnerabilities	Recommendations
Physical layer	Switching operations	Switch status changes are configured properly
	Generation uncertainty	Stochastic-based reliability models of sources
	Network, generator faults	Coordinated protection system
Control layer	Cyber vulnerabilities at sensors and actuators	Use of vulnerability database, intrusion detection system
	Control/Measurement service status	Host logs, regular checks
Communication layer	Network security, network topology	Attack graph metrics, communication path trace and redundancy
	Communication quality	Use of communication performance metrics

at these layers are broadly categorized as (i) deception/man-in-the-middle (MITM)/false data injection (FDI) attacks and (ii) denial-of-service (DoS) attacks. These attacks are potent enough to disrupt the MG operation, can cause system instability, and sometimes lead to a complete shutdown. The different types of attacks in a distributed controlled MG system and their impact on the test system shown in Figure 6.3 are detailed in the following sections.

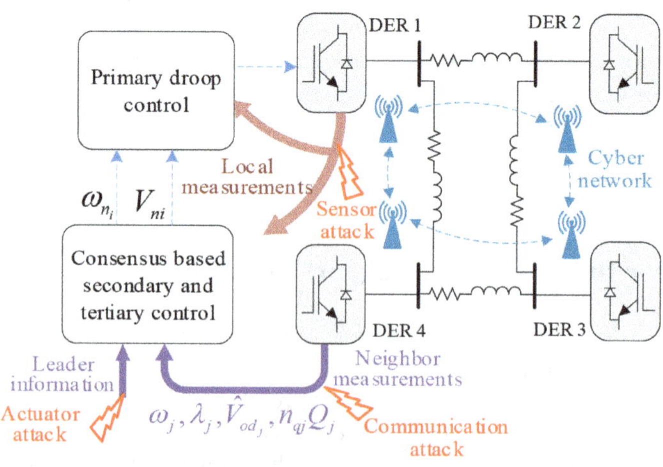

Figure 6.3 MG test system with the cyber layer and possible cyberattacks

6.3.1 Types of cyberattacks

6.3.1.1 FDI attack

In an FDI attack, the attacker gains access to the distributed communication network of the system and manipulates the measurement data, thereby modifying the results of state estimation. FDI attack can be effectively acknowledged by an MITM attack, in which the perpetrator places himself between multiple parties and eavesdrops or modifies the information that is being exchanged between the trusted parties. This attack permits the intruder to acquire, modify, and send data without the knowledge of trusted parties about the attacker. FDI attacks can be classified into two types: one with prior knowledge of the system and other the without any knowledge or having partial information about the system. As an example, Khalil [20] explains the construction of an FDI attack based on information, such as system topology, line parameters, and Jacobian matrix. In [23], a sparse attack strategy is proposed, in which state variables of a cluster are attacked by their own vector. Keeping in mind the real-world situations, where it is difficult to obtain the complete system information, Esmalifalak *et al.* [19] describe a stealthy false data injection attack based on independent component analysis where the attacker has no knowledge about the system. Authors in Reference 24 constructed a principal component analysis (PCA)-based blind FDI attack, but the presence of gross errors makes PCA-based strategies ineffective. Typically, the FDI attack [25] on frequency controller (in Figure 6.2) at ith DER is modeled as

$$\tilde{\omega}_i = \omega_i + \rho_i \omega_i^a \tag{6.32}$$

Here, ρ_i indicates the attack status, $\rho_i = 1$ reflects the presence of attack, and $\rho_i = 0$ reflects no attack. ω_i^a indicates the injected false data by the attacker, which can be a constant, time-varying bounded function, or time-varying unbounded function. ω_i is the original data transmitted to the controller, whereas $\tilde{\omega}_i$ is the corrupted information read by the controller. A similar attack can also be feasible on other variables λ_i, $\hat{V}_i$, Q_i that are used to achieve full distributed control in the AC MG shown in Figure 6.2. To evaluate the impact of the FDI attack, the test system shown in Figure 6.3 is simulated under FDI attacks. In Figure 6.4, initially, the system is operated with full distributed control without any attacks. At $t = 1$ s, the communication link 3–4 is hijacked and a false data of $\omega_3^a = 1.6$ are injected into the DER-3 frequency signal. Due to this attack, the frequency of the entire MG has deviated from the nominal values and the system is no longer operating as desired. It should be noted that constant bounded injection in frequency did not lead to any instabilities. However, at $t = 3$ s, false data $\hat{V}_3^a = 5$ are injected at the same DER, which led to a constant decrease in the voltage magnitude (Figure 6.4d) and led to the collapse of the MG. Similarly, time-varying bounded ($\omega_3^a = 2\sin 31.4t + 2.5\cos 31.4t$) and time-varying unbounded ($\omega_3^a = 0.1\exp(0.3t)$) data signals are injected into DER-3 and their corresponding responses are shown in Figures 6.5 and 6.6, respectively. With time-varying bounded attack, the MG shows an oscillatory instability whereas with time-varying unbounded attack, the system is asymptotically unstable.

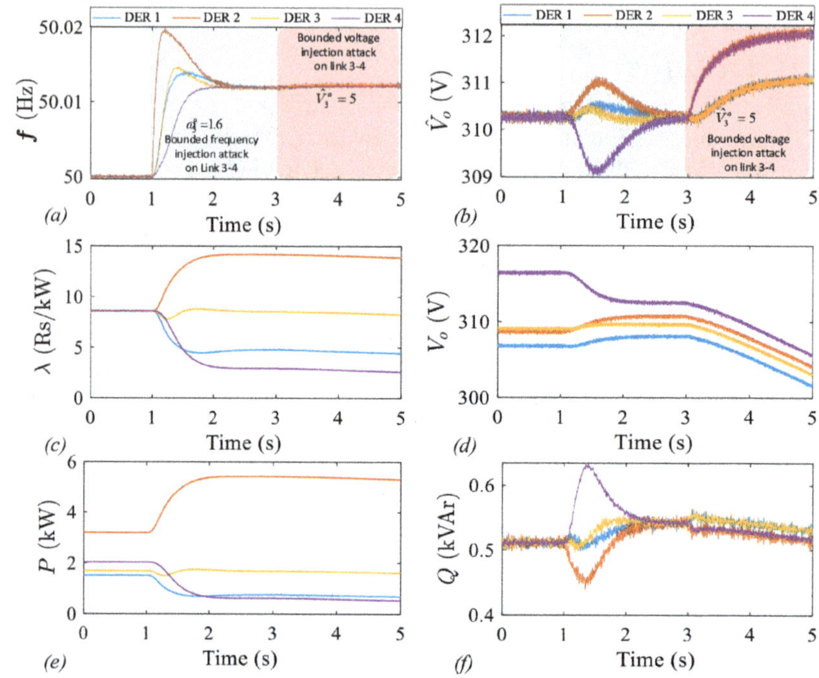

Figure 6.4 *Bounded constant FDI attack on the communication link 3–4 in AC MG*

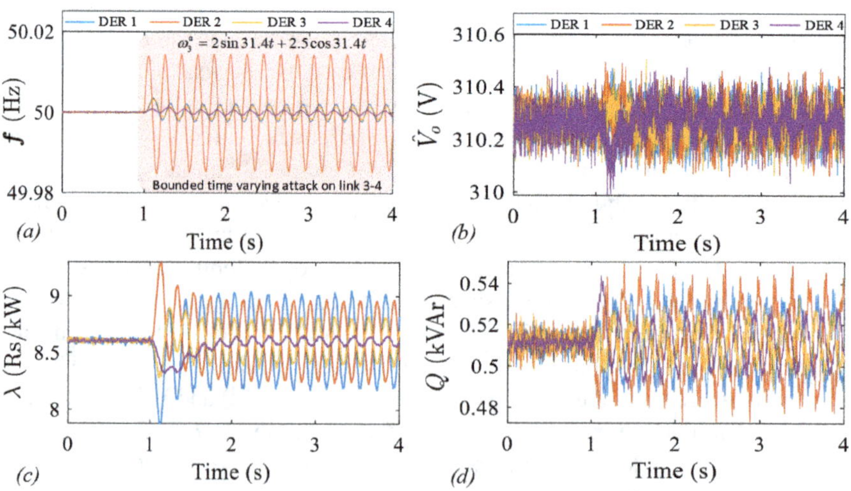

Figure 6.5 *Bounded time varying FDI attack on the communication link 3–4 in AC MG*

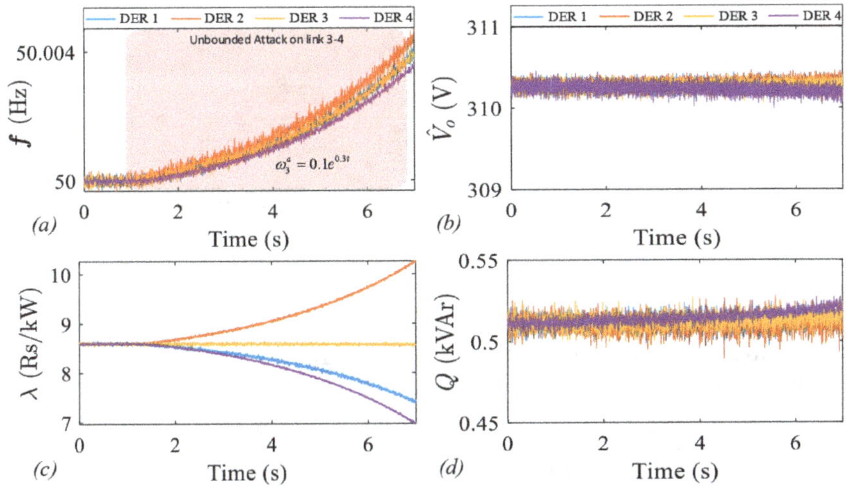

Figure 6.6 *Unbounded time varying FDI attack on the communication link 3–4 in AC MG*

6.3.1.2 DoS attack

Mostly communication inefficiencies arise due to DoS attacks. DoS attacks lead to unavailability or overloading of communication links, routers, servers, and so on, which interrupt or delay normal authorized functions. In this attack, the target system is flooded with unnecessary requests to make it overloaded, thereby delaying or preventing authorized users to access data in a timely manner. DoS attacks from outside the system can be prevented by proper design of firewall systems but in-house attacks cannot be prevented. The impacts of DOS attacks are more severe in the case of power system faults. DoS attacks can also be launched in a distributed manner, known as distributed denial-of-service (DDoS), for disabling MG functionalities by manipulating distributed sources, such as smart meters, Phasor measurement units (PMUs), and Intelligent electronic devices (IEDs). The DDoS attacks are initiated from multiple agents simultaneously, in which, the shutdown of a single agent does not stop the attack [26]. Compared to wired networks, wireless networks are more prone to DoS attacks [27]. They have additional means for launching DoS attacks, emitting radio frequency signals to jam authentic data, or misleading signals by counterfeiting a networking device. Any DoS attack on a communication, actuator or sensor, blocks the sharing of information leading to loss of data for the neighbor or its own controller. However, in classical CPSs, the packet losses can be reasonably modeled as random events, a stochastic characterization of the DoS attacks that inherently behave as a packet loss makes the system fail in capturing the malicious and intelligent nature of an attacker. Especially, in a classical sampled-CPS with continuous-time feedback loop, an attacker with some random strategy, can interrupt communication channels from sensors and neighbors. In the continuous time, the timing of attacks [16] is characterized as follows. Let $\{a_k \geq 0\}_{k\in\mathbb{Z}^+}$ represent

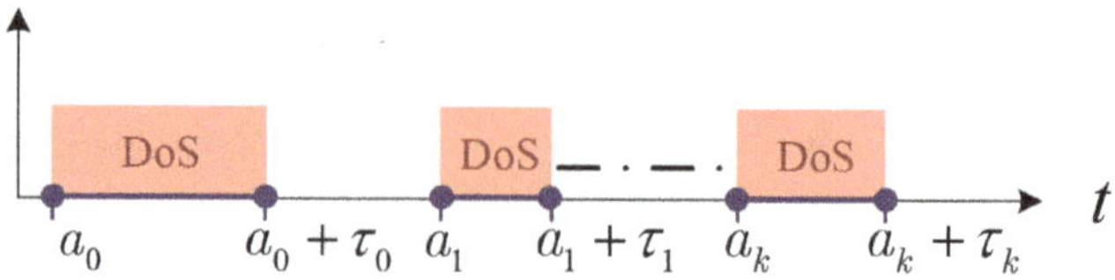

Figure 6.7 *DoS attack sequence on communication transmission*

the attack start instant, and $\{\tau_k \geq 0\}_{k\in\mathbb{Z}^+}$ indicate the attack duration. As shown in Figure 6.7, the kth attack starts at instant a_k for a duration of τ_k time. It is should be noted that to avoid attack overlapping, $a_{k+1} > a_k + \tau_k$.

If the packet transmission time, t, overlaps with any attack interval, $\bigcup_{k\in\mathbb{Z}^+} [a_k, a_k+\tau_k]$, then it causes the packet transmission failure. For any time interval $[t, t'] \subset [0, \infty)$ with $t, t' \in \mathbb{R}^+$, the set of DoS attack time instants, $\Xi(t, t')$, is defined as

$$\Xi(t, t') \triangleq \bigcup_{k\in\mathbb{Z}^+} [a_k, a_k + \tau_k] \bigcap [t, t']. \tag{6.33}$$

If the DoS attacks cover the entire time span of communication, then the desired objectives of distributed control cannot be achieved. However, the attackers may have limited resources or information about the system that prevents them from an all-time attack. The DoS attack on distributed communication channel is mathematically formulated as

$$\tilde{\omega}_i(t_k) = \Gamma_i(t_k)\omega_i(t_k), t_k \in [t, t'] \tag{6.34}$$

where $\tilde{\omega}_i(t_k)$ is the ith DER information received by its neighbor at kth instant between $[t, t']_{\mathbb{R}^+}$, and $\Gamma_i(t_k)$ is defined as

$$\Gamma_i(t_k) = \begin{cases} 0 & \text{if } t_k \in \Xi(t, t') \\ 1 & \text{otherwise} \end{cases} \tag{6.35}$$

The impact of the DoS attack is evaluated on the test system, as shown in Figure 6.3, by making the ω_3 unavailable at its neighbor DER-4 at $t = 1$ s, as shown in Figure 6.8. The DoS attack led to oscillatory instability of MG.

6.3.1.3 Replay attack

In replay attacks [28], the attacker aims to create an anomaly in the operation of an MG that operating in a steady state may inject undetectable exogenous inputs. To achieve this, the attacker hijacks the communication channel, actuators or sensors, observe, and record the data for certain duration when system is in normal operation. The attacker uses these recorded data afterward to complete a stealthy attack. This is most commonly used attack by the attacker who has minimum knowledge about the MG operation and dynamics but knows the fact that the MG is expected to be in steady state during the attack. For example, if there are surveillance cameras in a particular area, an intruder could hijack the surveillance system and record the footages when no one passed in the surveillance zone. Now, these footages are used to broadcast to the real-time monitoring system while he trespasses the surveillance zone safely with no one noticing. This attack is called a replay attack. It is assumed that the attacker can record sensor measurements, communication, and actuator information and choose the replay window size $[t, t']$, and at each timestep the attacker can send the correct or delayed information. The replay attacks are usually realized in two phases [21, 29]. In the first phase, the attacker performs a disclosure attack from t_0 to t_r time duration by recording the sequence of data r_k and formulated as

$$\text{Phase 1}: \begin{cases} \omega_{a_k}(t) = 0 \\ r_k = r_{k-1} \bigcup \omega_i(t_k) \end{cases} \quad \forall t \in [t_0, t_r];\ r_0 = \varnothing. \tag{6.36}$$

In the second phase, the attacker replays the recorded data from t_{r^+} to t_f, which are framed as

$$\text{Phase 2}: \begin{cases} \omega_{a_k}(t) = r_k \\ r_k = r_{k-1} \end{cases} \quad \forall t_k \in [t_{r^+}, t_f]; k = 1, 2, \ldots l_r, \tag{6.37}$$

where, l_r is the length of the recorded data set.

6.3.1.4 Configuration change attack

One of the familiar protection schemes in the distribution automation system is protection coordination between overcurrent relay (OCR), recloser, and fuse. Proper setting of protection coordination is very important to limit the outages. An attacker may change these settings to cause unwanted outages during normal operating periods or during faults on the feeder. Normally, there are two distinct approaches to

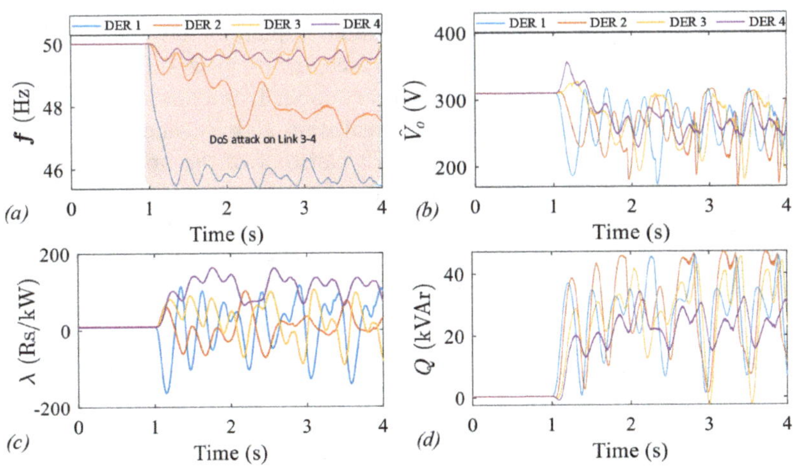

Figure 6.8 DoS attack on the communication link 3–4 in AC MG

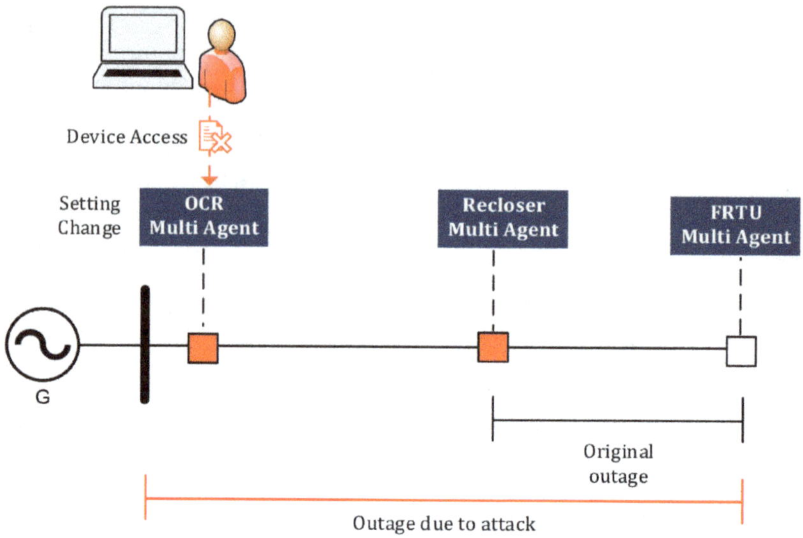

Figure 6.9 Protection coordination for recloser and OCR

change the protection setting of field devices: (i) the vendor's engineering tool or (ii) the front panel of the protective device. In the event that the attacker knows the ID and password of the user interface (UI) in the distribution operation center, he can execute the engineering tool for protective devices. At that point, they can change the security settings of devices with a default password. In the event that the changed pickup setting esteem is lower than load current, the protective device quickly trips the circuit breaker though the changed setting time delay value of the back-up device is quicker than primary protection's time delay esteem, it will trip the whole feeder and make undesirable blackout territory during the power system fault. Figure 6.9 shows a model of protection coordination between a recloser and OCR protective devices. To limit the outage area, the recloser's time delay needs to be quicker than an OCR to check a temporary fault. If the fault is permanent, the recloser will be locked-out subsequent to reclosing activity, and the fault will be cleared (territories from R1 to GS1) without blackout of the upside of the feeder (territories from CB1 to R1). Nonetheless, if an intruder compromises the engineering tool and changes the time defer setting of the OCR, regions from CB1 to R1 will have undesirable outages during the deficiency. This is on the grounds that the OCR will trip the circuit breaker (CB1) quicker than the recloser (R1).

6.4 Multi-agent–based cyberattack detection algorithms

Although there exists immense threat in the distributed control-based cyber-physical MGs, there are multiple defense mechanisms at the cyber, control, and physical layers that are developed to reduce the risk associated with cyberattacks. These defense mechanisms [30] are broadly classified as (i) prevention, (ii) resilience, and (iii) attack detection and isolation.

6.4.1 Security at cyber layer

In the cyber layer, the most common defense mechanism falls into the prevention category in which the inception of the attacks is usually postponed or blocked by the defense system. One such prevention technique is the use of *message authentication code* (MAC) algorithms [31, 32] while carrying the communication between different agents. In this, the key distribution center sends the same privacy key to the sender and receiver. Algorithms such as Galois Message Authentication Code or Hash Message Authentication Code are used for message authentication. Enhanced security may be achieved by using complex algorithms for MAC-based authentication, but it introduces time delay in the existing protection and control functions. Synchronization number is defined to check the updated messages, and Sync is increased by '1' for each transmission, so the new Sync ought to be higher than the Sync number received earlier. A comparison of these numbers reveals information on whether the received messages are reused by an intruder. The sender will produce an MAC tag utilizing the MAC calculation along with the shared key, time, synchronization number, and message body. The MAC tag produced is affixed to the body of the message and then sent to the receiver. When the receiver gets the message with the MAC tag, it will ascertain the MAC tag again utilizing the shared key, synchronization number, time, and conveyed message body. The message delivered can be considered to be attack free if the calculated and delivered MAC are matched, the sync number is as expected, and the delivered time is faster than the previous time. Time out can be considered as an additional verification, where a threshold of 3 ms can be set and any message delayed more than 3 ms may be discarded and a fresh message request may be sent to sender.

Similarly, logical connectivity among the agents (DERs) with a secured routing protocol might be used on top of a network topology for communication to improve the security. On top of these, software-defined networking (SDN) can also be used for improving resiliency against cyber threats. SDN is an open-source platform used to design, build, and manage the communication networks [33]. The SDN reduces the burden of decision functions from the centralized controller(s). The SDN layer will be an additional layer in the conventional system as it decouples the network control functions in communication devices. The SDN provides global visibility and dynamic programmability, which can detect the anomaly and mitigate its implications while preserving the MG observability.

6.4.2 Security at physical and control layer

In the physical and control layers, mostly the attack detection and isolation approach is used in which the source of the attack is identified and isolated from the system while restoring non-compromised units. In the attack detection and isolation approach, the states of the physical system (such as currents, voltages, frequencies, and powers) are closely observed continuously using which any abnormality or anomality is detected in the early stages of the attack. In Reference 34, the FDI is detected by checking the changes in sets of inferred *invariants* using Hynger prototype tool. In this, *invariants* are defined within bounds of the output voltage and current for individual DERs. A parametric time–frequency logic framework-based FDI attack detection method [35] can be used. In this, the time–frequency content of the current, voltage, or frequency is analyzed using Morlet (morl) wavelet and compared with the predefined time–frequency properties to check for any anomalies.

An active synchronous detection method (ASDM) [36] can also be used to detect deception attacks. In ASDM, first, probing signals are created by the centralized controller in which transmitted individual DERs, then the responses of these DERs, are monitored and compared with the predetermined values to check if the DERs are compromised. The probing signals are modified regularly to make ASDM a more effective defensive scheme. Similarly, a cooperative vulnerability factor framework [37] can be used for accurately identifying the attacked agent(s) through a cooperative approach among the connected DERs. A model-based anomaly detection [38] is developed for a partial primal-dual (PPD) optimization-based distributed control in MGs. The anomaly detection is developed by exploring the dual variable-related metrics, in which only the local information and its neighboring dual variable information are used to detect the attack.

In addition to the detection and isolation approaches, the resilient control approaches are also studied intensively as these approaches reduce the impact of the attacks by some controller or sensor actions and operate CPS normally, even during attacks. A distributed observer-based resilient cooperative synchronization protocol [39] is used to reduce the trust of the compromised DERs and regain normalcy in MG operation. In the approach, each DER observes its neighbor's information, updates its local confidence and trust factors, and shares the confidence factor with its neighbor. The trust factors are obtained by passing the received neighbors' information through the distributed observer. By this, the DERs in close proximity to the attack point reach to lower confidence factor and lower trust factor compromised DER. Thus, the spread of corrupted data is discarded completely and brings MG operation to normalcy. Similarly, a weighted mean subsequence reduced algorithm [40] is used to eliminate the corrupted information by managing the connectivity of communication graphs and making MG operations resilient.

6.5 Summary

In this chapter, a brief understanding of distributed cyber-physical MGs, the cyber risk associated with them, and a few existing possible security measures are presented. Specifically, the impact of MG operations with different FDI attacks such as time-invariant, bounded time-variant, and unbounded attacks is analyzed. In addition to this, deceptive attack such as DoS and replay attacks impacts on cyber-physical MG is presented. Furthermore, different existing security measures applicable at the cyber layer, physical layer, and control layer are also detailed in this chapter. However, cyber security in MGs remains an ongoing research, with potential future research scope in the direction of the distributed attack detection techniques and security measures for multiple coordinated stealthy attacks.

References

[1] 'IEEE recommended practice for the planning and design of the microgrid'. *IEEE Std 20309-2019*. 2019:1–46.

[2] Lasseter R.H. 2002 IEEE Power Engineering Society Winter Meeting. (Cat. No.02CH37309); 2002. pp. 305–8.

[3] Olivares D.E., Mehrizi-Sani A., Etemadi A.H., *et al.* 'Trends in microgrid control'. *IEEE Transactions on Smart Grid*. 2014;5(4):1905–19.

[4] Guerrero J.M., Vasquez J.C., Matas J., de Vicuna L.G., Castilla M. 'Hierarchical control of droop-controlled AC and DC microgrids—a general approach toward standardization'. *IEEE Transactions on Industrial Electronics*. 2011;58(1):158–72.

[5] Bidram A., Davoudi A. 'Hierarchical structure of microgrids control system'. *IEEE Transactions on Smart Grid*. 2012;3(4):1963–76.

[6] Ilic M., Liu S. *Hierarchical power systems control: its value in a changing industry. advances in industrial control*. London: Springer; 2012.

[7] Lewis F.L., Zhang H., Hengster-Movric K. *Cooperative Control of Multi-Agent Systems: Optimal and Adaptive Design Approaches*. Springer; 2014.

[8] Simpson-Porco J.W., Dörfler F., Bullo F. 'Synchronization and power sharing for droop-controlled inverters in Islanded microgrids'. *Automatica*. 2013;49(9):2603–11.

[9] Bidram A., Davoudi A., Lewis F.L., Qu Z. 'Secondary control of microgrids based on distributed cooperative control of multi-agent systems'. *IET Generation, Transmission & Distribution*. 2013;7(8):822–31.

[10] Wang X., Zhang H., Li C. 'Distributed finite-time cooperative control of droop-controlled microgrids under switching topology'. *IET Renewable Power Generation*. 2017;11(5):707–14.

[11] Dehkordi N.M., Baghaee H.R., Sadati N., Guerrero J.M. 'Distributed noise-resilient secondary voltage and frequency control for Islanded microgrids'. *IEEE Transactions on Smart Grid*. 2018;10(4):3780–90.

[12] Zuo S., Davoudi A., Song Y., Lewis F.L. 'Distributed Finite-time voltage and frequency restoration in Islanded AC microgrids'. *IEEE Transactions on Industrial Electronics*. 2016;63(10):5988–97.

[13] Zhao T., Ding Z. 'Distributed agent consensus-based optimal resource management for microgrids'. *IEEE Transactions on Sustainable Energy*. 2018;9(1):443–52.

[14] Pullaguram D., Rana R., Mishra S., Senroy N. 'Fully distributed hierarchical control strategy for multi-inverter-based AC microgrids'. *IET Renewable Power Generation*. 2020;14(13):2468–76.

[15] Sahoo S., Dragičević T., Blaabjerg F. 'Cyber security in control of grid-tied power electronic Converters–Challenges and vulnerabilities'. *IEEE Journal of Emerging and Selected Topics in Power Electronics*. 2019:1–1.

[16] Cetinkaya A., Ishii H., Hayakawa T. 'An overview on denial-of-service attacks in control systems: attack models and security analyses'. *Entropy*. 2019;21(2):210.

[17] Nutkani I.U., Loh P.C., Blaabjerg F. 'Droop scheme with consideration of operating costs'. *IEEE Transactions on Power Electronics*. 2014;29(3):1047–52.

[18] Wood A.J., Wollenberg B.F., Sheblé G.B. *Power Generation, Operation, and Control*. 3. Hoboken, NJ: Wiley; 2003.

[19] Esmalifalak M., Nguyen H., Zheng R. 'Stealth false data injection using independent component analysis in smart grid'. 2011 IEEE International Conference on Smart Grid Communications (SmartGridComm); 2011. pp. 244–8.

[20] Khalil H.K. *Nonlinear systems. The Book Can Be Consulted By Contacting: PH-AID: Wallet, Lionel*. 3. Upper Saddle River, NJ: Prentice Hall; 2002.

[21] Teixeira A., Shames I., Sandberg H., Johansson K.H. 'A secure control framework for resource-limited adversaries'. *Automatica*. 2015;51(6):135–48.

[22] Venkataramanan V., Hahn A., Srivastava A. 'CP-SAM: Cyber-Physical security assessment metric for monitoring microgrid resiliency'. *IEEE Transactions on Smart Grid*. 2020;11(2):1055–65.

[23] Ozay M., Esnaola I., Vural F.T.Y., Kulkarni S.R., Poor H.V. 'Sparse attack construction and state estimation in the smart grid: centralized and distributed models'. *IEEE Journal on Selected Areas in Communications*. 2013;31(7):1306–18.

[24] Yu Z.-H., Chin W.-L. 'Blind false data injection attack using PCA approximation method in smart grid'. *IEEE Transactions on Smart Grid*. 2015;6(3):1219–26.

[25] Mustafa A., Poudel B., Bidram A., Modares H. 'Detection and mitigation of data manipulation attacks in AC microgrids'. *IEEE Transactions on Smart Grid*. 2019;11(3):2588–603.

[26] Chang R.K.C. 'Defending against flooding-based distributed denial-of-service attacks: a tutorial'. *IEEE Communications Magazine*. 2002;40(10):42–51.

[27] Raymond D.R., Midkiff S.F. 'Denial-of-service in wireless sensor networks: attacks and defenses'. *IEEE Pervasive Computing*. 2008;7(1):74–81.

[28] Mo Y., Sinopoli B. 'Secure control against replay attacks'. *47th Annual Allerton Conference on Communication, Control, and Computing (Allerton)*; Allerton House, UIUC, IL, 30 Sep–2 Oct; 2009. pp. 911–8.

[29] Sánchez H.S., Rotondo D., Escobet T., Puig V., Quevedo J. 'Bibliographical review on cyber attacks from a control oriented perspective'. *Annual Reviews in Control*. 2019;48(15):103–28.

[30] Deng C., Wang Y., Wen C., Xu Y., Lin P. 'Distributed resilient control for energy storage systems in cyber–physical microgrids'. *IEEE Transactions on Industrial Informatics*. 2020;17(2):1331–41.

[31] Hussain S.M.S., Farooq S.M., Ustun T.S. 'Analysis and implementation of message authentication code (MAC) algorithms for goose message security'. *IEEE Access*. 2019;7:80980–4.

[32] Power systems 'Power systems management and associated information exchange - data and communications security–part 6: security for IEC 61850' in *International electrotechnical commission*. Geneva: IEC; 2020.

[33] Jin D., Li Z., Hannon C., *et al*. 'Toward a cyber resilient and secure Microgrid using Software-Defined networking'. *IEEE Transactions on Smart Grid*. 2017;8(5):2494–504.

[34] Beg O.A., Johnson T.T., Davoudi A. 'Detection of false-data injection attacks in cyber-physical DC microgrids'. *IEEE Transactions on Industrial Informatics*. 2017;13(5):2693–703.

[35] Beg O.A., Nguyen L.V., Johnson T.T., Davoudi A. 'Cyber-physical anomaly detection in microgrids using time-frequency logic formalism'. *IEEE Access*. 2021;9:20012–21.

[36] Li Y., Zhang P., Zhang L., Wang B. 'Active synchronous detection of deception attacks in microgrid control systems'. *IEEE Transactions on Smart Grid*. 2016;8(1):373–5.

[37] Sahoo S., Mishra S., Peng J.C.-H., Dragicevic T. 'A stealth cyber-attack detection strategy for DC microgrids'. *IEEE Transactions on Power Electronics*. 2018;34(8):8162–74.

[38] LY L., Liu H.J., Zhu H. 'Intrusion detection in distributed frequency control of isolated microgrids'. *IEEE Transactions on Smart Grid*. 2019;10(6):6502–15.

[39] Abhinav S., Modares H., Lewis F.L., Ferrese F., Davoudi A. 'Synchrony in networked microgrids under attacks'. *IEEE Transactions on Smart Grid*. 2017;9(6):6731–41.

[40] Duan J., Chow M.-Y. 'A resilient consensus-based distributed energy management algorithm against data integrity attacks'. *IEEE Transactions on Smart Grid*. 2019;10(5):4729–40.

Chapter 7

Communication-assisted protections for DC microgrids and their performance analysis during cyberattacks

Rabindra Mohanty[1], *Subham Sahoo*[2], *Ashok Kumar Pradhan*[3], *and Frede Blaabjerg*[2]

In recent years, fossil fuel-based electric power sources are a major concern due to environmental issue. Depletion of such energy sources is also a matter of worry. The power delivery networks in a power system have been stressed due to aging of transmission and distribution infrastructure and the growing demand of electricity. Integration of renewable energy sources (RESs) to the power grid using advanced power electronic converters is a promising solution for enhanced reliability of supply, clean energy and improved power quality [1]. High penetration of RESs in a grid causes voltage rise and protection issues threatening grid security and reliability. The concepts of 'Microgrid' and 'Smart grid' are emerging in power system as viable solution to these problems.

7.1 Introduction

Microgrid can be categorized into AC or DC depending on the type of common bus to which sources and loads are connected. Hybrid AC-DC microgrid is another type where both AC and DC buses are present to interface AC and DC sources and loads with less conversion stages. RESs like PV arrays, fuel cells and storage devices produce DC. Even, interfacing of small wind turbines to DC bus is simpler and economical compared to AC. Number of conversion stages needed with DC is lesser than AC. LEDs, fans and modern electronic loads consume DC. Several stand-alone applications like vehicular power system, telecommunication center, data center, marine and aerospace need DC supply for reliable, economic and efficient operation. DC microgrid, therefore, is a feasible solution for electricity supply.

[1]Department of Electrical and Electronics Engineering, BITS Pilani Hyderabad campus, India
[2]Department of Energy Technology, Aalborg University, 9220 Aalborg, Denmark
[3]Department of Electrical Engineering, Indian Institute of Technology Kharagpur, India

Operation, control and protection of AC microgrid are well established compared to DC. Power electronic converters cannot bear overloading for several milliseconds beyond its rating in a DC microgrid. With the presence of power electronic converters having current control strategy fault, current does not exceed twice of the rated current, and therefore, current magnitude-based protection techniques are not a solution for DC microgrid [2]. Furthermore, DC microgrid has protection challenges due to difficulties in extinguishing arc, accurate short-circuit current calculation, fault detection and isolation within the minimum time frame [3]. In addition, low cable impedance in DC causes high rate of rise of current during fault. Available protection schemes for AC cannot be applied directly to DC as the voltage and current patterns in DC during faults are different than AC. A DC microgrid requires reliable protection schemes for safe and secured operation.

DC microgrid is a better solution for energy management due to DC producing RESs, storage devices, advances in power electronic devices, modern control strategies and DC consuming loads. Lot of literature is available on DC system architecture, applications, standard and benefits [4, 5]. The incorporation of power electronic converters with current control strategy in DC microgrid limits the current during fault. Current magnitude-based protection scheme cannot identify the fault accurately. Detection of high resistance fault is also a problem in DC microgrid. Rate of change of DC current is high due to less cable resistance and inductance. Fast and accurate protection scheme is required for reliable operation. This section details the literature survey of DC microgrid protection and fault location.

This chapter introduces new algorithms for protection and fault location for DC microgrid. Using local voltage and current data, a least square-based protection scheme is proposed. The method estimates the seen inductance at intelligent electronic devices (IEDs) and the protection decision is derived using seen inductance at both ends of a line. Frequency of DC current oscillation and associated transient power in the first cycle of oscillation are used to develop another protection algorithm. A unit protection scheme is proposed for DC microgrid using superimposed fault current at both ends of the line segment. Correct faulted section is identified in a DC distribution system using consumer-end smart meter data. An algorithm is derived for exact fault distance calculation in an isolated faulted section using damped resonant frequency and attenuation of injected current from a probe power unit. The algorithms are tested for different systems (radial and ring configurations) using data from simulation and validated with data from scaled-down laboratory setup. The protection and fault location schemes supplemented by these algorithms will be able to provide improved and reliable operation of DC microgrid.

7.1.1 Related works

For a DC microgrid, one of the important challenges is the lack of effective protection solution [6]. Detection and clearance of DC fault are difficult because of the presence of power electronic converters and associated DC circuit breakers [7]. Absence of zero crossing and association of more aggressive arc cause longer time to clear a DC fault. Furthermore, the relatively small resistance and inductance in cables, their limited impact on DC faults and the sensitivity of converters to faults make the realization of protection selectivity difficult [6]. The DC network fault can be cleared by protection arrangement in AC side where isolation of entire DC system leads to the issue of system reliability and security [8].

Concept of a microgrid development leads to sufficient sensors and communication infrastructure within a distribution network to derive improved network monitoring and management [9]. Such an arrangement provides an opportunity for unit protection in a microgrid. Current differential scheme using magnitude of differential current is proposed in [8, 10–12]. Such scheme is less vulnerable to the effect of variable fault levels. Unlike the operating time of current differential protection in AC system (20 ms), that in DC network is around 2 ms [12]. The factors that prevent the reduction in operating time in AC network are the requirement of individual phase currents and subsequent phasor comparison. For DC system, only current magnitude needs to be compared. High resistance fault and synchronization error are the issues in such approach. A high resistance fault results low magnitude fault current, and therefore, available current differential techniques find problem in identifying such faults. In [6], a fast and selective protection scheme based on direction of current information from both ends is proposed for DC network. Such a method has also limitation in identifying direction of fault during high resistance fault as prefault current may be more than fault component of current.

Based on the connection of equipment types, networks in the microgrid can be AC, DC or a combination of the two [13]. DC network is more feasible for a demarcated power system, for example, rural power systems, office buildings and ships where the majority of loads are sensitive electronic equipment and electric vehicles. Further advantages with DC microgrid are high efficiency [2], easy connection of sources to DC bus [14], negligible transmission loss due to small and localized system [11], enhanced power transfer capacity [15] and interfacing through more efficient power electronic devices [14]. However, quick detection of a fault, DC arc breaking, DC protective equipment and lack of standards, guidelines and experience are the major issues in DC network protection [14]. As there are differences in patterns of voltage and current during fault in DC systems compared to AC, the methods available for protection of AC network cannot be copied for DC systems directly.

Power electronic converters are required to connect both AC and DC sources and loads to a common bus (AC or DC type) in a microgrid system. Moreover, DC network uses less stages of conversion [3, 16]. Internal faults in converter include failure of switches such as insulated-gate bipolar transistor (IGBT). Such faults are difficult to be cleared, and for a fault tolerant converter system, redundant devices are suggested [3]. Fault in DC network with parallel converters is the more severe one. The DC link capacitor of converter and small cable impedance cause severe overcurrent and undervoltage during a fault.

In a DC network, the common fault is of pole-to-ground type, and this is because of physical damage, aging or severe electrical stress in cables [17, 18]. Differential current-based protection schemes for cable fault in DC network are proposed in [12, 19] that require reliable communication channel for instantaneous data transfer between protective devices placed at both ends of the cable. Chances of communication failure and loss of data result in performance limitation of differential scheme. The cost of such protection scheme is also a concern in microgrid.

In [10] and [20], unit and non-unit protection schemes are investigated for DC network faults. High rate of rise of fault current and its large steady state value demand for fast operation of DC network protection. The challenge is to avoid damage to the power electronic devices and keep the fault current within interruptible limit; the fault must be cleared in a timely manner. In [21], a non-unit protection scheme is proposed for DC network fault detection and the method disconnects all the converters in AC side. The capacitor present in the DC side of the converters supplies current for short duration during a fault in the line. The main drawback of this method is that it shuts down the system completely for a fault. A fast and selective protection scheme based on direction of current information only for DC distribution network is proposed in [6]. During high resistance fault, flow of current direction may not be indicative of fault direction for a protection system. With increased number of sensors and communication infrastructure, the concept of smart microgrid are being realized for automatic network monitoring and management, increase the usage of intermittent sources and decrease the network congestion [10, 22]. Protection action can be performed utilizing this advanced communication infrastructure of future smart DC microgrid. Furthermore, the distributed generation (DGs) in a microgrid operate in different modes, like grid supporting, grid following and maximum power point tracking (MPPT) [23].

7.1.2 Chapter focus

Power electronic converters with different control strategies modulate the voltage current during fault in a DC microgrid. Current magnitude-based protection schemes overcurrent, change in impedance find difficulty to identify the fault [19]. Presence of DGs, power electronic devices controlling current and voltage and bidirectional power flow in today's DC distribution system make the protection issues more challenging. Thus, there is a scope for the development of improved protection techniques for DC microgrid.

The objectives set for the DC microgrid protection are to develop

- a unit protection scheme using both ends current of a line segment to be protected.
- centralized protection of DC microgrid robust to communication delay and link failure.
- performance analysis of the protection schemes for cyberattack.

7.2 Fault analysis in DC microgrids

A short-circuit fault in DC network is a severe condition for converters and the IGBTs can be blocked for self-protection letting reverse diodes exposed to overcurrent [24]. During fault, the network passes through two states. First state is the natural response of the RLC circuit due to discharge of DC link capacitor just after the occurrence of a fault, and second state starts when the fault current reaches the peak and capacitor voltage becomes less than the maximum of input voltage, that results in the participation from converter interfaces sources and loads (if applicable) to the fault current [3, 20]. Both states are further analyzed below.

7.2.1 Analysis of capacitor discharge

With the onset of a fault, the capacitor starts discharging through cable impedance as in the equivalent circuit shown in Figure 7.1. The peak current resulting from the discharging of DC link capacitor is shown in Figure 7.1 and the magnitude goes around ten times of the rated current in a system of 100 kW, 350 V DC, without the operation of any IEDs. Considering the fault inception at time t_0, the RLC circuit response in frequency domain can be written as [24]:

$$I(s) = \frac{\frac{V_0}{L} + I_0 s}{s^2 + \frac{R}{L}s + \frac{1}{LC}} \tag{7.1}$$

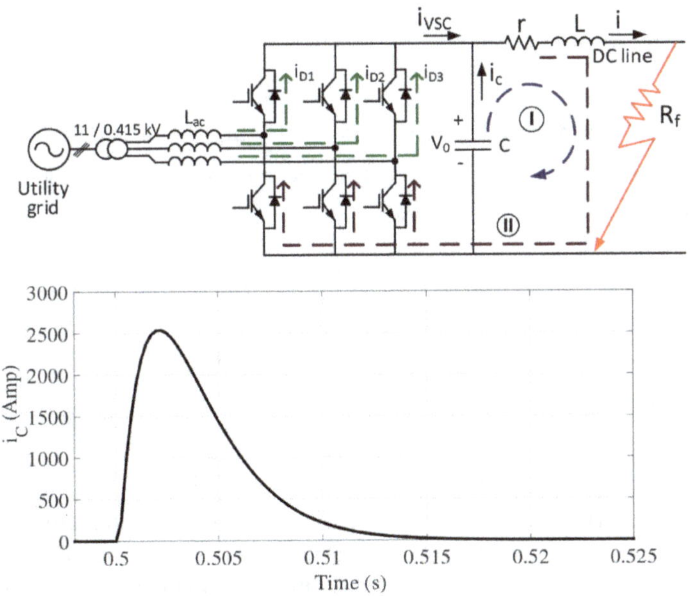

Figure 7.1 (a) Equivalent faulted network of a DC short-circuit fault at VSC terminal (b) Capacitor discharging current during short-circuit fault

where $v_c(t_0) = V_0$, $i(t_0) = I_0$ and r are the inductance and resistance of the cable from converter terminal to fault point. R is the sum of r and fault resistance. In (7.1), the contribution of converter current to the fault is not considered due to the slow response of the converter controllers [25]. The voltage across the capacitor and fault current in time domain will be

$$v_c(t) = \frac{V_0\omega_0}{\omega}e^{-\alpha t}sin(\omega t + \beta) - \frac{I_0}{\omega C}e^{-\alpha t}sin(\omega t) \tag{7.2}$$

$$i(t) = C\frac{dv_c(t)}{dt} = -\frac{I_0\omega_0}{\omega}e^{-\alpha t}sin(\omega t - \beta) + \frac{V_0}{\omega L}e^{-\alpha t}sin(\omega t) \tag{7.3}$$

where $\alpha = R/2L$, $\omega = \sqrt{\frac{1}{LC} - \left(\frac{R}{2L}\right)^2}$, $\omega_0 = \sqrt{\alpha^2 + \omega^2}$, and $\beta = \tan^{-1}\left(\frac{\omega}{\alpha}\right)$. The form of current response is defined by the relationship between magnitudes of α and ω_0^2. The fault current response will be over, under and critically damped for $\alpha^2 > \omega_0^2$, $\alpha^2 < \omega_0^2$, and $\alpha^2 = \omega_0^2$ respectively. With high DC link capacitance and low inductance of the cable, the predominant part of any underdamped fault current shown in (7.3) is due to the voltage of initially charged capacitor. Therefore, the expression for fault will now reduce to

$$i(t) \simeq \frac{V_0}{\omega L}e^{-\alpha t}sin(\omega t) \tag{7.4}$$

For highly underdamped response (α^2 and ω), (7.14) can be further simplified to

$$i(t) \simeq \frac{V_0}{Z_0}e^{-\alpha t}sin(\omega_0 t) \tag{7.5}$$

where Z_0 is the surge impedance of cable up to fault point and is represented as $Z_0 = \sqrt{\frac{L}{C}}$. Time taken by the fault current to reach its peak during underdamped and overdamped cases is as in [24], respectively,

$$t_{pund} = \frac{1}{\omega_0}tan^{-1}\frac{\omega_0}{\alpha} \tag{7.6}$$

$$t_{povr} = \frac{ln(n_2/n_1)}{n_1 - n_2} \tag{7.7}$$

where $n_1 = -\alpha + \sqrt{\alpha^2 - \omega_0^2}$ and $n_2 = -\alpha - \sqrt{\alpha^2 - \omega_0^2}$. Using cable parameters and DC link capacitor (in Table 7.1), time to reach peak current during fault is calculated for both extreme cases (solid fault, $R_f = 0\ \Omega$ and high resistance fault, $R_f = 1\ \Omega$) and found that the allowed fault detection time before load voltage collapses in DC system is much lesser compared to AC system.

Table 7.1 *DC microgrid components and parameters*

System voltage	**350 V DC**
Base power	500 kW
Battery	96 V, 0.4 kAh
Battery DC-DC converter	125 kW
Solar panel	V_{mp}= 54.7V, I_{mp} = 5.58A at STC
PV converter	125 kW
VSC connecting to utility grid	250 kW
Load	Constant impedance load, 500 kW
Fault resistance	0–10Ω
DC link capacitor	25 mF
Cable parameters:	
Cross-section area	240 mm^2
Resistance	0.125 Ω/km
Inductance	0.232 mH/km
Capacitance	0.67 μF/km
Cable length in the ring	300–500 m

7.2.2 *Freewheeling diode operation*

The second step of fault current analysis is with the presence of freewheeling diode in parallel with active devices in the converter [20, 21]. This phase starts when the voltage across DC link capacitor drops to zero or becomes negative, and it is after t_{pund} or t_{povr} as defined in (7.6) and (7.7). As an effect, the voltage at converter terminal gets reversed and the diode starts conducting, as shown in Figure 7.2. The diode D represents the equivalent of any conducting leg of converter [26]. The current

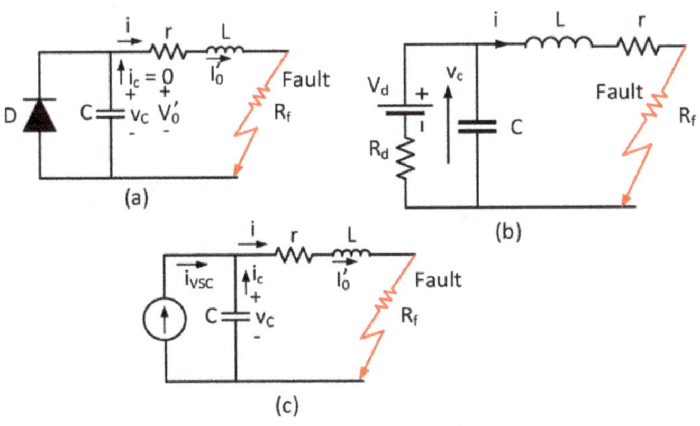

Figure 7.2 *(a) Freewheeling diode operation (b) equivalent circuit of conducting freewheeling diode and (c) current from converter*

response changes due to this alternative path irrespective of the IGBT conducting state. Figure 7.2c shows the equivalent circuit of faulted network during conducting freewheeling diode. The reverse polarity circulating current can be expressed as

$$i(t) = \frac{v_d}{R_d + r} + A_1 e^{n_1 t} + A_2 e^{n_2 t} \tag{7.8}$$

where A_1, A_2 are the coefficients that depend on initial conditions and n_1, n_2 are the roots as defined in (7.7), in which the damping factor $\alpha = \frac{r}{2L} + \frac{1}{2R_dC}$ and the resonant frequency $\omega_0 = \frac{1}{\sqrt{LC}} + \frac{r}{\sqrt{R_dLC}}$. The voltage across the parallel branch diode-capacitor will be $v(t) = i(t)r + L\frac{di}{dt}$. Substituting $i(t)$, the voltage can be expressed as

$$v(t) = \frac{v_d}{R_d + r} r + A_1 e^{n_1 t}(r + n_1 L) + A_2 e^{n_2 t}(r + n_2 L) \tag{7.9}$$

Current through cable (i_L) and voltage across DC link capacitor (v_C) during diode freewheeling are shown in Figure 7.3. In the figure, $I_0' = i_L(t_{C0})$ and t_{C0} is the time when DC link capacitor becomes zero.

The contribution of converter to the fault current, as shown in Figure 7.1, can be obtained by three-phase short-circuit analysis. The grid side phase current is [26]

$$i_g^a = I_g sin(\omega_s t + \alpha - \phi) + [I_{g0} sin(\alpha - \phi_0) - I_g sin(\alpha - \phi)]e^{-t/\tau} \tag{7.10}$$

where $\phi = tan^{-1}[\omega_0(L_{ac} + L)/R]$, $\tau = (L_{ac} + L)/R$, I_{g0} and ϕ_0 are the initial current and phase angle, and L_{ac} is the grid side inductance. The current i_g^a (positive part only) flows through diode D_1; similarly i_g^b and i_g^c flow through D_2 and D_3, respectively and total i_{VSC} is the summation of all three-phase currents.

$$i_{VSC} = i_{D1} + i_{D2} + i_{D3} = i^a_{g(>0)} + i^b_{g(>0)} + i^c_{g(>0)} \tag{7.11}$$

The complete response of the non-linear circuit during fault is analyzed in two different steps. The protection system is expected to operate within the peak reaching time (t_{pund} or t_{povr}) for the sake of converter and other components.

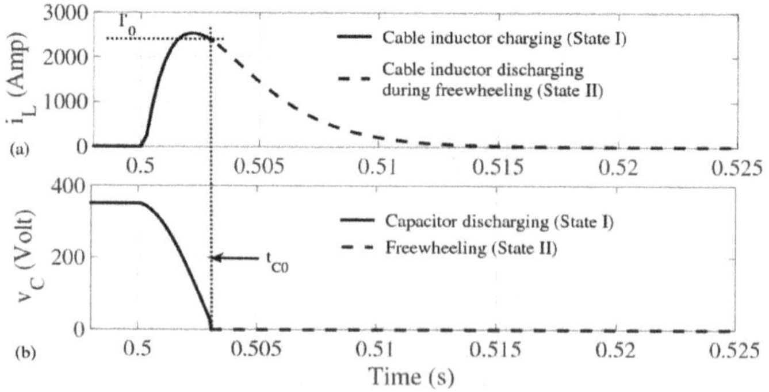

Figure 7.3 During diode freewheeling (a) cable inductor current i_L (b) DC link capacitor voltage v_C

7.3 A Unit protection scheme for DC microgrid

In a DC microgrid, power electronic converters retain current control schemes to avoid possible damage during fault. Therefore, current magnitude-based protection schemes find limitation in such systems. With enhanced processing capacity and communication infrastructure available in smart grid platform, unit protection scheme is a promising technique. High resistance fault and time synchronization error are the problems associated with current differential and current direction-based unit protection solution. In this paper, a protection scheme capable of fast fault detection and reliable selectivity for DC microgrid is proposed. The method uses superimposed currents of both ends of a line segment during fault to derive protection decision. In the proposed method, low magnitude superimposed current during high resistance fault is still informative on Δi-plane for proper protection decision. Using PSCAD/EMTDC simulations, performance of the proposed method is evaluated and compared with available techniques. With a hardware setup in the laboratory, the performance of the protection algorithm is also tested and found to be accurate.

7.3.1 The unit protection methodology

There are protection methods available that use voltage and current (both sensors) and two IEDs for a line segment to be protected in DC microgrids. In this chapter, a protection scheme for DC microgrid is proposed reducing the number of sensors and IEDs in a line segment. Usually, protection methods require data communication as both ends current information is used for protection. Analysis reveals that the superimposed currents of both ends correctly discriminate internal and external faults of a line in a DC microgrid. Small line resistance and inductance and charged DC link capacitor cause fast rate of change of current during fault in such a system.

A 500 kW, 350 V DC microgrid, as shown in Figure 7.4 is considered consisting of DC-DC boost converters at PV array terminals, a battery with bidirectional DC-DC buck-boost converter and a front-end voltage source converter (VSC) connected to utility grid. Both AC and DC loads in the microgrid are supplied by local PV arrays and when necessary, grid feeds power. Battery acts as a backup to manage during islanded mode of operation. The DG operations in a microgrid can be categorized into grid-forming and grid-following modes [27]. In this case, the battery operates as grid-forming source, regulating voltage and acting as a slack terminal in the islanded mode. The front-end VSC maintains voltage at the point of common coupling for grid connected mode of operation. The PV arrays operate in grid-following mode and injecting power using MPPT control strategy. The details on different functions and control of DG units are provided in Table 7.2.

The DC microgrid of Figure 7.4 is considered. The microgrid is divided into different protection segments. For each such segment, a solid-state circuit breaker (SSCB) is provided at each end. An IED is installed for each segment to derive protection decision using currents of both ends. Data are collected at a rate of 4 kHz

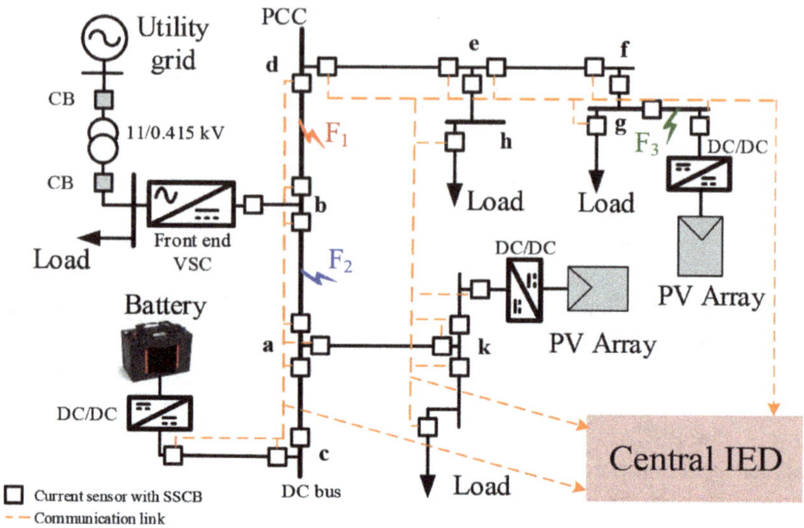

Figure 7.4 Single-line diagram of the DC microgrid

Table 7.2 Assigned functions and control of sources in DC microgrid

Types	Sources	Functions	Control strategies
Grid-forming	Battery	Energy storage and voltage control	PI controllers
	Front-end VSC	Slack terminal	
Grid-following	PV arrays	Power injection	MPPT

by the IEDs. Once a fresh sample of current is available, a disturbance index (g) is calculated using following relation:

$$g = \frac{1}{N\Delta t}\left(\sum_{j=1}^{N}|i_{j+1} - i_j|\right) \tag{7.12}$$

where i_j corresponds to sample value of current, Δt is sampling interval and number of samples in the window $N = 4$. When g exceeds a threshold ξ, the main algorithm for protection decision is activated. For disturbance like load change, the rate of rise of current is slower, and g does not exceed ξ. For 4 kHz sampling, the threshold ξ = 5000 A/s with change of current over the period of $N\Delta t$ is 5 A. This provides high security in protection decision. Currents of both ends of a line segment are communicated to the IED (Figure 7.5 and Figure 7.6.) and the sign of the superimposed current at each end is obtained.

The line current during an internal fault consists of two components, i.e., fault component of current and prefault current.

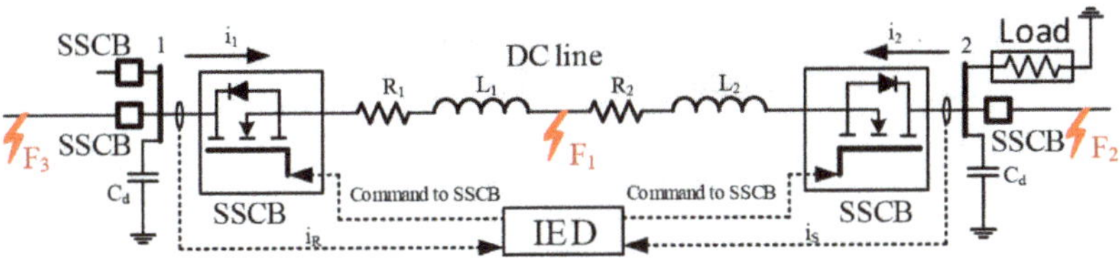

Figure 7.5 An internal fault in the line

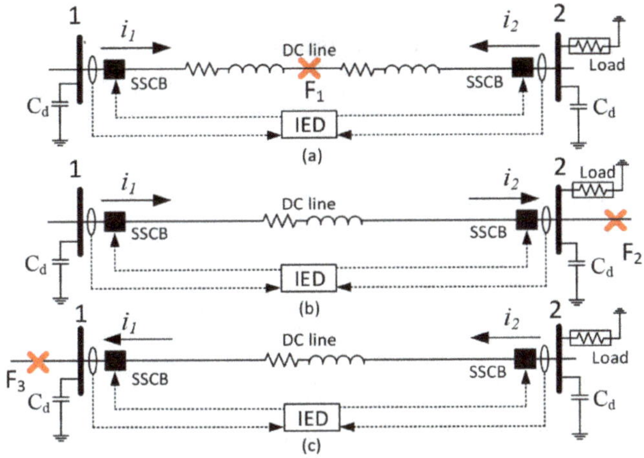

Figure 7.6 Internal and external faults for a DC line

$$i_1 = i_1^{F1} + i_{load} \tag{7.13}$$

In state-1 fault analysis, considering switches being blocked and a small fault loop resistance, the current during fault at F_1 will be [17]

$$i_1^{F1}(t) \simeq \frac{V_{dc0}}{\omega_1 L_1} e^{-\alpha_1 t} sin(\omega_1 t + \phi) \tag{7.14}$$

where $\alpha_1 = R_1/2_{L1}$, $\omega_1 = \sqrt{\frac{1}{L_1 C_d} - (\frac{R_1}{2L_1})^2}$ and $\omega_0 = \sqrt{\alpha_1^2 + \omega_1^2}$. Fault component of current is the resultant of current fed by source and equivalent capacitance in the fault loop.

$$i_1^{F1}(t) = \frac{V_{dc0}}{R_1}(1 - e^{-\frac{R_1}{L_1}t}) + \frac{V_{dc0}}{Z_{01}} e^{-\alpha_1 t} sin(\omega_1 t + \phi) \tag{7.15}$$

where Z_{01} is the surge impedance of cable up to fault point represented by $Z_{01} = \sqrt{\frac{L_1}{C_d}}$. In this case, the superimposed current at bus 1 for fault in line segment 1–2 of Figure 7.5, Δi_1 is obtained as

$$\Delta i_1(t) = i_1^{F1}(t) - i_1^{pre} \tag{7.16}$$

where $i_1^{pre} = \frac{V_{dc0}}{Req}$ and $R_{eq} = R_{line} + R_{load}$. It is clear from (7.16) that Δi_1 is positive during an internal fault. In a similar way at bus 2,

$$i_2^{F1}(t) = \frac{V_{dc0}}{R_2}(1 - e^{\frac{R_2}{L_2}t}) + \frac{V_{dc0}}{Z_{02}} e^{-\alpha_2 t} sin(\omega_2 t + \phi) \tag{7.17}$$

and as convention on direction of currents, as shown in Figure 7.5:

$$i_2^{pre} = -i_1^{pre} \tag{7.18}$$

where $\alpha_2 = R_2/2L_2$, $\omega_2 = \sqrt{\frac{1}{L_2 C_d} - (\frac{R_2}{2L_2})^2}$, $\omega_0 = \sqrt{\alpha_2^2 + \omega_2^2}$ and Z_{02} is the impedance of cable up to fault point and is represented as $Z_{02} = \sqrt{\frac{L_2}{C_d}}$. Furthermore, from Figure 7.7. it is clear that for the internal fault, i_2^{F1} and i_2^{pre} are in opposition; thus

$$\Delta i_2(t) = i_2^{F1}(t) - i_2^{pre} \tag{7.19}$$

Thus, Δi_2 is also positive during an internal fault case.

In case of an external fault F_2, the current seen at bus 1 is,

$$i_1^{F2}(t) = \frac{V_{dc0}}{R_{eq}}(1 - e^{-\frac{Req}{Leq}t}) + \frac{V_{dc0}}{Z_{eq}} e^{-\alpha_1 t} sin(\omega_1 t + \phi) \tag{7.20}$$

where $\alpha_1 = R_{eq}/2L_{eq}$, $\omega_1 = \sqrt{\frac{1}{LeqC_{deq}} - (\frac{Req}{2Leq})^2}$ and $\omega_0 = \sqrt{\alpha_1^2 + \omega_1^2}$, L_{eq} and C_{deq} are the equivalent resistance, inductance and capacitance up to the fault F_2 point, respectively, and $Z_{eq} = \sqrt{\frac{Leq}{C_{deq}}}$. In this case, $i_2^{F2} = - i_1^{F2}$. The superimposed currents at both ends of the line segment during the external fault F_2 are,

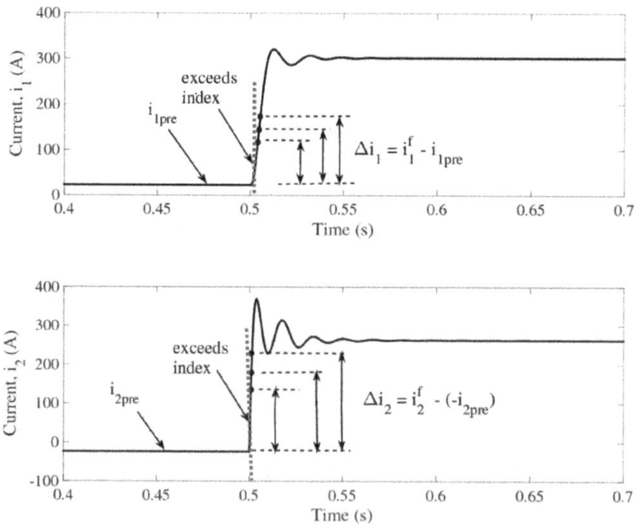

Figure 7.7 Current response during an internal fault (a) sensor at bus 1 (b) sensor at bus 2

$$\Delta i_1(t) = i_1^{F2}(t) - i_1^{pre} \qquad \Delta i_2(t) = -i_1^{F2}(t) - (-i_1^{pre}) \tag{7.21}$$

It is clear that for external fault (F_2) case, the sign of Δi_1 is positive and negative for Δi_2.

External fault at F_3 is analyzed in a similar way, when Δi_1 becomes negative and Δi_2 is positive. Thus, an internal fault is confirmed when both Δi_1 and Δi_2 are positive. Otherwise, it is a case of external fault. Note that from the sign of Δi_1 and Δi_2, even the external fault position with respect to the line segment can also be identified. The changes in currents of both sides for different situations can be represented in a four-quadrant plane, as shown in Figure 7.8. The small region around the origin is left out to avoid errors due to measurements, etc. and is taken care with as threshold. The calculation steps of proposed protection scheme are provided in Figure 7.9. The consistency of the sign of superimposed current is checked till three consecutive values should be either positive or negative. This avoids any spurious decision for spikes or so.

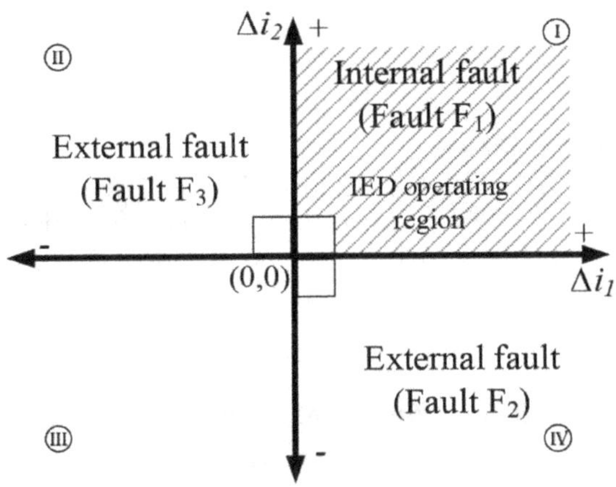

Figure 7.8 Δi_2 vs. Δi_1 in four-quadrant

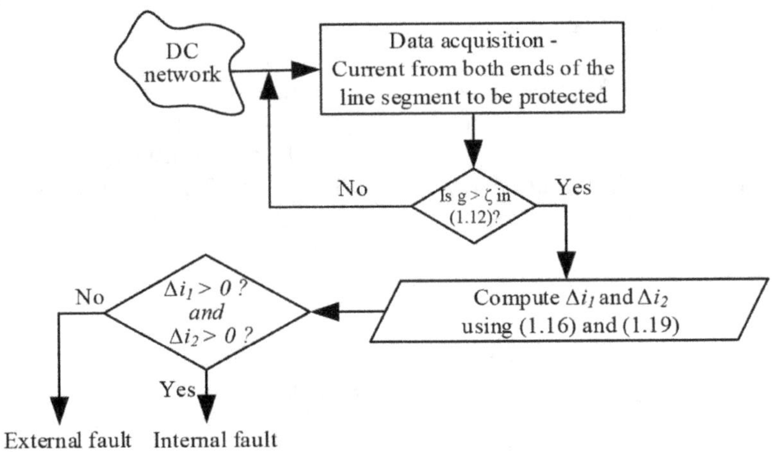

Figure 7.9 Flow diagram of proposed protection algorithm

7.3.2 Operation issues and economic aspects

The design criteria of any protection system needs to address its operability and associated cost. For the proposed protection scheme, these aspects are addressed below.

7.3.2.1 Synchronization error

Correct operation of current differential protection requires time synchronized measurements [28]. It is difficult to achieve this in case of high *di*/*dt* (often during a fault in DC system) [12]. There are different causes of time synchronization error, out of which timing error between communication devices [29] and non-synchronous current sampling are common [12]. The problems associated with differential sum due to synchronization error are analyzed in [12] and a threshold (current in Ampere) is set considering time synchronization error for high *di*/*dt*. However, identification of high resistance internal fault is a concern with such threshold.

Change in currents at both ends of the line segment during fault including time synchronization error can be expressed as

$$\Delta i_1(t_k) = i_1^f(t_k) - i_{pre} \quad (7.22)$$

$$\Delta i_2(t_k) = i_2^f(t_k + \epsilon_t) - i_{pre} \quad (7.23)$$

where ϵ_t is the time synchronization error (in this case one sampling interval, 250 μs). Proposed method uses superimposed currents with sign rather than their magnitudes and the coordinates of Δi-plane for decision. Therefore, from (7.22) and (7.23) it can be inferred that synchronization error is not a problem to identify an internal fault with proposed method.

7.3.2.2 Communication channel requirement

Deployment of digital microwave and direct fiber optic connections along with synchronous optical network or synchronous digital hierarchy systems provides readily available communication channels and expands the application of differential protection solutions. Today's current differential scheme works with 64 kbps channels [30]. The amount of data that needs to be transferred with 4 kHz sampling rate for the application of proposed method can be realized as 16 bits per each sample (in a quarter of ms). With the communication protocol in IEC 61850 that requires high speed peer-to-peer communication capability [31] and infrastructure, this is not difficult to achieve in a smart DC microgrid. For short distances within 2 km, channel delays for fiber and microwave are 4.9 μs and 3.3 μs, respectively [32]. Synchronization error up to 250 μs considered in the above case and is not a problem for proposed Δi-plane-based protection solution.

7.3.3 Economics

Installation of current and voltage sensors including IEDs in each end of a line segment for a simple protection system increases the implementation cost [33]. Cost of power supply interruption, increased network loss, DG unit outage and penalty for power quality or contract/regulatory are the important consideration for distribution system operator and DG owners. The economic evaluation of protection system must include the aforementioned operational and maintenance measures that can be due to protection operation [34]. Proposed method uses only one IED and two current sensors for each line segment. Compared to available protection schemes [7, 35–37] that use both voltage and current data, investment on proposed scheme is economical.

7.3.4 Case study

The proposed algorithm is tested for a monopolar 350 V DC microgrid shown in Figure 7.4. The system components and their parameters are provided in Table 7.1. Using PSCAD/EMTDC simulation, faults are created in different line segments and data are acquired at a sampling rate of 4 kHz. The performance of the method is tested for different cases like internal/external faults, high resistance fault and time synchronization error issue.

7.3.4.1 Internal and external fault cases

A pole-to-ground internal fault (F_1) with a fault resistance of 0.2 Ω is simulated at 0.5 seconds in the middle of the line segment 1–2 (Figure 7.5). The currents seen by the sensors at bus 1 and 2 are shown in Figure 7.7. and Figure 7.7., respectively. The superimposed currents of both ends (Δi_1 and Δi_2) are found to be positive. Similarly, external faults F_2 and F_3 as in Figure 7.5 with a fault resistance of 0.1 Ω are created. For external fault F_2 (right to line segment 1–2), current seen by sensor at bus 2 is shown in Figure 7.10. where the superimposed current Δi_2 is negative. The trajectories of current changes during F_1, F_2 and F_3 faults are found in first, fourth and second quadrant, respectively, as shown in Figure 7.11. The performance

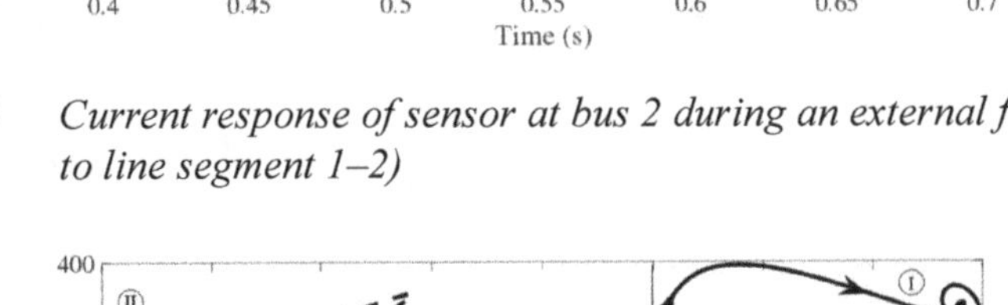

Figure 7.10 Current response of sensor at bus 2 during an external fault (right to line segment 1–2)

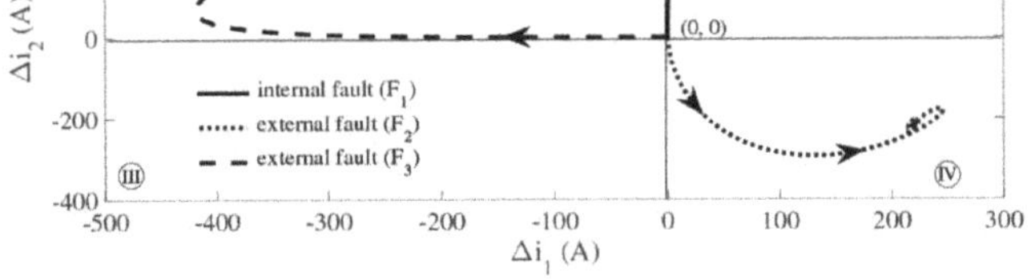

Figure 7.11 Internal and external faults in Δi-plane

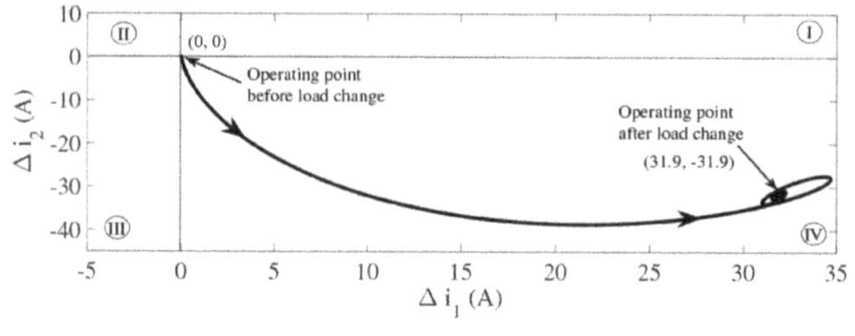

Figure 7.12 Shifting of operating point for sudden load change

of the method is also tested for sudden change in load case. The load is varied from 60% to 110% of rated value at bus 2. The trajectory for the case in Δi-plane settles in fourth quadrant, as shown in Figure 7.12, confirming not an internal fault.

7.3.4.2 Comparative assessment

The performance of the proposed method is compared with available current differential and current direction information-based protection schemes during high resistance fault and time synchronization error.

7.3.4.3 Performance during high resistance internal fault

High resistance internal fault is a problem for current differential [12] and current direction-based protection schemes [6]. The operating current in the differential technique goes below the set threshold for fault resistance more than 5.5 Ω, as evident from Figure 7.13. Similarly, the direction of current has not changed during the internal fault, as shown in Figure 7.14 and the fault cannot be identified correctly. Using the trajectory of change in current instead of simply that of current has the advantage during high resistance internal fault. Furthermore, in Figure 7.15, it is evident that the trajectory of currents of both ends is found in fourth quadrant even during high resistance internal fault, whereas the trajectory of superimposed currents is found in first quadrant that identifies the high resistance internal fault correctly.

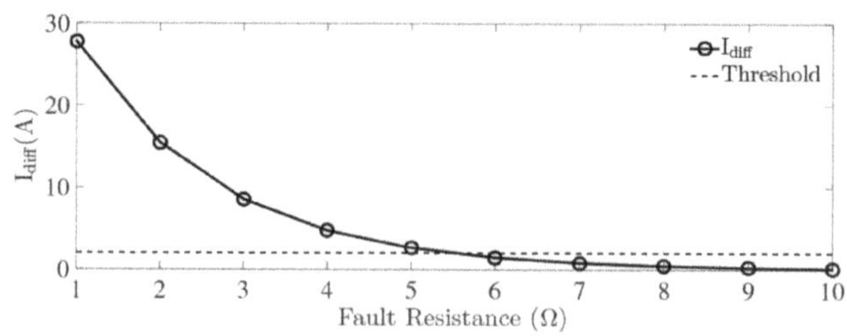

Figure 7.13 Current difference during high resistance internal fault

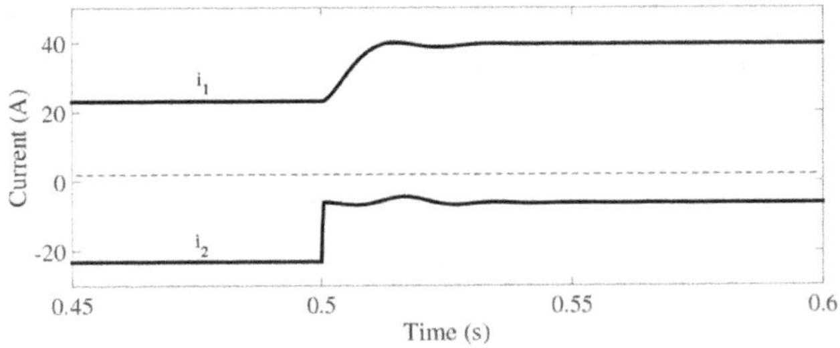

Figure 7.14 Currents of both ends during high resistance internal fault

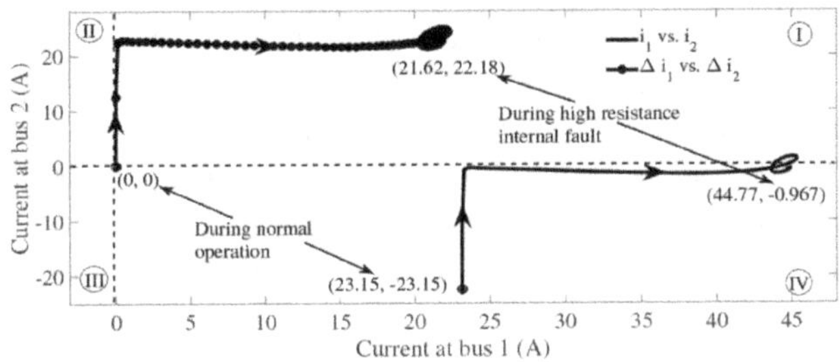

Figure 7.15 Currents and superimposed currents during high resistance internal fault

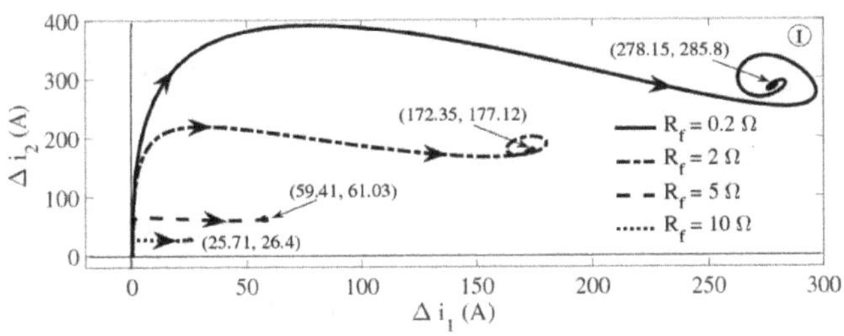

Figure 7.16 Performance of proposed method during high resistance internal fault

Internal faults for different fault resistances are simulated for pole-to-ground fault at 0.5 seconds of simulation run in the line segment 1–2. Performance of the method is evaluated for the cases, and the changes in current trajectories are plotted in Figure 7.16. All the trajectories settle in first quadrant in Δi-plane and the method identifies the internal faults correctly.

7.3.4.4 Performance with time synchronization error

Time synchronization error (ϵ_t) results in maloperation of current differential scheme during an external fault. Pole-to-ground fault is created beyond the line segment 1–2 and in line 2–3 with $\epsilon_t = 100\ \mu s$. For the differential scheme, the current difference exceeds the threshold and enters into operating region, as shown in Figure 7.17. Thus, method in [12] will malfunction in such a case. The proposed algorithm is tested for the case and change in current is found in fourth-quadrant of Δi-plane, as shown in Figure 7.18, ensuring an external fault. This confirms that proposed method is correct in the decision.

7.3.5 Performance with noisy signals

Noise in signal causes non-zero current difference at both ends of a line segment during normal operation. Available differential protection is vulnerable in such a situation. The performance of proposed method is tested for signals contaminated with noise. Internal and external faults, F_1, F_2 and F_3 are created in the line segment 1–2, and the current signals are measured at both ends and contaminated by a uniform distribution noise with zero mean and a standard deviation of 2.5% [38]. The changes in currents are plotted in Δi-plane, as shown in Figure 7.19. Proposed method correctly identifies fault at F_1 as internal and faults F_2 and F_3 as external faults, respectively.

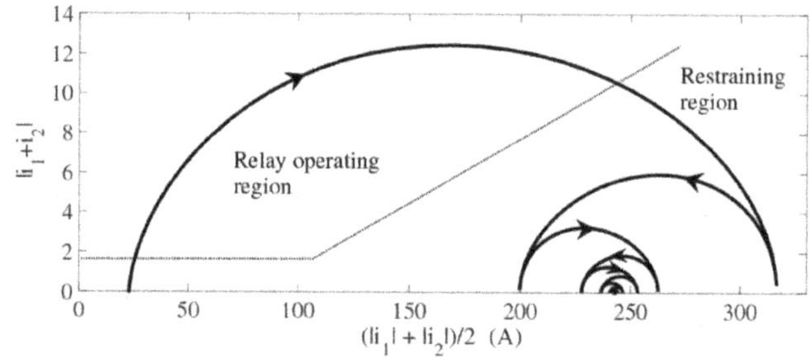

Figure 7.17 Current differences with synchronization error during an external fault

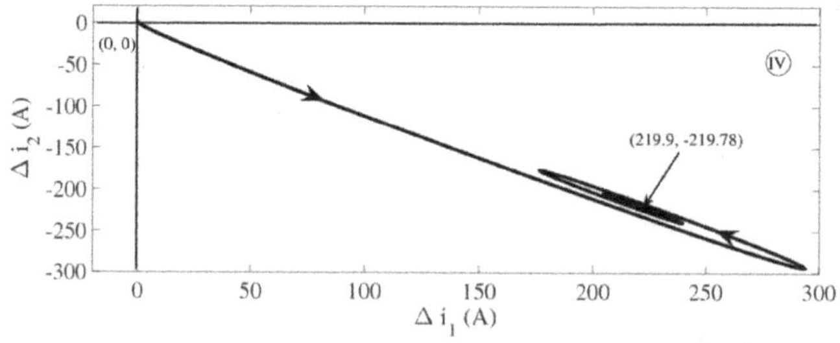

Figure 7.18 Change in current during an external fault with time synchronization error

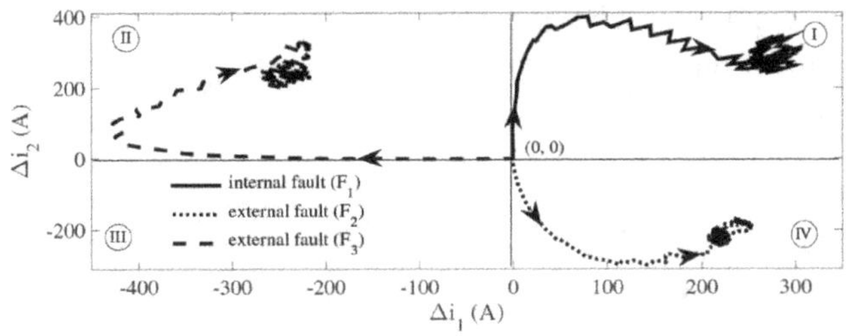

Figure 7.19 Internal and external faults representation in Δi-plane with noise in the current signals.

7.4 Centralized protection of DC microgrids

Unit protection schemes exploit data from both ends of the line to be protected by overcoming the abovementioned issues. Regardless of its advantages, high resistance fault could be an issue for current differential [11, 12] and current directional [6]-based unit protection schemes. The performance of the differential scheme is highly dependent on the current threshold, since the operating current may not exceed it during high resistance fault. Similarly, the direction of fault current during high resistance fault does not alter, as the pre-fault current can be more than the faulted current under heavy loading condition. In addition, communication delay and communication link failure limit the application of unit protection schemes.

IEDs are used for protection and automation with the support of IEC 61850 protocol [39]. Since IEC 61850 is known for its flexibility due to its capability of decoupling domain-specific models from the communication stack, it has been adopted as an international communication medium to monitor, control and measure the physical processes in a microgrid. A centralized protection scheme compatible with IEC 61850 protocol has the capability to provide promising solution for microgrids that can deal with the above-stated issues associated with both non-unit and unit protection schemes. In [40–42], similar protection schemes have been proposed for DC microgrids. A current derivative-based protection algorithm is proposed in [40]. However, the selection of threshold in a multi-converter system can become a challenging task as low impedance of DC line segment can offer high rate of change of current even during non-faulted conditions. Furthermore, a differential current-based protection approach is used in [41, 42], where the accuracy of these schemes cannot be guaranteed for high resistance faults. Moreover, the effects of communication delay and link failure on the performance of these methodologies are not studied.

To address these issues, a centralized protection scheme based on the similarity of current change over a window of both ends of a line segment is proposed for DC microgrids. The cosine similarity index (*CSI*) is used to differentiate the internal and external faults. Unlike sample-to-sample-based approach [43], a window of data set is considered to overcome the effect of communication delay in the protection scheme. The current from each line segment is communicated to a central IED using IEC 61850 protocol for DC microgrid. The central IED overcomes the issue associated with communication link failure and delay from any location using data of adjacent line segments and generates a trip signal accurately for the corresponding breaker in the faulted segment. Robustness in the performance of the proposed method is evaluated for various cases including high resistance fault, communication delay and link failure.

7.4.1 The centralized protection methodology

The DC microgrid considered in this work is shown in Figure 7.4. The microgrid is divided into different protection segments, and each segment consists of an SSCB at each end. A central IED is installed for the entire microgrid. The cyber layer, shown in Figure 7.4, is realized using *SimEvents*-based realistic communication model to simulate all the intermittent parameters associated with the communication network [44]. Currents of both ends of a line segment are communicated to the IED (highlighted in Figure 7.4) and the rate of change of current at each end is obtained.

7.4.1.1 Fault detection

The collected line currents at IED are sampled at a rate of 4 kHz. A disturbance index (h) is calculated using current information from the following relation

$$h = \frac{1}{N\Delta t}\left(\sum_{n=1}^{N} |i_{n+1} - i_n|\right) \tag{7.24}$$

where i_n corresponds to sample value of current for nth instant with $N = 4$ as four samples in 1 ms, Δt is the sampling interval. The absolute value of the current difference is considered in (7.24) to monitor the magnitude change only. When $h > \xi$ (threshold), a fault is ensured; else, it suggests a normal condition. It is to be noted that the disturbance index becomes high even for load change or any other switching phenomena. This is similar to a fault detector in available numerical relays that triggers the next step on the decision process.

7.4.1.2 Current response during an internal fault

In Stage 1 of fault analysis, the RLC response of the DC line current i_{db} at bus d in frequency domain during F_1 can be written as:

$$i_{db}(s) = \frac{\frac{V_{dd}(0)}{L_1} + i_{db}(0)}{s^2 + \frac{R_1}{L_1}s + \frac{1}{L_1 C_d}} \tag{7.25}$$

where $V_{dd}(0)$ is the voltage across the C_{dd} at bus d and $i_{db}(0)$ is the current through L_1 at the instant of fault inception. Fault component of current in time domain is given by:

$$i_{db}^{F1}(t) = \frac{V_{dd}(0)}{L_1(m_2 - m_1)}(e^{-m_1 t} + e^{-m_2 t}) + \frac{i_{db}(0)}{m_2 - m_1}\left(-m_1 e^{-m_1 t} + m_2 e^{-m_2 t}\right) \tag{7.26}$$

where $m_1 = -\alpha_1 + \sqrt{\alpha_1^2 - \omega_0^2}$, $m_2 = -\alpha_1 - \sqrt{\alpha_1^2 - \omega_0^2}$ and, $\alpha_1 = R_1/2L_1$, $\omega_1 = \sqrt{\frac{1}{L_1 C_{dd}} - (\frac{R_1}{2L_1})^2}$ and $\omega_0 = \sqrt{\alpha_1^2 + \omega_1^2}$, where, Z_{01} is the impedance of cable up to fault point represented by $Z_{01} = \sqrt{\frac{L_1}{C_{dd}}}$. The time derivative of (7.26) at $t = 0^+$at the instant of fault inception will be:

$$\frac{di_{db}(t)}{dt} = \frac{V_{dd}(0)}{L_1} - 2\alpha_1 i_{db}(0) \tag{7.27}$$

Using (7.27), it can be concluded that the rate of change of current depends on the initial conditions of voltage across the DC link capacitor and line current just before the fault.

In a similar way at bus b, for the fault F_1, the rate of change of line current i_{bd} will be:

$$\frac{di_{bd}(t)}{dt} = \frac{V_{db}(0)}{L_2} + \frac{R_2}{L_2} i_{bd}(0) \tag{7.28}$$

Remark I: *As a convention with the direction of currents, as shown in Figure 7.4, it is clear that for internal fault* F_1, $\frac{di_{db}(t)}{dt}$ *and* $\frac{di_{bd}(t)}{dt}$ *are positive.*

7.4.1.3 Current response during an external fault

Remark II: *In case of an external fault* F_2, *as shown in Figure 7.4,* $\frac{di_{db}(t)}{dt}$ *and* $\frac{di_{bd}(t)}{dt}$ *are equal in magnitude as well as their rate of changes. However, the sign will be opposite, e.g.,* $\frac{di_{db}(t)}{dt} = -\frac{di_{bd}(t)}{dt}$.

Thus,

$$\frac{di_{bd}(t)}{dt} = \frac{d(-i_{db}(t))}{dt} = \frac{-di_{db}(t)}{dt} = -\frac{V_{dd}(0)}{L_1} + \frac{R_1}{L_1} i_{db}(0) \tag{7.29}$$

7.4.2 Cosine similarity index (CSI)-based protection decision

In the above subsections, the direction of current during transient is analyzed. A simple threshold-based protection decision is challenging for power-electronic converters due to distortion and high-frequency ripples during transients. Therefore, a window of 2 ms (i.e., 8 samples$\times\frac{1}{4000}$s) is considered, as shown in Figure 7.20. The average of the rate of change of current in a window is obtained at each end of a line segment. The *alikeness* of two data windows of both ends of the line segment is measured using cosine similarity. The idea of cosine similarity used in the proposed method is analogous to the *CSI* of two non-zero vectors in a dot product, which has been widely used in data mining applications [45, 46].

Using two consecutive samples, the rate of change of current is obtained in a window of 2 ms; 7 number of $\frac{di}{dt}$ samples in a window (in Figure 7.20), where

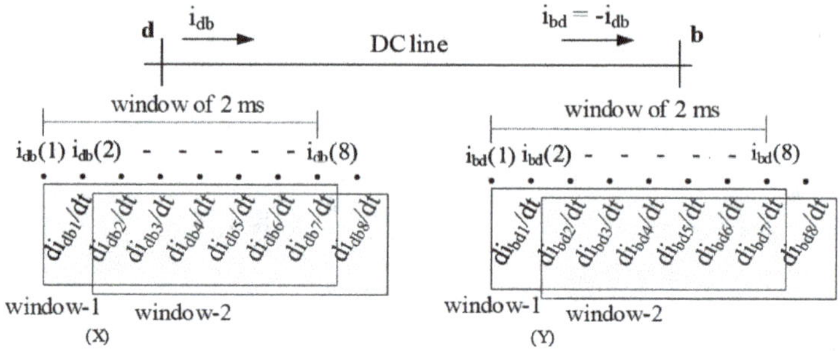

Figure 7.20 Independent data windows of both ends of the line segment b–d

$di_{db}(t)/dt = (i_{db}(t_2) - i_{db}(t_1))/(t_2 - t_1)$. For a window of data set, $X = \{x_1, x_2, x_3, ...x_7\} = \left\{\frac{di_{db1}}{dt}, \frac{di_{db2}}{dt}, \frac{di_{db3}}{dt}, ...\frac{di_{db7}}{dt}\right\}$ and $Y = \{y_1, y_2, y_3, ...y_7\} = \left\{\frac{di_{bd1}}{dt}, \frac{di_{bd2}}{dt}, \frac{di_{bd3}}{dt}, ...\frac{di_{bd7}}{dt}\right\}$, CSI can be calculated as below:

$$CSI = \frac{X.Y}{|X|\,|Y|} = \frac{\frac{1}{7}\sum_{n=1}^{7} x_n . \frac{1}{7}\sum_{n=1}^{7} y_n}{\left|\frac{1}{7}\sum_{n=1}^{7} x_n\right| \left|\frac{1}{7}\sum_{n=1}^{7} y_n\right|} = \frac{\sum_{n=1}^{7} x_n . \sum_{n=1}^{7} y_n}{\left|\sum_{n=1}^{7} x_n\right| \left|\sum_{n=1}^{7} y_n\right|} \tag{7.30}$$

For an external fault F_2, $X \approx \sum_{n=1}^{7} x_n$ and $Y \approx \sum_{n=1}^{7} y_n$ in the line segment d–b are positive and negative, respectively, using Remark II. However, for internal fault F_1, both X and Y are positive using Remark I. The similarity index output can be summarized by [47]:

$$\text{CSI}=\begin{cases} +1, & \textbf{if}, \text{internal fault;} \\ -1, & \textbf{if}, \text{external;} \end{cases} \tag{7.31}$$

The similarity of change in current at both ends of the line segment, X and Y in (7.30) is shown in Figure 7.21. For the internal fault (F_1), CSI comes out to be +1

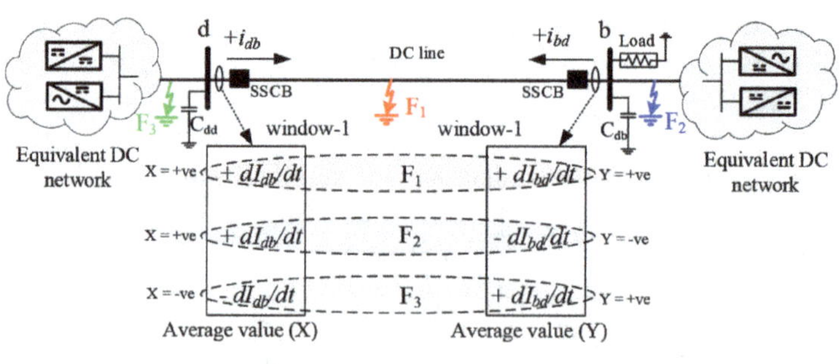

Figure 7.21 Similarity of windows during internal and external faults

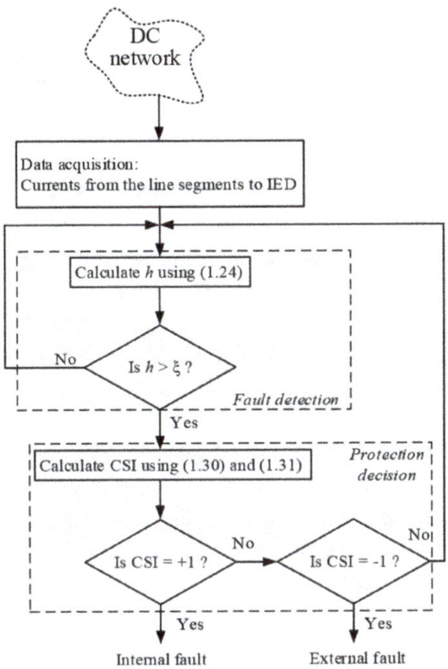

Figure 7.22 Flow diagram of proposed protection algorithm

as both X and Y are positive. However, *CSI* is –1 for the external faults (F_2 and F_3) as X and Y are opposite in sign. A detailed flowchart of the proposed protection scheme is provided in Figure 7.22.

7.4.2.1 Communication link failure from one end of the faulted segment

The fault F_1 cannot be confirmed from X and Y of line segment d–b if i_{bd} is not communicated due to communication link failure, as shown in Figure 7.23. In this case,

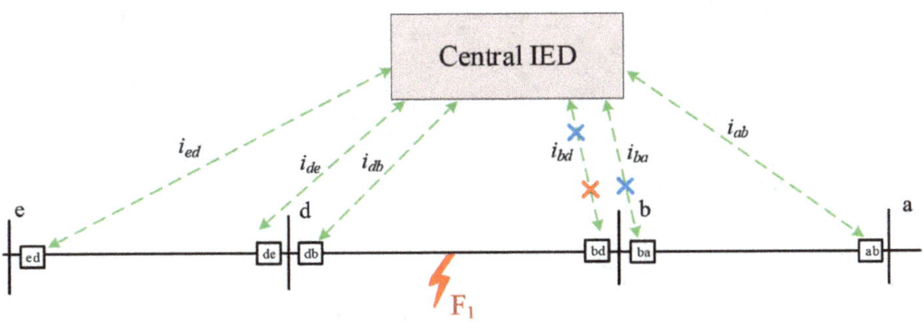

Figure 7.23 Performance during communication links failure

the IED uses the similarity of i_{db} and i_{ba} to generate a trip decision. Subsequently, the trip signal will be generated if $|i_{db}| - |i_{ba}| > |i_{load_b\_rat}|$, where $i_{load_b\_rat}$ is the rated load current at bus b. Thus, the proposed method performs satisfactorily, even if communication link fails from one end of a line segment.

7.4.2.2 Performance during multiple link failure

In case of multiple communication link failure, a fault can be identified using the available information of rate of change of current. For example, fault F_1 can be identified using the *CSI* of i_{db} and i_{ab} in case of communication failure of i_{bd} and i_{ba} (refer Figure 7.23). Under such circumstances, the adjacent segment including the faulted one will be isolated. The trip decision will be derived based on *CSI* =+1 and for the fault F_1, the line segments d–b and a–b will be isolated.

It is worth notifying that the proposed method identifies the exact faulted segment when one communication channel fails in the microgrid or two channels fail (one from each non-adjacent segment). Further, when two or three channels fail from adjacent segments, two segments are identified as possible faulted segments. The selectivity of proposed method during communication link failure is listed in Table 7.3.

7.4.2.3 Performance during communication delay

There are different causes of communication delay, out of which time-out error between communication devices [29] and non-synchronous current sampling are common [12]. The change in currents at both ends of the line segment during fault in the presence of communication delay can be expressed as

$$X = \frac{1}{7}\sum_{n=1}^{7} x_n \quad \text{and,} \quad Y = \frac{1}{7}\sum_{n+\epsilon_n=1}^{7+\epsilon_n} y_n \tag{7.32}$$

where ϵ_n denotes the delayed sample number (in this case, one lost sample corresponds to a delay of 250 μs). The proposed method uses the sign similarity of average rate of change of current in a window of 2 ms rather than their magnitudes to generate the trip decision. It can be inferred that the proposed strategy is resilient to communication delay.

Table 7.3 Selectivity with communication link failure

Number of communication channel failure	Segment isolation
No channel	Only the faulted segment
One channel	
Two channels	Faulted and the adjacent one
Three channels	

7.4.3 Case study

The proposed algorithm is tested for the 350 V DC microgrid shown in Figure 7.4. The system components and their parameters are provided in Table 7.1. Using PSCAD/EMTDC simulation, faults are created in different line segments, and data are acquired at a sampling rate of 4 kHz. The performance of the method is tested for several cases, e.g., fault at different locations, communication link failure, high resistance fault and communication delay.

7.4.3.1 Fault identification and protection decision

A pole-to-ground fault (F_1) with a fault resistance of 0.1 Ω is simulated at 0.5 seconds in the middle of the line segment b–d (Figure 7.4). The currents and the rate of change of current seen by the sensors at SSCB-db and SSCB-bd are shown in Figure 7.24(a) and Figure 7.24(b), respectively. As shown in Figure 7.24(c), the average in the change of current in a window at both ends (X and Y) is found to be positive, which results in *CSI* to be +1 using (7.30).

Similarly, an external fault F_2 (in the system in Figure 7.4) with a fault resistance of 0.1 Ω is created at 0.5 seconds. The currents seen by sensors at SSCB-db and

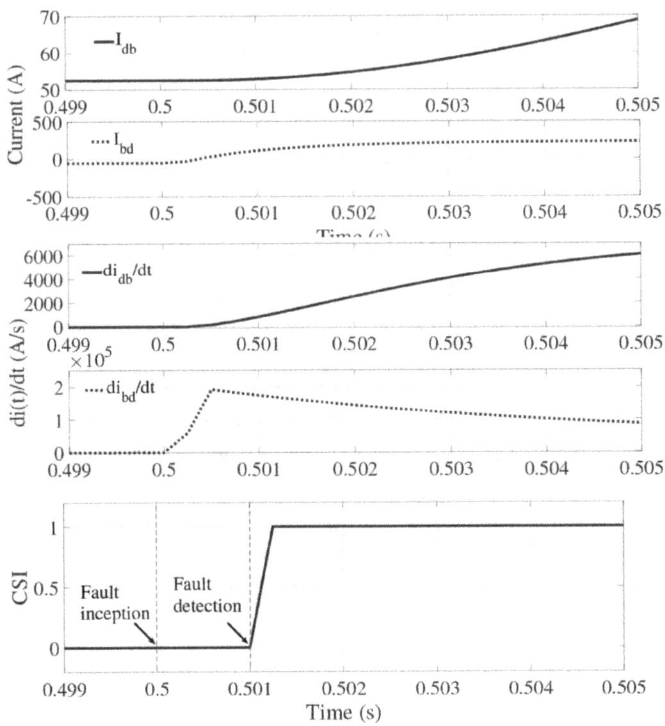

Figure 7.24 For the internal fault F_1 *in the line segment d–b (a) current at d and b sides (b) rate of change in current (c) the correspondingCSI*

SSCB-bd are shown in Figure 7.25(a). Furthermore, it can be seen in Figure 7.25(b) that the signs of change in current over a window at both ends of the line segment d–b are moving in opposite directions. As a result, it is confirmed from Figure 7.25(c) that *CSI* is −1.

The performance of the method is also tested for sudden change in loading condition. The load is varied from 60 to 110% of rated value at bus b. The *CSI* is obtained as −1, as shown in Figure 7.26. Therefore, no trip signal will be generated.

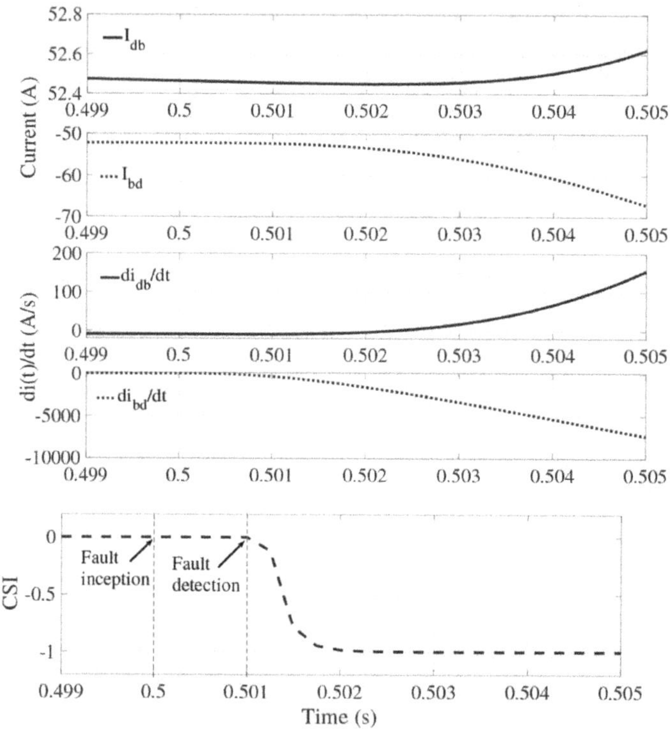

Figure 7.25 For the external fault F$_2$ *in the line segment a–b (a) current at d and b sides (b) rate of change in current (c) the correspondingCSI*

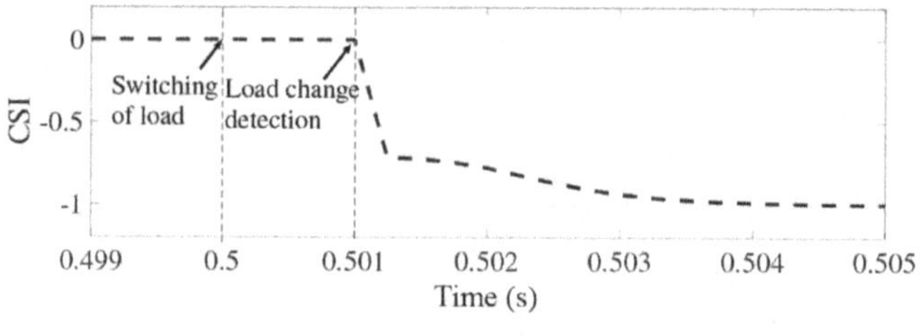

Figure 7.26 CSI for sudden load change

Table 7.4 Selectivity during communication failure for fault F_1

Events	Line segment bd	At central IED	Remarks
No communication failure	i_{db} is available i_{bd} is available	X and Y are positive and CSI =+	Trip SSCB-db and SSCB-bd
One end communication failure	i_{db} is not available i_{bd} is available	Apply KCL at bus d	Trip SSCB-db and SSCB-bd
Both ends communication failure	i_{db} is not available i_{bd} is not available	Apply KCL at bus d and b	Trip SSCB-db and SSCB-bd
Both ends and one end from adjacent lines fails	i_{db} is not available i_{bd} and i_{ba} are not available	Apply KCL at bus d and a	Trip SSCBs at segment bd and ab
Load change in bus b	i_{db} is available i_{bd} is available	i_{db}, iab are positive and i_{bd}, i_{ba} are negative, CSI = −1	No trip signal

7.4.3.2 Performance during communication link failure

In case of one or two communication link failure, the proposed method is able to derive trip decision without losing proper selectivity. The influence of communication link failure is analyzed for various cases and the selectivity is provided in Table 7.4.

7.4.4 Discussion

In the proposed method, the fault detection does not take more than 1 ms as it uses four samples at a sampling frequency of 4 kHz (using (7.24)). A window of 2 ms is used to calculate *CSI*, thus the protection decision can be taken in 3 ms of the fault inception, which is evident from the results. The average of current change over a window is considered to avoid the issues associated with communication delay and any spurious data in the signal. As an advantage, the proposed method identifies the exact faulted segment during a communication channel fails. Furthermore, the faulted segment along with an adjacent one are isolated in case two or three channels fail. The method can be applied to any DC microgrid including ring bus, grounded and ungrounded systems as it uses current signal of each line segment.

7.5 Performance of communication-based protections during cyber-attacks

Communication-assisted protection schemes are prone to cyberattacks where the relays maloperate even during no-fault situations.

7.5.1 Performance of unit protection for cyberattack

Current differential technique is a widely used unit protection in practice. The differential quantity is the sum of both end currents (Fig. 7.5):

$$I_{diff} = I_1 + I_2 \quad (7.33)$$

For normal condition, I_{diff} and for an internal fault $I_{diff} > \alpha$, where α is a protection reliability co-efficient (ranges between 0.06 and 0.2) and I_{nom} is the nominal current of the line segment.

In case if one end current is subjected to cyberattack, as shown in Figure 7.27., the differential relay maloperates even if there is no internal fault, as shown in Figure 7.28. At 0.4 s, false data are injected to the current I_1, which leads to the current difference.

7.5.2 Performance of centralized protection for cyberattack

The performance of the centralized protection scheme as discussed in 7.4.2 is studied for cyberattack in a communication channel. For example, the current data i_{32}

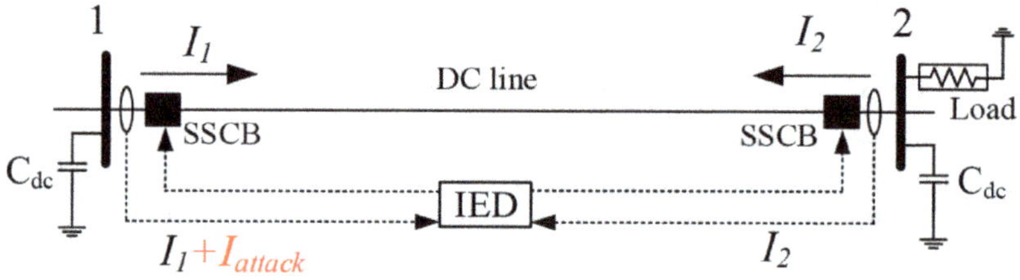

Figure 7.27 Differential protection under cyberattack

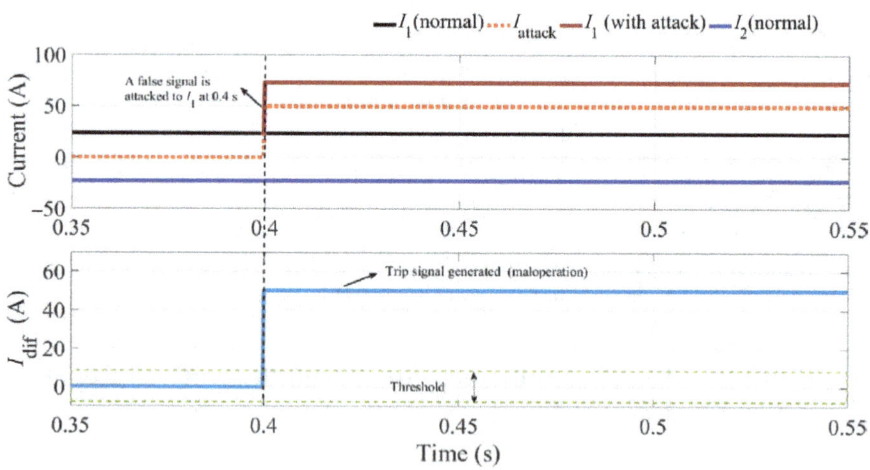

Figure 7.28 Trip signal during cyberattack

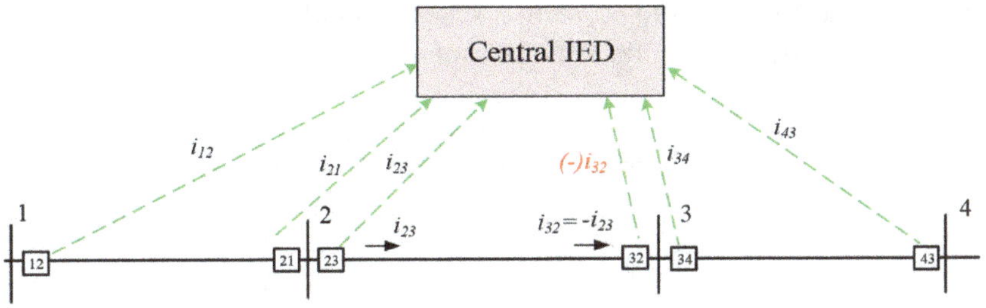

Figure 7.29 *Cyberattack in a channel associated with centralized protection*

communicated to central IED is multiplied by a factor −1 (attack), as shown in Figure 7.29.

$$CSI = \frac{\left(\frac{di_{23}}{dt}\right).\left(\frac{-di_{32}}{dt}\right)}{\left|\frac{di_{23}}{dt}\right|\left|\frac{di_{32}}{dt}\right|} = +1 \tag{7.34}$$

The corresponding CSI is obtained as +1, which leads to the generation of trip signal even with no fault in the system.

7.6 Conclusion

In this chapter, a protection scheme is proposed for DC microgrid using superimposed fault currents from both ends of a line segment. The method is able to distinguish internal and external faults correctly using corresponding coordinates of superimposed currents in the Δi-plane. Fast and selective protections are the features of proposed method. It is found that the method has advantage for high resistance internal fault compared to current differential and direction-based protection schemes. The performance of the proposed method is tested for time synchronization error, signals with noise and high fault resistance cases. For external faults, proposed method can identify the fault position with respect to the sensor positions. Proposed algorithm is also tested with data from laboratory setup and found to be accurate.

A centralized DC microgrid protection scheme is proposed. The similarity of rate of current change over a window of both ends of the line segment identifies an internal fault. IEC 61850 protocol is realized for the communication in DC microgrid application using *SimEvents* blocks. The performance of the protection scheme is tested for high resistance fault, communication delay and link failure. The protection decision is taken within 3 ms or even faster in case of no delayed signals. This protection algorithm aids in power electronic-based systems by providing an auxiliary signal to block the switching pulses being supplied to converter. As compared to available protection schemes that use both voltage and current data, the proposed protection scheme entails an economic alternative, which employs only one central IED for entire microgrid and one current sensor for each end of a line segment.

The performance of the communication-assisted protection schemes is also verified under cyberattack. It is seen that such protections are prone to cyberattacks that lead to IED maloperation even during no-fault situations. There is a scope of development of cyber-secured protection schemes in communication-assisted protections in future power networks integrated with renewable sources.

References

[1] Kumar D., Zare F., Ghosh A. 'DC microgrid technology: system architectures, AC grid interfaces, grounding schemes, power quality, communication networks, applications, and standardizations aspects'. *IEEE Access*. 2017;5:12230–56.

[2] Dragicevic T., Lu X., Vasquez J.C., Guerrero J.M. 'DC microgrids—Part II: a review of power architectures, applications, and standardization issues'. *IEEE Transactions on Power Electronics*. 2016;31(5):3528–49.

[3] Cuzner R.M., Venkataramanan G. 'The status of DC micro-grid protection'. *Proceedings of the IEEE Industry Applications Society Annual Meeting*. 2008:1–8.

[4] Salonen P., Kaipia T., Nuutinen P., Peltoniemi P., Partanen J. 'An LVDC distribution system concept'. *Nordic Workshops on Power and Industrial Electronics*; 2008. pp. A3–1A316.

[5] Kwasinski A. 'Quantitative evaluation of DC microgrids availability: effects of system architecture and converter topology design choices'. *IEEE Transactions on Power Electronics*. 2011;26(3):835–51.

[6] Emhemed A.A.S., Fong K., Fletcher S., Burt G.M. 'Validation of fast and selective protection scheme for an LVDC distribution network'. *IEEE Transactions on Power Delivery*. 2017;32(3):1432–40.

[7] Dhar S., Patnaik R.K., Dash P.K. 'Fault detection and location of photovoltaic based DC microgrid using differential protection strategy'. *IEEE Transactions on Smart Grid*. 2017;9(5):4303–12.

[8] Yuan C., Haj-ahmed M.A., Illindala M.S. 'Protection strategies for medium-voltage direct-current microgrid at a remote area mine site'. *IEEE Transactions on Industry Applications*. 2015;51(4):2846–53.

[9] Venkataramanan G., Marnay C. 'A larger role for microgrids'. *IEEE Power and Energy Magazine*. 2008;6(3):78–82.

[10] Fletcher S.D.A., Norman P.J., Galloway S.J., Crolla P., Burt G.M. 'Optimizing the roles of unit and non-unit protection methods within DC microgrids'. *IEEE Transactions on Smart Grid*. 2012;3(4):2079–87.

[11] Park J.-D., Candelaria J. 'Fault detection and isolation in low-voltage DC-bus microgrid system'. *IEEE Transactions on Power Delivery*. 2013;28(2):779–87.

[12] Fletcher S.D.A., Norman P.J., Fong K., Galloway S.J., Burt G.M. 'High-speed differential protection for smart DC distribution systems'. *IEEE Transactions on Smart Grid*. 2014;5(5):2610–17.
[13] Dong J., Gao F., Guan X., Zhai Q., Wu J. 'Storage-reserve sizing with qualified reliability for connected high renewable penetration micro-grid'. *IEEE Transactions on Sustainable Energy*. 2016;7(2):732–43.
[14] Salomonsson D., Soder L., Sannino A. 'Protection of low-voltage DC microgrids'. *IEEE Transactions on Power Delivery*. 2009;24(3):1045–53.
[15] Starke M., Li F., Tolbert L.M., Ozpineci B. 'AC vs. DC distribution: maximum transfer capability'. *Proceedings of the IEEE Power and Energy Society General Meeting – Conversion and Delivery of Electrical Energy in the 21st Century*. 2008:1–6.
[16] Saeedifard M., Graovac M., Dias R.F., Iravani R. *DC power systems: challenges and opportunities*. IEEE PES General Meeting; 2010. pp. 1–7.
[17] Meghwani A., Srivastava S.C., Chakrabarti S. 'A non-unit protection scheme for DC microgrid based on local measurements'. *IEEE Transactions on Power Delivery*. 2017;32(1):172–81.
[18] Das J.C., Osman R.H. 'Grounding of AC and DC low-voltage and medium-voltage drive systems'. *IEEE Transactions on Industry Applications*. 1998;34(1):205–16.
[19] Park J.-D., Candelaria J., Ma L., Dunn K. 'DC Ring-Bus microgrid fault protection and identification of fault location'. *IEEE Transactions on Power Delivery*. 2013;28(4):2574–84.
[20] Baran M.E., Mahajan N.R. 'Overcurrent protection on voltage-source-converter-based multiterminal DC distribution systems'. *IEEE Transactions on Power Delivery*. 2007;22(1):406–12.
[21] Tang L., Ooi B.-T. 'Locating and isolating DC faults in multi-terminal DC systems'. *IEEE Transactions on Power Delivery*. 2007;22(3):1877–84.
[22] Wang B., Sechilariu M., Locment F. 'Intelligent DC microgrid with smart grid communications: control strategy consideration and design'. *IEEE Transactions on Smart Grid*. 2012;3(4):2148–56.
[23] Che L., Khodayar M., Shahidehpour M. 'Only connect: microgrids for distribution system restoration'. *IEEE Power and Energy Magazine*. 2014;12(1):70–81.
[24] Fletcher S.D.A., Burt G.M., Galloway S.J., Norman P.J. 'Determination of protection system requirements for DC unmanned aerial vehicle electrical power networks for enhanced capability and survivability'. *IET Electrical Systems in Transportation*. 2011;1(4):137–47.
[25] Berizzi A., Silvestri A., Zaninelli D., Massucco S. 'Short-circuit current calculations for DC systems'. *IEEE Transactions on Industry Applications*. 1996;32(5):990–7.
[26] Yang J., Fletcher J.E., O'Reilly J. 'Short-circuit and ground fault analyses and location in VSC-Based DC network cables'. *IEEE Transactions on Industrial Electronics*. 2012;59(10):3827–37.

[27] Katiraei F., Iravani R., Hatziargyriou N., Dimeas A. 'IEEE power and energy magazine'. *Microgrids Management*. 2008 May;6(3):54–65.
[28] Network protection and automation guide, ch. 10 [online]. Available from http://www.alstom.com/grid.
[29] Villamagna N., Crossley P.A. 'A symmetrical component-based GPs signal failure-detection algorithm for use in feeder current differential protection'. *IEEE Transactions on Power Delivery*. 2008;23(4):1821–8.
[30] Miller H., Burger J., Fischer N., Kasztenny B. 'In: 63rd annual conference for protective relay engineers'. *Modern Line Current Differential Protection Solutions*. 2010:1–25.
[31] Cintuglu M.H., Ma T., Mohammed O.A. 'Protection of autonomous micro-grids using agent-based distributed communication'. *IEEE Transactions on Power Delivery*. 2017;32(1):351–60.
[32] Adamiak M. Digital communication for relay protection. Available from http://wwwpes-psrcorg/Reports/Digital%20communications%20for%20re-laying%20(H9)pdf.
[33] Nikolaidis V.C., Papanikolaou E., Safigianni A.S. 'A communication-assisted overcurrent protection scheme for radial distribution systems with distributed generation'. *IEEE Transactions on Smart Grid*. 2016;7(1):114–23.
[34] Penuela Meneses C.A., Mantovani J.R.S. 'Improving the grid operation and reliability cost of distribution systems with dispersed generation'. *IEEE Transactions on Power Systems*. 2013;28(3):2485–96.
[35] Salomonsson D., Soder L., Sannino A. 'Protection of low-voltage DC micro-grids'. *IEEE Transactions on Power Delivery*. 2009;24(3):1045–53.
[36] Orji U., Schantz C., Leeb S.B., *et al.* 'Adaptive zonal protection for ring Microgrids'. *IEEE Transactions on Smart Grid*. 2017;8(4):1843–51.
[37] Etemadi A.H., Iravani R. 'Overcurrent and overload protection of directly Voltage-controlled distributed resources in a microgrid'. *IEEE Transactions on Industrial Electronics*. 2013;60(12):5629–38.
[38] Mohanty R., Balaji U.S.M., Pradhan A.K. 'An accurate noniterative fault-location technique for low-voltage DC microgrid'. *IEEE Transactions on Power Delivery*. 2016;31(2):475–81.
[39] Alndreev A.N., Patrashkin E.A., Zheltov A.S. 'Centralized relay protection of power plants using IEC-61850'. *International Multi-Conference on Industrial Engineering and Modern Technologies*. 2018:1–3.
[40] Augustine S., Reno M.J., Brahma S.M., Lavrova O. 'Fault current control and protection in a standalone DC microgrid using adaptive droop and current derivative'. *IEEE Journal of Emerging and Selected Topics in Power Electronics*. 2020;9(3):2529–39.
[41] Nougain V., Mishra S., Pradhan A.K. 'MVDC Microgrid protection using a centralized communication with a localized backup scheme of adaptive parameters'. *IEEE Transactions on Power Delivery*. 2019;34(3):869–78.
[42] Monadi M., Gavriluta C., Luna A., Candela J.I., Rodriguez P. 'Centralized protection strategy for medium voltage DC Microgrids'. *IEEE Transactions on Power Delivery*. 2017;32(1):430–40.

[43] Jia K., Wang C., Bi T., Zhu R., Xuan Z. 'Transient current waveform similarity based protection for flexible DC distribution system'. *IEEE Transactions on Industrial Electronics*. 2019;66(12):9301–11.

[44] Rana R., Sahoo S., Mishra S., Peng J.C.H 'Performance validation of cooperative controllers in autonomous ac microgrids under communication delay'.*2019 IEEE Power & Energy Society General Meeting (PESGM)*; Atlanta, GA, USA, 2019. pp. 1–5. Available from https://ieeexplore.ieee.org/xpl/mostRecentIssue.jsp?punumber=8960953

[45] Singhal A. 'Modern information retrieval: a brief overview'. *IEEE Engineering Bulletin*. 2001;24(4):35–43.

[46] Tan P.N., Stenbach M., Kumar V. 'Introduction to data mining'. *IEEE Engineering Bulletin*; MA, 8, 2001. pp. 500.

[47] Mohanty R., Sahoo S., Pradhan A.K., Blaabjerg F. 'A cosine similarity-based centralized protection scheme for DC Microgrids'. *IEEE Journal of Emerging and Selected Topics in Power Electronics*. 2021;9(5):5646–56.

Chapter 8

Cyber-physical microgrids: toward flexible energy districts

Argo Rosin[1], Roya Ahmadiahangar[1], Joao Martins[2], Subham Sahoo[3], and Frede Blaabjerg[3]

8.1 Introduction

Flexibility can be defined as the power system's ability to respond to both expected and unexpected changes, either in demand or on the supply side. This concept contributes to improving the grid's stability allowing a higher penetration of renewable energy sources (RESs). A real-time flexibility utilization system can be viewed as a cyber-physical system where the communication component is cyber, whereas the control components have physical effects. In this chapter, we discuss security properties and challenges that must be considered in the flexibility utilization of energy districts.

8.2 Impact of nearly zero energy districts

The concept of nearly zero energy buildings (nZEB) is not a new concept. It has been highly adopted either by the scientific community or by the building industry itself. The European Union (EU), through its Energy Performance of Buildings Directive 2010/31/EU (EPBD), defines nZEB as a "building that has a very high energy performance," where "the nearly zero or very low amount of energy required should to a very significant extent be covered by energy from renewable sources, including renewable energy produced on-site or nearby" [1]. This nZEB definition is quite flexible, as EU member states are responsible for defining and implementing the nZEB concept in their respective countries. Regardless of regulatory issues, there is a scientific common understanding to consider nZEB as a building with a neutral energy balance during a certain period of time (usually one year), taking into account that efficiency measures were implemented [2].

[1]Tallinn University of Technology, Estonia
[2]NOVA School of Science and Technology Universidade NOVA de Lisboa & CTS/UNINOVA, Portugal
[3]Department of Energy, Aalborg University, Denmark

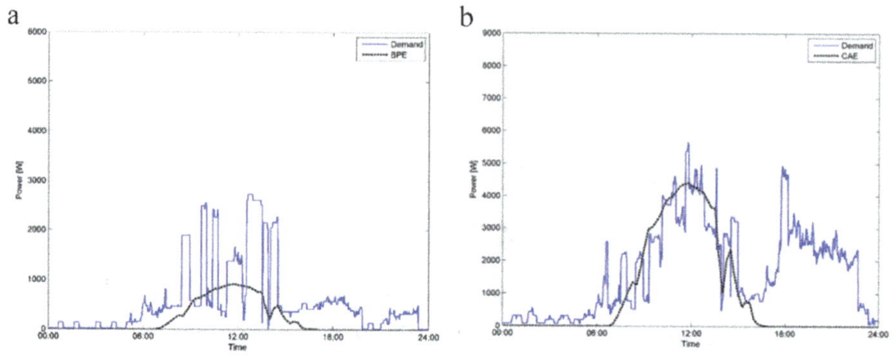

Figure 8.1 Load and on-site PV-based production profiles for (a) single nZEB and (b) aggregated nZEBs

nZEB benefits are widely described in the literature [3]. Their flexibility capabilities have been already addressed, through the use of heat storage or load shifting [4], and can be further enhanced by using energy storage systems [5]. These benefits will be scaled up if the aggregation of nZEB is to be considered. In [2], it was shown that the load-cover factor (defined as the ratio of the electrical demand that is covered by the on-site nZEB electricity generation) can be multiplied by three when evolving from one single nZEB into an aggregation of five nZEB. Figure 8.1 presents the load and on-site photovoltaic (PV)-based production profiles for the aforementioned cases, when load flexibility control is considered. In this example, it is clear that scaling up nZEBs will allow the exploitation of load diversity between buildings and renewable resource usage in such ways that would be impossible for single buildings.

There are several ways to consider the aggregation of several nZEBs. Current trend is the establishment of energy communities, which are defined by two separate directives of the European Commission Clean Energy Package (Renewable Energy Directive (EU) 2018/2001 and Electricity Market Directive (EU) 2019/944) as way of coordinating collective citizen activities in the energy system [6]. The establishment and operation of energy communities, which will bring advantages either to consumers or to the energy grid itself, should be based on four principles: flexibility, interconnectivity, bi-directionality, and complementarity [7]. This will increase the number of distributed energy resources and the grid flexibility to cope with the mismatching between the intermittent renewable production and the time-varying load. The potential of energy communities to provide ancillary services to the grid has also been recognized in the literature [8]. Energy communities can be established inside a housing company, crossing property boundaries, or even take a distributed configuration [9].

A natural evolution of energy communities will be the establishment of flexible energy districts, maximizing building efficiency, renewable energy potential, and load control [10]. This is, definitely, a hard and huge task that will lead to the fulfillment of the Paris Climate Agreement. The EC report "From nearly-zero energy buildings to net-zero energy districts" gives an interesting overview of the lessons learned from existing EU projects, where besides technical issues, strong political commitment and inhabitant's engagement emerge as key factors for the success of flexible energy districts [11].

Despite the obvious grid advantages of flexible energy districts, some questions may be posed. Are there any negative technical impacts on the grid or on grid equipment, such as power transformers? Are current standards suitable for the deployment of flexible energy districts? What, and how, assets (flexible loads, storage systems) can be used to mitigate possible negative impacts? It has been demonstrated that uncontrolled deployment of nZEBs into existing low-voltage distribution grids will have negative impacts on distribution transformers [12]. The mismatch between PV-based renewable production and building loads leads to high reverse power flows through the distribution transformer, thus largely increasing its aging due to prohibitive values of peak loads and maximum hot-spot temperatures. Figure 8.2 presents the load profile of a distribution transformer when, within a community of 19 buildings, the number of buildings converted to nZEB goes from 0 (none of them is nZEB) to 19 (all of them are nZEB).

It is interesting to note that the rated power of the transformer is reached in the "all nZEB" limit case. Figure 8.3 presents the increasing equivalent aging of the distribution transformer with the increase of the number of buildings converted to nZEB. When 1 to 15 buildings were converted to nZEBs, the annual equivalent aging actually decreases, with a minimum value of 7.4 days/year when the number of converted buildings is 8. However, when more than 15 buildings are converted, the aging increases exponentially reaching 2278 days, which is clearly unsustainable. To mitigate this situation, flexibility measures should be considered, using either controllable home loads or energy storage systems (two distinct forms of building flexibility).

Building flexibility and self-consumption can be highly optimized when using on-site energy storage systems. Building owners will benefit because their energy bills will be lower as long as their self-consumption levels increase. From the grid

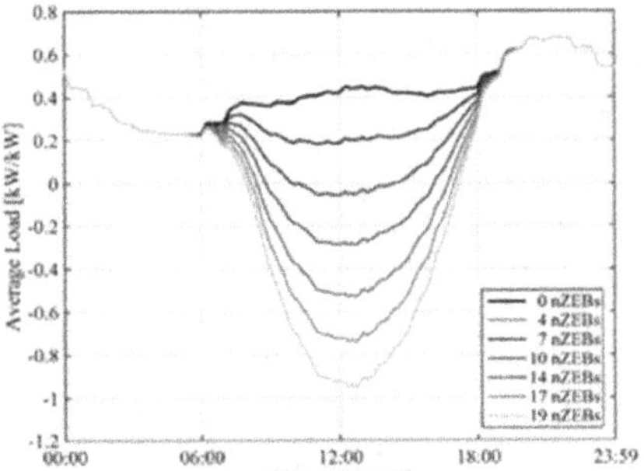

Figure 8.2 Daily average load profiles for different levels of buildings converted to nZEB normalized to distribution transformer rated load

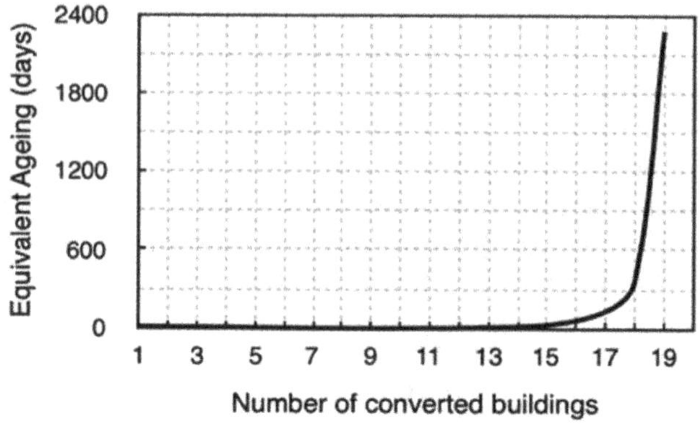

Figure 8.3 *Distribution transformer equivalent aging for different levels of buildings converted to nZEB*

point of view, transmission and distribution costs will decrease, and new investments can be deferred. It is also important to note that the use of grid-based flexible storage systems is becoming a viable technology for grid ancillary services delaying grid expansion and reinforcement [13].

On-site nZEB storage systems are a valuable mitigation strategy to prevent the precocious aging of distribution transformers. In [14] a battery-based Transformer Anti-Aging Protection System, to cope with the nZEB negative impacts, is presented. It uses a combination of battery storage and PV curtailment, leading to a Pareto front presented in Figure 8.4. For each case, this trade-off should be carefully evaluated together with extensive economic analysis.

Positive energy districts are the perfect platform to upscale the nZEB flexibility concept to the urban context. Although there are numerous advantages, this upscaling should be carefully planned from the social and technical point of view.

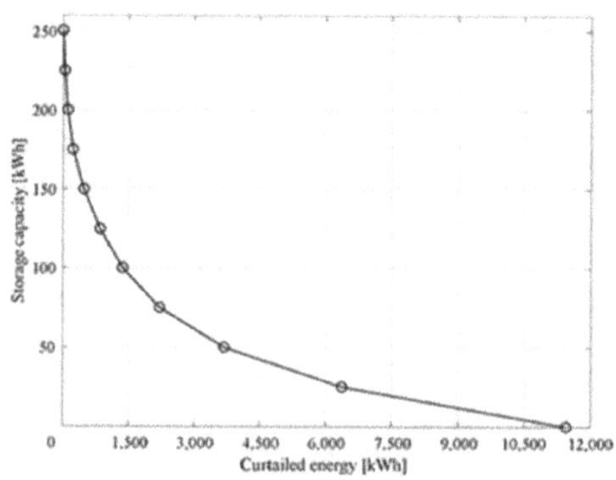

Figure 8.4 *Pareto front relating battery-based storage capacity and on-site PV curtailed energy*

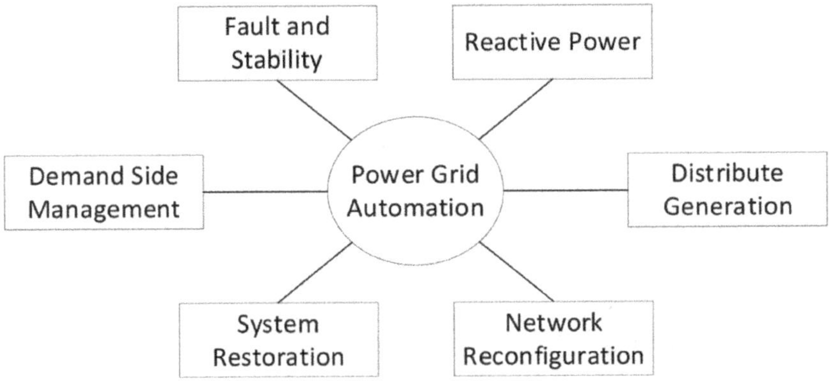

Figure 8.5 Smart grid intelligent automation functions [15]

8.3 Methods for improvement of energy flexibility in energy districts

The smart grid is planned to have the following key characteristics, like self-healing, consumer involvement, physical and cybersecurity, power quality, a wide variety of generation technologies, high optimization, and digital maturity level of assets. So, the main goals are to increase the grid observability; improve the controllability of assets, enhance performance and security; and reduce costs of operation, maintenance, and system planning.

Communications pitfalls among system operators as well as lack of predictive real-time management tools, predictive optimal power flow control software and hardware, and distribution management system (DMS) are not well developed for systems operations. Advance interactions of agents such as telecommunication, sensing, control, and optimization will help to achieve adaptive interaction, adaptability, self-healing, efficiency, and reliability of the electrical grid. The related goals (Figure 8.5) of intelligent functions in smart grid design include [15]:

1. Real-time angle and voltage stability; collapse detection and prevention via intelligent data.
2. Reactive power control based on intelligent coordination controls.
3. Fault analysis and reconfiguration schemes based on intelligent switching operations.
4. Power generation and load balance with the use of intelligent switching operation to energize loads while controlling frequency and oscillations, and minimize demand interruption.
5. Distributed generation and demand side management (DSM) via demand response strategy for peak shaving. These also include increased proliferation and control of renewable energy resources.

Flexibility is the ability to respond in a timely manner to variations in electricity supply and demand and has been officially recognized by organizations like IEA [16], NERC [17], and IRENA [18]. Flexibility can be categorized as flexibility for power (short-term equilibrium between the power supply and power demand), flexibility for energy (medium- to long-term), flexibility for transfer capacity, and flexibility for voltage [19–21].

With the evolution of the smart grid and electricity markets, the responsibility of balancing between generation and demand is shifting toward the demand side [22–24]. Although existing smart appliances, energy storage systems, including batteries, flywheels, electric vehicles (EV), supercapacitors, etc. provide a significant potential for flexibility, it is not clear what the best approach is for unlocking this potential, particularly within an energy district. For the demand-side flexibility estimation, some estimation methods have been reported in the literature [25] considering profitability assessment of flexibility, estimation of the flexibility of EVs charging stations [26, 27], and common buildings. Increased demand for smart solutions related to disturbance and flexibility management on district level is described in IEA-EBC Annex 82, "Energy flexible buildings towards resilient low carbon energy systems," and Annex 83, "Positive Energy Districts."

Smart buildings are key enablers of future energy systems as they will help to facilitate a larger share of renewables, distributed supply, and demand-side energy flexibility. In the revised EPBD, one of the focal points is to improve the realization of this potential through smart ready technologies in the buildings sector. Therefore, the revised EPBD requires the development of a voluntary European scheme for rating the smart readiness of buildings: the "Smart Readiness Indicator" (SRI). An SRI for buildings provides information on the technological readiness of buildings to interact with their occupants and the energy grids, and on their capabilities for more efficient operation and improved performance through the use of information and communication technologies (ICT) [28].

The possible methods for improvement (utilization) of energy flexibility in energy districts are [29]:

1. delivering customized designs, innovative technologies, and decision support strategies and tools (i.e., flexibility evaluation tools) to maximize buildings flexibility (i.e., demand response capabilities)
2. encouraging community engagement and empowering user's control/participation;
3. unlocking the potential of neighborhoods as flexibility providers;
4. improving characterization of energy flexibility in microgrids (incl. nanogrids);
5. providing advanced control systems/schemes in buildings and improving control strategies and algorithms, i.e., AI algorithms, model predictive control for integration into building inertia as flexible load;
6. providing big data-based infrastructure management (i.e., governance and processing, advanced computing);
7. providing smart networks and cybersecurity of devices/systems in energy districts;
8. demonstrating new designs and efficient operation of plus energy microgrids/districts.

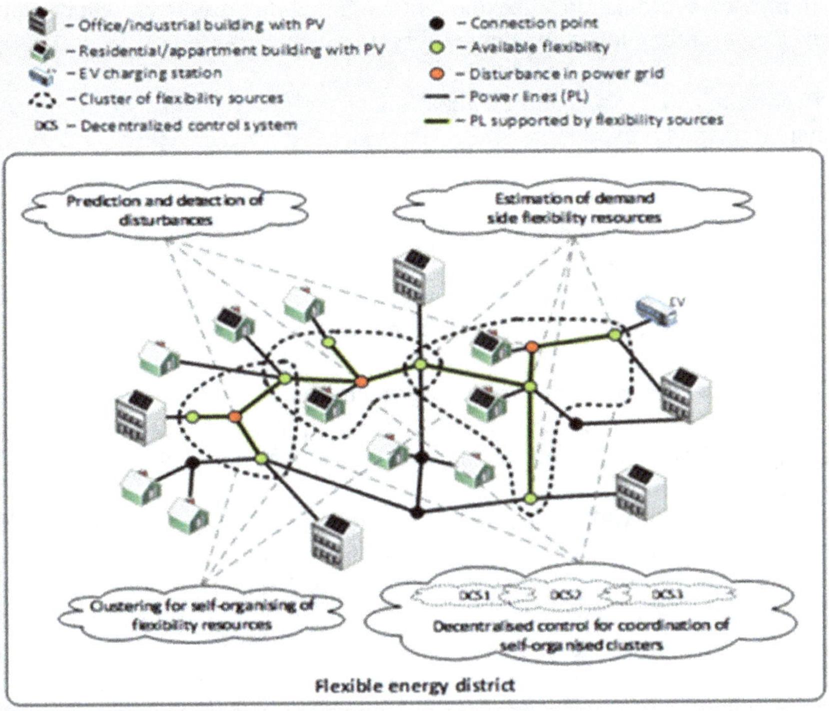

Figure 8.6 Energy management of flexible energy districts

According to previous description, the improvement of energy flexibility should consider technological, social, and economic methods. Social methods will consider mainly users' or prosumers' behavior and measures (i.e., regulations, advertising) to affect these behaviors. Economic methods mainly consider costs and regulations. Social and economic methods are mainly market and policy driven. Technology methods consider mainly the utilization of novel design, technologies, and methodologies. From the cyber security point of view, the most important are technology-based methods, which should ensure the flexibility, security, and power quality of energy district. Here, the improvement of cyber-physical flexibility management systems has important role. The main challenges of improved cyber-physical flexibility management systems on the district level are as follows (Figure 8.6):

1. Increasing the prediction, detection, and localization accuracy of disturbances [30].
2. Increasing estimation accuracy of available demand-side flexibility resources [31–33].
3. Minimizing load shedding and costs in on- and off-grid modes by using clustering for self-organizing the flexibility sources [34, 35].

4. Improving coordination and exploitation of flexibility resources with decentralized control for coordination of self-organized clusters [36–39].

Prediction, **detection**, and **localization** of the different type of disturbances estimation methods cover:

1. Voltage and frequency disturbances in the off-grid mode caused by RESs, prosumers stochastic behavior, state of charge of energy storage systems, and power line contingency in district.
2. Active and reactive power disturbances in the on-grid mode caused by RESs, prosumers stochastic behavior, contingency of power grid, and market signals.

Here different artificial intelligence algorithms for the prediction and determination of disturbances improve the performance of V/f and P/Q regulation algorithms in energy districts, considering weather conditions, prosumers stochastic behavior, power grid constraints.

Demand-side flexibility estimation methods in energy districts (i.e., positive energy districts) will commonly consider weather conditions, power generation of renewable energy systems, prosumers stochastic behavior, the status of energy storage systems, the effect of power grid disturbances, and changes in market signals. These methods increase the forecast accuracy of available flexibility resources and the profit for flexibility service providers, considering prosumers behavior, share, and characteristics of different flexibility resources.

Clustering for self-organizing of flexibility resources (particularly energy storage systems) will minimize load shedding and generation cost in the off-grid mode, considering the voltage and frequency changes; and on-grid costs, considering the market signals, prosumers' stochastic behavior, and power grid disturbances. The self-organizing clusters of flexibility resources are capable to compensate or balance disturbances inside the clusters, preventing spreading of the problem to other parts of the grid, by providing required active or reactive power to the grid. Thoughtful clustering approaches will open the potential of utilizing available flexibility to maintain most power quality problems automatically in a more profitable manner by reduction of investment on special equipment in the main grid.

Decentralized control methods can improve coordination and exploitation of flexibility resources to adjust frequency and voltage in the off-grid mode and optimizing active and reactive power exchange with the main grid in the on-grid mode. The agent-based decentralized control methods would be the best solution for coordinating the self-organizing clusters of flexibility resources. Decentralized control with minimal communication requirements is the best solution from the perspective of practical implementation, security, and privacy (amount of information exchanged; local control decisions; independent and distributed control technologies). In combination of optimal clustering, it will improve the voltage and frequency regulation accuracy in off-grid mode and can enhance the main grid capacity support in on-grid mode.

8.4 Cybersecurity of flexible energy districts

Electrical network of Flexible energy districts (FED) consists of different domains, including nZEB, electric grid (usually distribution grid, operation and service provider, aggregator and market), as shown in Figure 8.7. Domains are interconnected by two-way energy flow and two-way information flow. This two-way flow of energy and data will enable new functionality in nZEBs and FED and even between the FED and the main grid.

FED is supposed to have several smart functionalities, e.g., dynamic or real-time pricing [40], demand response [41], dynamic billing, self-healing and smart outage management, microgrids islanding detection, theft detection, etc. Smart meter is the key to accomplish these functionalities. A large number of heterogeneous devices (e.g., smart meters, sensors) have been deployed between nZEBs, operators and aggregator, and form the metering system as one of the main infrastructures in FED [42]. Smart metering system is "an intelligent electronic device that measures energy usage data, with more precise information than a traditional meter, and sends and receives data via two-way communication" [43]. Therefore, ICT provided by smart meters allows the aggregator to access necessary data for prediction and utilization of potential energy flexibility in FED from each nZEB. All these data are sent to the grid operators and aggregator at regular intervals of 5–15 minutes based on the grid code.

However, on the one hand, the energy companies can be predominantly subjected to targeted attacks due to the mass dependence on ICT and smart metering network technologies. This kind of attack could potentially lead to loss of power for

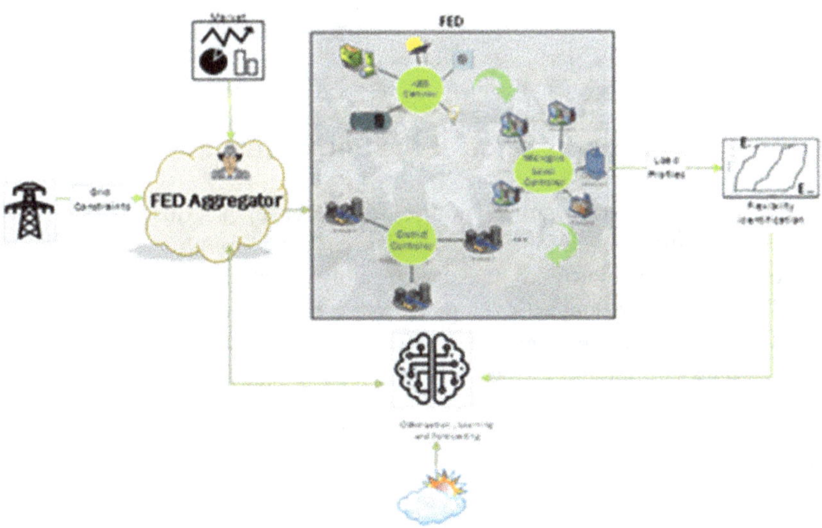

Figure 8.7 Conceptual cyber-physical model of FED

some of nZEBs or even blackout a part or whole FED. The attack could also affect infrastructures such as the aggregator control center or distribution substations. On the other hand, as shown in Figure 8.7, in order to forecast and estimate individual nZEB flexibility and also the aggregated flexibility in the aggregator control center, a large amount of consumption data is needed [44–46]. It should be noted that with an access to individual nZEB metering data, it is possible to apply non-intrusive load monitoring techniques to analyze consumption patterns and infer user activities [47]. A relatively low load consumption for several days may indicate that households are not around; while a drop of consumption power late at night may be a sign that the households go to sleep [48]. The determination of personal behavior patterns is a serious privacy concern in smart grid defined by Electronic Privacy Information Center [49]. Therefore, a cybersecurity threat may raise privacy issues for the nZEB and its households.

Following the aforementioned issues, several cyber security standards and guidelines have been developed, including:

- ISO/IEC 27000-series (27001, 27002, 27005...) for general IT security
- NIST SP 800–82 (rev. 2): Guide to Industrial Control Systems (ICS) Security
- IEC 62351 for communication protocols
- NIST Framework for Improving Critical Infrastructure Cybersecurity
- NERC CIP (Critical Infrastructure Protection)
- NISTIR 7628 (rev. 1): Guidelines for Smart Grid Cyber Security

However, none of them is particularly considering the case of FED and achieving privacy-friendly aggregation. Vulnerabilities of FED are mainly due to complexity, multiple layer control and communication, and interdependencies and it is significantly important to resist sophisticated attacks from intelligent and capable threat agents.

8.5 Impact of cyberattacks on flexibility utilization

To increase grid resilience and reliability, networked microgrids are being investigated as a promising solution. Networked microgrids are clusters of geographically close, islanded microgrids that can flexibly function as a single, aggregate island. This flexibility enables customer-level resilience and reliability improvements during extreme event outages and reduces utility costs during normal grid operations. The coordinated cluster of microgrids would increase the number of available resources and extend service availability; this coordination can be performed through an energy management system (EMS), DMS, or a central microgrid controller [50].

While there are various challenges to address for a full, comprehensive implementation of networked microgrids, a prominent concern that needs to be addressed from the design stage is networked microgrid cybersecurity. The networking of the microgrids is dependent on the operation of communicative microgrid controllers that enable the quick reconfiguration of the aggregate system. Furthermore, individual microgrids could employ communication-assisted control and protection schemes that expand operational capabilities but also broaden the cyberattack landscape [51]. Additional complexity is derived from the potential for each microgrid to have different levels of cybersecurity defenses; an integral question is if the connection of a cyber-secure microgrid to a microgrid with less or non-existent cybersecurity decreases overall networked security and increases cascading impacts.

For networked microgrids, vulnerability analysis of the external connections and devices that enable those connections is needed. In this document, we will not conduct a full vulnerability analysis but provide an example of the types of vulnerabilities that could arise. For this purpose, we will examine communications over external connections, protocol translators, and microgrid controllers. These communications could stem from operational functions such as inter-microgrid transactions as well as protection schemes (either individual microgrid communication-assisted protection or exchange of protection settings with networked microgrids).

Communications can occur every 1 s to 100 ms and include protocols such as IEC 61850, Modbus (RS485/TCP/IP), IEC 60870-5-104, OPC, Ethernet, and DNP3 [52]. Overall, microgrid controllers can detect disturbances, such as points out of range, digital state changes, open/close breaker status, and drop time-outs, and implement response to maintain stability. These controllers can be stand-alone devices or a function within real-time automation controllers (RTACs).

Considering microgrid controllers, either stand-alone or within an RTAC, the vulnerabilities that should be considered include:

- Open, unused ports
- Software/firmware updating procedure
- Internet/Wi-Fi connection capabilities
- Use of third-party software
- Hardware supply chain

These example vulnerabilities are general for grid end-devices, especially those with additional communication capabilities. These advanced functions enable the microgrid controller to communicate quickly and host processes that would have otherwise been too burdensome (e.g., exchanging pricing information). These added functions can be achieved with third-party software, but without proper security mechanisms and analysis, the third-party software can broaden the vulnerability landscape. For example, the third-party software could include backdoor access or software Trojans, if not properly examined prior to installation (even if end-device is from a trusted source).

To test the impact of cyberattacks on flexibility utilization, we consider a test case study of central power management in the microgrid, as shown in Figure 8.8. This microgrid can be divided into four important parts: A diesel generator, acting as the base power generator; A PV farm combined with a wind farm, to produce renewable energy; a V2G system installed next to the last part of the system, which is the load of the grid. The size of the microgrid represents approximately a community

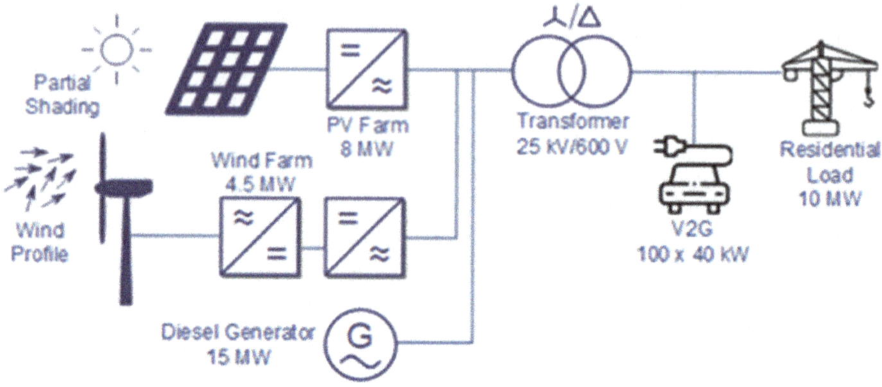

Figure 8.8 *System under consideration to investigate the impact of cyberattacks on flexibility of renewable energy resources*

of a thousand households during a low consumption day in spring or fall. There are 100 EVs in the base model, which means that there is a 1:10 ratio between the cars and the households. The PV farm produces energy proportional to three factors: the size of the area covered by the PV farm, the efficiency of the solar panels, and the irradiance data. A simplified model of a wind farm produces electrical power following a linear relationship with the wind. When the wind reaches a nominal value, the wind farm produces the nominal power. The wind farm trips from the grid when the wind speed exceeds the maximum wind value, until the wind gets back to its nominal value.

The load comprises residential load and an asynchronous machine that is used to represent the impact of an industrial inductive load (like a ventilation system) on the microgrid. The residential load follows a consumption profile with a given power factor. The asynchronous machine is controlled by a square relation between the rotor speed and the mechanical torque. The EVs are managed in such a way that their charging is controlled, where the available power is used to regulate the grid. However, based on the associated vulnerabilities outlined above, the coordination among different fleet of EVs responsible for grid regulation can be simply interrupted through a denial-of-service (DoS) attack. This is easily possible initially by hijacking the firmware and then fusing it with multiple random packets causing unnecessary traffic.

By doing so, it can be seen in Figure 8.9 that EVs regulate and support the grid demand by providing auxiliary power instead of charging. However, when the DoS attack is launched around 12 hours, as shown in Figure 8.10, it can be seen that the flexibility of the system drops significantly as any intermittent behavior of renewable sources has to be matched by the diesel generator. In the worst case, if the load goes above the diesel generation, the system operator must ultimately resort to load shedding. This affects the system reliability due to the presence of cyberattacks.

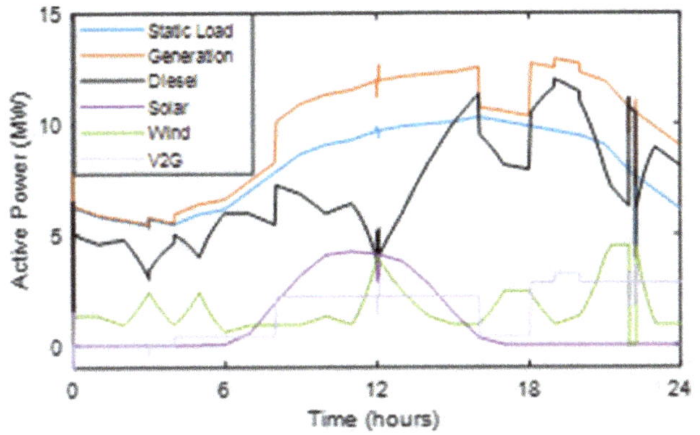

Figure 8.9 *Active power profile of flexible allocation in the system in Figure 8.8 – without any cyberattacks*

Hence, the cybersecurity principles for networked microgrids will vary depending on operator and aggregator policies to ensure maximum flexibility. Nonetheless, networked microgrid cybersecurity should be assessed from multiple perspectives (from both cyber- and physical-sides) and a defense-in-depth approach should be implemented.

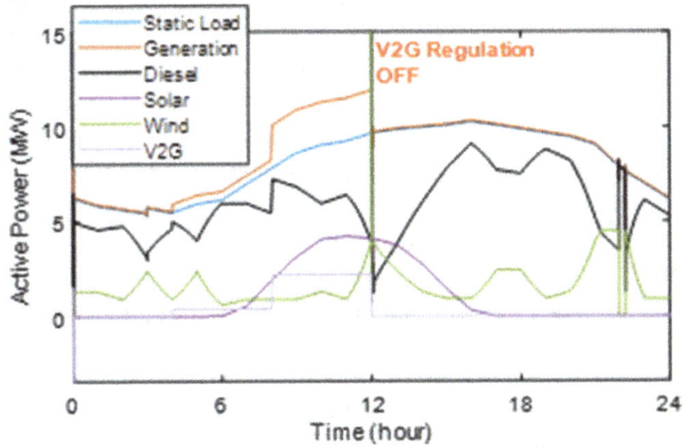

Figure 8.10 *Active power profile of flexible allocation in the system in Figure 8.8 – with a denial-of-service attack in the charging stations around 12th hour.*

8.6 Conclusion

Designing a practical penalty signal relies on two main factors: (i) energy consumption behavior (consumer's load, generation, and the SOC level of the batteries) and (ii) analyzing a large volume of data. Since conventional methods are inefficient to handle the mentioned factors properly, we propose a new clustering-based method to analyze these datasets and extract different levels for load and PV generation and quantify them. In addition, we have documented how cyberattacks can easily manipulate the penalty signal, thereby restricting optimal usage of RESs. These cyberattacks could also affect the flexible usage of RESs and hamper the system's objectives. Hence, advanced and resilient technologies and mitigation measures will be developed as future scope of work and implemented at every level to ensure the secure energy flexibility of microgrids. Moreover, an appropriate optimization framework needs to be co-designed in parallel to the resilience feasibility and flexibility of RESs as a future scope of work.

References

[1] European Commission. *Energy Performance of Building Directive (EPBD) [online]*. 2014. Available from https://ec.europa.eu/energy/en/topics/energy-efficiency/energy-performance-ofbuildings/overview# content-heading-0 [Accessed 27 Jan 2021].

[2] Lopes R.A., Martins J., Aelenei D., Lima C.P. 'A cooperative net zero energy community to improve load matching'. *Renewable Energy*. 2016;93(1):1–13.

[3] Le Dréau J., Heiselberg P. 'Energy flexibility of residential buildings using short term heat storage in the thermal mass'. *Energy*. 2016;111(Part A):991–1002.

[4] Jensen S.Ø., Madsen H., Lopes R. Energy flexibility as a key asset in a smart building future—contribution of annex 67 to the european smart building initiatives [online]. Position Paper of the IEA-EBC Annex 67 Energy Flexible Buildings. Available from http://www.annex67.org/publications/position-paper/ [Accessed 27 Jan 2021].

[5] Xiaosong H., Changfu Z., Zhang Caiping L.Y. 'Technological developments in batteries'. *IEEE Power Energy Magazine*. 2017;15(5):20–31.

[6] Roberts J., Frieden D., Gubina A. *Energy community definitions, compile project: integrating community power in energy islands*. EU project website; 2019.

[7] Lowitzsch J., Hoicka C.E., van Tulder F.J. 'Renewable energy communities under the 2019 European clean energy package – governance model for the energy clusters of the future?' *Renewable and Sustainable Energy Reviews*. 2020;122:109489.

[8] Stadler M., Cardoso G., Mashayekh S, *et al*. 'Value streams in microgrids: a literature review'. *Applied Energy Elsevier*. 2016;**162**(4):980–89.

[9] Pahkala T., Uimonen H., Väre V. *Flexible and ccustomer-ccentred electricity system; final report of the smart grid working group*. Finland: Ministry of Economic Affairs; 2018.

[10] Polly B., Kutscher C., Macumber D., Schott M. 'From zero energy buildings to zero energy districts'. *ACEEE Summer Study on Energy Efficiency in Buildings*; 2016. pp. 1–16.

[11] Saheb Y., Shnapp S., Paci D. *From Nearly-Zero Energy Buildings To Net-Zero Energy Districts—Lessons Learned From Existing EU projects, EUR 29734 EN*. Luxembourg: Publications Office of the European Union; 2019.

[12] Lopes R.A., Magalhães P., Gouveia J.P., Aelenei D., Lima C., Martins J. 'A case study on the impact of nearly zero-energy buildings on distribution transformer aging'. *Energy*. 2018;157(27):669–78.

[13] Garcia-Garcia L., Paaso E.A., Avendano-Mora M. 'Assessment of battery energy storage for distribution capacity upgrade deferral'. *IEEE Power Energy Society Innovative Smart Grid Technologies Conference (ISGT)*; 2017. pp. 1–5.

[14] Queiroz H., Amaral Lopes R., Martins J. 'Automated energy storage and curtailment system to mitigate distribution transformer aging due to high renewable energy penetration'. *Electric Power Systems Research*. 2020;182:106199.

[15] Momoh J.A. 'Smart grid design for efficient and flexible power networks operation and control'. *IEEE/PES Power Systems Conference and Exposition*; Seattle, WA; 2009. pp. 1–8.

[16] I.E.A. *Status of Power System Transformation 2018: Advanced Power Plant Flexibility*. Paris: IEA; 2018.

[17] Corporation, North American Electric Reliability. Accommodating high levels of variable generation. Tech. Rep. Princeton, NJ: NERC; 2009.

[18] International Renewable Energy Agency (IRENA). *Power System Flexibility for the Energy Transition, Part 1: Overview for Policy Makers*. Abu Dhabi: IRENA; 2018.

[19] Zolfaghari M., Ahmadiahangar R., Gharehpetian G.B., Rosin A., Plaum F. 'Using V2G technology as virtual active power filter for flexibility enhancement of HVDC systems'. IEEE Power Engineering Conference; Portugal; 2020.

[20] Zolfaghari M., Ahmadiahangar R., Gharehpetian G.B., Häring T., Rosin A. 'Flexibility enhancement of hybrid microgrids using optimal H filtering-based fuzzy control of UIPC'. *IEEE Power Engineering Conference*; Portugal; 2020.

[21] Ponocko J., Milanovic J.V. 'Forecasting demand flexibility of aggregated residential load using smart meter data'. *IEEE Transactions on Power Systems*. 2018;33(5):5446–55.

[22] Ahmadiahangar R., Rosin A., Palu I., Azizi A. 'New approaches for increasing demand-side flexibility'. *Demand-side Flexibility in Smart Grid. Springer Briefs in Applied Sciences and Technology*. Singapore: Springer; 2020.

[23] Ahmadiahangar R., Rosin A., Palu I., Azizi A. 'On the concept of flexibility in electrical power systems: signs of inflexibility'. *Demand-side Flexibility in*

Smart Grid. SpringerBriefs in Applied Sciences and Technology. Singapore: Springer; 2020.

[24] Simonov M., Bertone F., Goga K., Tibiletti L. 'Demand-side flexibility measurement and estimation for transactive energy control'. *IEEE International Telecommunications Energy Conference*; Torino, Italy, 2018.

[25] Mishra S., Koduvere H., Palu I., Kuhi-Thalfeldt R., Rosin A. 'Assessing demand side flexibility with renewable energy resources'. *IEEE 16th International Conference on Environment and Electrical Engineering (EEEIC)*; Florence, Italy; 2016–1–6.

[26] Kütt L., Saarijärvi E., Lehtonen M., Rosin A., Mõlder H. 'Load shifting in the existing distribution network and perspectives for EV charging-case study'. *IEEE PES Innovative Smart Grid Technologies*; Istanbul, IEEE, 2014. pp. 1–6.

[27] Drovtar I., Rosin A., Landsberg M., Kilter J. 'Large scale electric vehicle integration and its impact on the Estonian power system'. *IEEE Grenoble Conference*; Grenoble, France; 2013. pp. 1–6.

[28] OVERVIEW | Energy Flexibility in buildings [online]. Available from https://www.buildup.eu/en/news/overview-energy-flexibility-buildings [Accessed 7 Feb 2021].

[29] *The reviewed EPBD: upcoming new targets for energy efficiency in buildings*. Available from https://www.buildup.eu/en/news/reviewed-epbd-upcoming-new-targets-energy-efficiency-buildings, [Accessed 7 Feb 2021].

[30] Amaripadath D., Roche R., Joseph-Auguste L, *et al.* 'Power quality disturbances on smart grids: overview and grid measurement configurations'. *2017 52nd International Universities Power Engineering Conference (UPEC)*; Heraklion, IEEE, 2017. pp. 1–6.

[31] Ahmadiahangar R., Häring T., Rosin A., Korõtko T., Martins J. 'Residential load forecasting for flexibility prediction using machine learning-based regression model'. *IEEE International Conference on Environment and Electrical Engineering*; 2019.

[32] Shabbir N., Ahmadiahangar R., Kütt L., Rosin A. 'Comparison of machine learning based methods for residential load forecasting'. *Electric Power Quality and Supply Reliability Conference (PQ)*; IEEE, 2019.

[33] Shabbir N., AhmadiAhangar R., Kütt L., Iqbal M.N., Rosin A. 'Forecasting short term wind energy generation using machine learning'. *IEEE 60th International Scientific Conference on Power and Electrical Engineering of Riga Technical University (RTUCON)*; Riga, Latvia; 2019. pp. 1–4.

[34] Minniti S., Nguyen P.H., Vo T., Haque A.N.M.M. 'Development of grid-flexibility services from aggregators a clustering algorithm for deploying flexible DERs'. *IEEE International Conference on Environment and Electrical Engineering and 2018 IEEE Industrial and Commercial Power Systems*; 2018. pp. 1–7.

[35] Zhu Y., Yi Z., Lu Q., Liu S., Xu Y., Wang K. 'Typical scene acquisition strategy for VPP based on multi-scale spectral clustering algorithm'. *2nd IEEE Conference on Energy Internet and Energy System Integration (EI2)*; 2018. pp. 1–5.

[36] Fallahzadeh-Abarghouei H., Hasanvand S., Nikoobakht A., Doostizadeh M. 'Decentralized and hierarchical voltage management of renewable energy resources in distribution smart grid'. *International Journal of Electrical Power & Energy Systems*. 2018;100(4):117–28.

[37] Armstorfer A., Biechl H., Rosin A. 'Energy scheduling of battery storage systems in micro grids'. *Electrical, Control and Communication Engineering*. 2017;12(1):27–33.

[38] Khodadoost Arani A.A., Gharehpetian G.B., Abedi M. 'Decentralised primary and secondary control strategies for islanded microgrids considering energy storage systems characteristics'. *IET Generation, Transmission & Distribution*. 2019;13(14):2986–92.

[39] Sen G., Baysal M. 'Multi agent-based decentralized control for improving the efficiency and voltage regulation in smart microgrid'. *International Conference on Smart Energy Systems and Technologies (SEST)*; Sevilla; 2018. pp. 1–6.

[40] Azizi E., M. Shotorbani A., Ahmadiahangar R., *et al*. 'Cost/comfort-oriented clustering-based extended time of use pricing'. *Sustainable Cities and Society*. 2021;66(102673):102673–6707.

[41] Dadashi-Rad M.H., Ghasemi-Marzbali A., Ahangar R.A. 'Modeling and planning of smart buildings energy in power system considering demand response'. *Energy*. 2020;213(3):118770–5442.

[42] Kumar P., Lin Y., Bai G., Paverd A., Dong J.S., Martin A. 'Smart grid Metering networks: a survey on security, privacy and open research issues'. *IEEE Communications Surveys & Tutorials*. 2019;21(3):2886–927.

[43] Directive 2012/27/EU of the European Parliament and of the Council of 25 October 2012 on Energy Efficiency. Amending Directives 2009/125/EC and 2010/30/EU and Repealing Directives 2004/8/EC and 2006/32/EC Text With EEA Relevance [online]. 2021. Available from https://eur-lex.europa.eu/legalcontent/EN/TXT/?uri=celex%3A32012L0027 [Accessed Feb 2021].

[44] Sanjareh M.B., Nazari M.H., Gharehpetian G.B., Ahmadiahangar R., Rosin A. 'Optimal scheduling of HVACs in islanded residential microgrids to reduce BESS size considering effect of discharge duration on voltage and capacity of battery cells'. *Sustainable Energy, Grids and Networks*. 2021;25(20):100424–4677.

[45] Rosin A., Ahmadiahangar R., Azizi E., *et al*. 'Clustering-based penalty signal design for flexibility utilization'. *IEEE Access*. 2020;8:208850–60.

[46] Ahmadiahangar R., Rosin A., Palu I., Azizi A. 'Forecasting available demand-side flexibility'. *Demand-side Flexibility in Smart Grid. SpringerBriefs in Applied Sciences and Technology*. Singapore: Springer; 2020.

[47] Mustafa M.A., Cleemput S., Aly A., Abidin A. 'A secure and privacy-preserving protocol for smart metering operational data collection'. *IEEE transactions on smart grid*. 2019;10(6):6481–90.

[48] Ni J., Zhang K., Lin X., Shen X.S. 'Balancing security and efficiency for smart metering against misbehaving collectors'. *IEEE Transactions on Smart Grid*. 2019;10(2):1225–36.

[49] EPIC. *The smart grid and privacy, electron.* Washington, DC: Privacy Inf. Center; 2015.

[50] Eddy J., Miner N.E., Stamp J. 'Sandia's microgrid design toolkit'. *The Electricity Journal*. 2017;30(4):62–7.

[51] Hossain-McKenzie S., Reno M.J., Quiroz J., Schulz P., Summers A., Carter C. Cyber security issues and solutions for protective relaying and local monitoring. Tech. Rep. SAND2018-0406. Sandia National Laboratories; 2018.

[52] Emerson. Ovation™ microgrid control [online]. Available from https//www.emerson.com/documents/automation/microgrid-control-system-en-1263362.pdf.

Chapter 9

Design and modeling approaches to cyberattacks for grid-tied PV systems

V. S. Bharath Kurukuru[1] *and Mohammed Ali Khan*[2]

9.1 Overview

The rise in the installation of a grid-connected photovoltaic system (GCPVS) has a major impact on the global energy market. More than 20% (200 GW) of growth in PV installation is expected in 2022 [1]. The PV inverter is a vital component for the operation of a PV plant as it not only converts power from DC to AC but also supports the utilities in adverse operating conditions by the implementation of advanced control algorithms. The advancement in technology over the years has enabled the integration of information and communication technology (ICT) with power electronics converters (PECs), which further enable real-time remote control. However, the increase in controllability and introduction of ICT have made the PECs more susceptible to cyberattacks.

The cyberattack can introduce false information or false state of operation in the power system, which may result in physical damage to the system [2] or may even cause monitory loss to the state operators [3]. The security breaches in PEC's communication line may result in unauthorized power delivery. Furthermore, an unauthorized change in the control parameters may cause fluctuations in voltage and current that are to be supplied to the consumer.

To keep the cyberattack threat in check, the protection scheme is part of the resilience strategy. To secure the system from cyberattacks, various security software is being developed by the DISCOMs to prevent false data injection into the system by the attacker. Multiple researchers are working on abnormality detection and identifying the cause of abnormality (system error or external intrusion).

[1]Advance Power Electronics and Research Laboratory, Department of Electrical Engineering, Jamia Millia Islamia (A Central University), New Delhi, India
[2]Department of Electrical Power Engineering, Faculty of Electrical Engineering and Communication, Brno University of Technology, Brno, Czech Republic

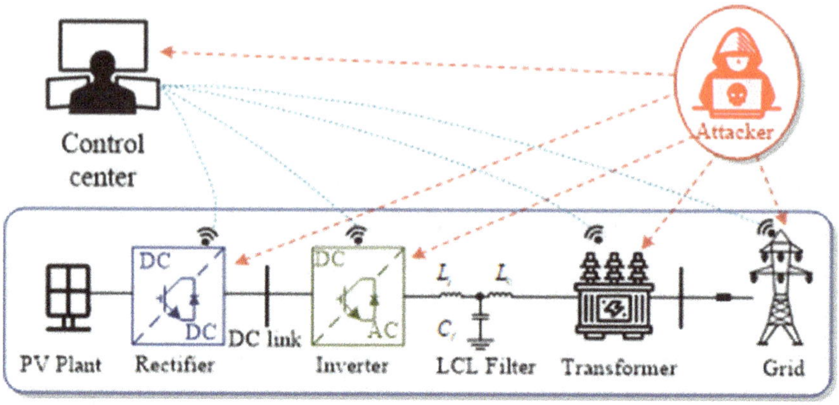

Figure 9.1 Attacker threatening the security of a grid-connected system

9.2 Cyberattack for grid-tied PV system

A few years back, the issue of cyber risk to the solar plants was relatively minor as most of the PECs did not communicate for the purpose of control or monitoring. The electrical grids are becoming digitalized for easy maintenance and management; however, with the integration between ICT and PECs, the risk of cyberattacks is increasing constantly. A brief representation of how an attacker can target the data transfer between a grid-connected system and the control center is shown in Figure 9.1.

The PECs that act as the interface between the panel and the grid are one of the most vulnerable components to the cyberattack. One of the attack options is concerning the inverter controller or the plant supervisory system, i.e., modification of the control software to access unit control or shutdown plan operation. Such attacks can take place at any instance while either plant monitoring, control or via internet-based communication. If the inverter controller is not secure against the cyberattack, the attacker can easily intercept and manipulate the data, which can cause a false alarm and may even let to black out. Even the attacker can introduce malware into the system, which may affect the power system at large. The attack on the supply change can be made possible by the sophisticated power electronic components that are used for designing the micro-controller and signal processing unit. If any of the components harbor any of the malware with them, then they may be able to attenuate the inverter operation and cause a system-level failure. Another concerning attack is related to the monitoring and protection aspect of the system. This attack is possible due to the digitalization of the plants, and if the attacker gets access to the Internet-based communication between the system operator and the system, false data can be injected to mimic the operation of a failure and terminate data integrity for falsifying the sensor measurement. An overview of various cyber-physical attacks that are targeted by the attacker is shown in Figure 9.2.

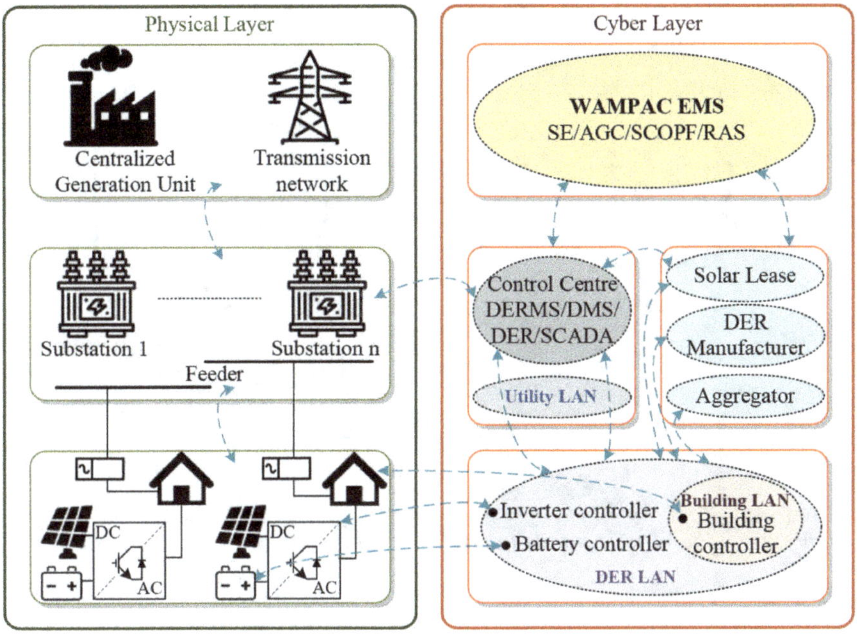

Figure 9.2 Domains for cyber-physical attacks in a GCPVS

To protect the inverters, many of the manufacturers are also focusing on cyber security and attempts are made by the manufacturers to provide an end-to-end encryption for all the data that are being sent by the device either to the consumer or among the devices themselves. The other form of attack focuses on the grid and impacts the power system by escalating issues such as disconnection of the plant from the grid or emerging false energy demand. The attacker may influence the operation of the inverter forcing the breaker to be tripped by causing concerns of voltage or frequency fluctuations. Few of the vulnerable attack points in the grid-tied PV system are represented in Figure 9.3.

The most common attacks on the cyber-physical system such as power grids and PECs are denial of services (DoS) attack, data integrity attack (DIA), replay attack, stealthy attack, and false data injection (FDI) attack. The attacker may spoof the sensor data, which may result in the falsification of power output in the PV plant. As per Figures 9.1 and 9.3, the vulnerable sensor data such as filter capacitor voltage (V_c), filter inductor current (I_L), DC link voltage (V_{dc}), grid current (I_g), and PV current (I_{pv}) are represented by $Y(t) = [V_c(t), I_L(t), V_{dc}(t), I_g(t), I_{pv}(t)]^T$, whereas the power control reference of the controller is represented as $S(t) = [P^*(t), Q^*(t)]^T$. The attack model based on the vulnerable components can be represented as

$$Y(t) = \alpha Y_f(t) + \beta Y_o(t - t_{delay}) \tag{9.1}$$

$$S(t) = \Upsilon S_f(t) + \phi S_o(t - t_{delay}) \tag{9.2}$$

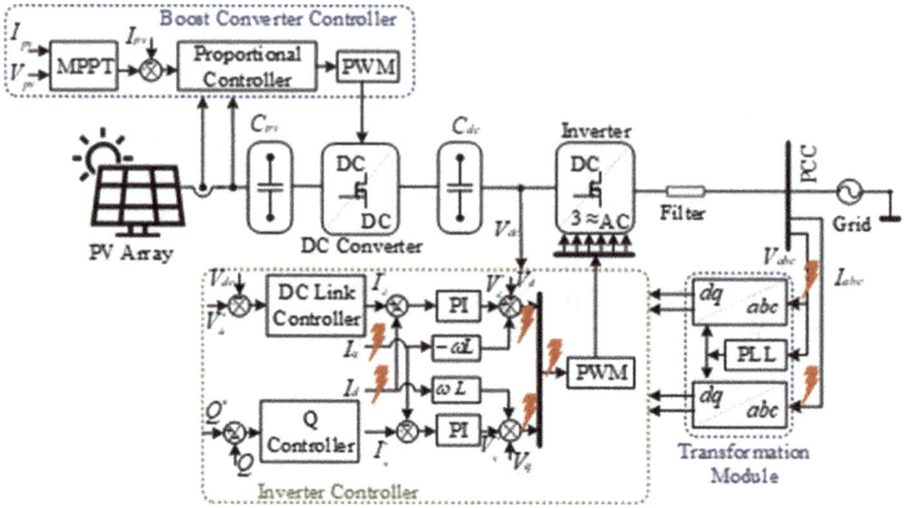

Figure 9.3 Vulnerable points in a GCPVS and its converter control structure

where S_o and Y_o represent the original measurement, and Y_F and S_F are the biased functions that can be either independent or can be a function of Y_o. The multiplying factors β and ϕ define the weight of the real vectors, whereas α and Υ are the multiplying factors that define the weight of the attacking vectors. The Y and S are the overall data vectors that are the inputs to the controller. The delay in communication is represented by t_{delay} that can be caused by cyberattack. The cyberattack can be represented as $t_a = [t_s, t_e]$, where the starting and the ending time of the attack are represented by t_s and t_e, respectively. Furthermore, the description of different types of attacks is presented in sections 9.2.1–9.2.5.

9.2.1 DoS attack

One of the wicked attacks that may take place on the power system is DoS as the communication channel is jammed by a large number of unauthorized packets of information. The transmission burden is heavy, and because of the excessive network bandwidth being consumed, the communication may be interrupted [4]. The DoS-based attacks are generally performed by manipulating the communication so that the control command or the measured data can be prevented from reaching the destination. In such attacks, the prior system stated knowledge is not utilized by the attacker; hence, the attacker has constrained knowledge about the controller [5]. The detection of DoS-based attacks is relatively easy than other attacks; hence, the attacker generally targets the system with constraints such as inadequate memory or a low-end processor [6]. In large distribution networks, the DoS can be effective as the communication delay can cause issues such as frequency instability [7].

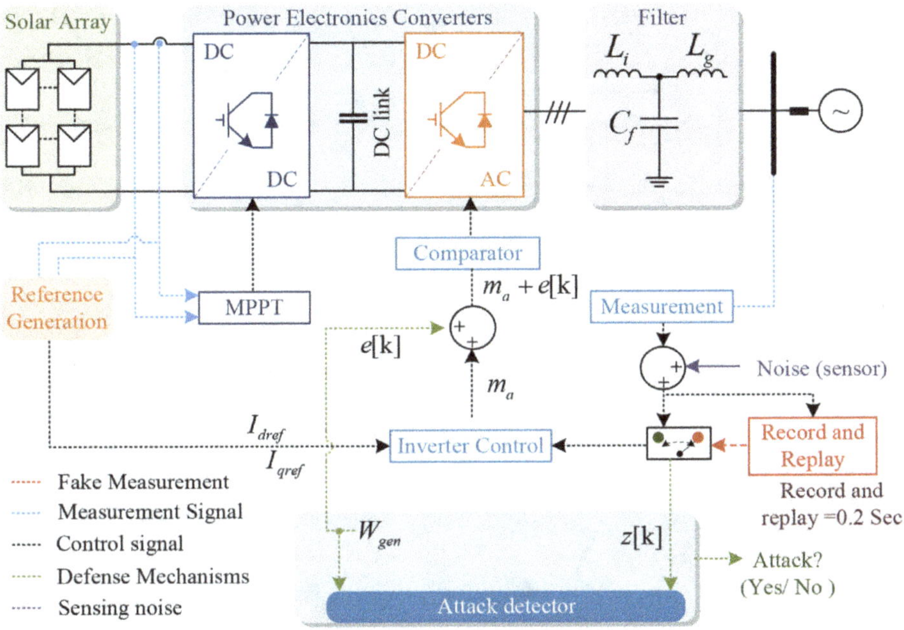

Figure 9.4 Implementation of replay attack in a GCPVS

9.2.2 Data integrity attack

The data integrity attack (DIA) is implemented by manipulating the control commands and measurement signals that are transmitted through the power system [8]. The manipulated data are injected by the attacker into the sensor channel by keeping the transmitted data within the limit. If the data exceed a specific limit, there is a chance that the attack may be detected easily. The attacker needs to follow the system operating principle to succeed [9].

9.2.3 Replay attack

In the replay attack shown in Figure 9.4, a disclosed attack is performed by the attacker to gather the data, and once the response data are gathered, the response is played continuously until the end of the attack [10]. The replay attack continuously replays the fault data to delay the authorized data transmission [4]. Like the DIA, this type of attack does not also require any prior knowledge of the system model, control, or design [8]. But with the advancement in the attack strategy and the involvement of physics-oriented attacks, the knowledge of the scenario is needed [5]. The attack can be further modified with a time-bound delay along with unknown information about the upper limit and change in rate [4]. The attacker can target the sensor by using the cryptography algorithm or by inducing false sensor data and manipulating the local conditions around it [11].

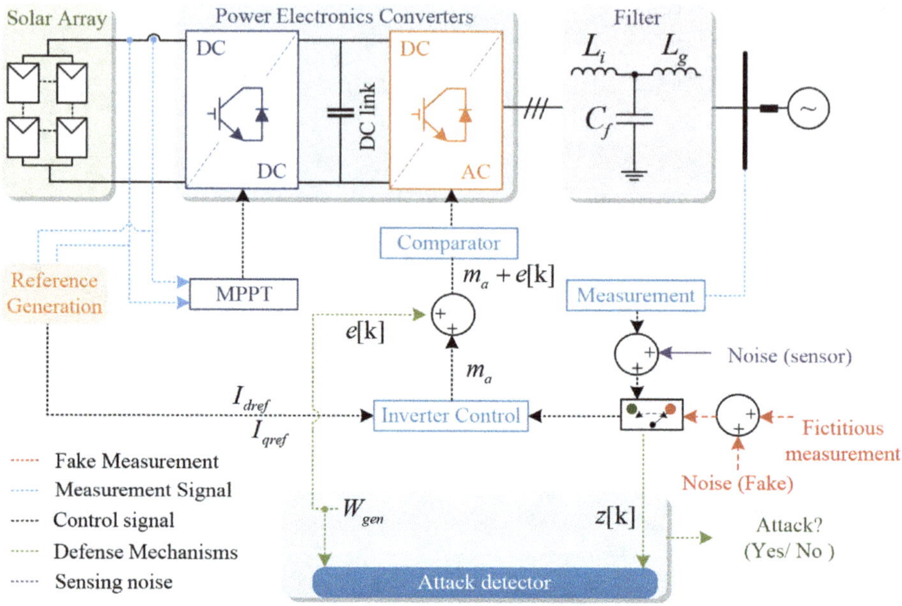

Figure 9.5 Implementation of stealth attack in a GCPVS

The attack at the sensor can be modeled as

$$Ay(t) = -Cx(t) + y(t - \tau) \tag{9.3}$$

where the sensor data are represented by $0 < \tau < t.y(t - \tau)$ through monitoring. The replay attack can be modeled based on the two stages involved [11, 12]:

Monitoring Phase

$$(0 < t < t_o)\begin{Bmatrix} y_a(t) = 0 \\ j(t) = \Gamma^y \cdot y(t) \end{Bmatrix} \tag{9.4}$$

Replay Phase

$$(t_o < t < 2t_o)\begin{Bmatrix} y_a(t) = j(t - t_o) \\ j(t) = j(t - 1) \end{Bmatrix} \tag{9.5}$$

9.2.4 Stealthy attack

The attacker feeds the fabricated measurements obtained by the simulated system, as illustrated in Figure 9.5. It can be observed that the noise is added to the system along with the fabricated measured value keeping the real-life system noise into consideration. The system noise can be caused because of interference from jitters and thermal and electromechanical parameters of the system. The fake noise makes the process of detection difficult. The watermarking signal $(e[k])$ along with the

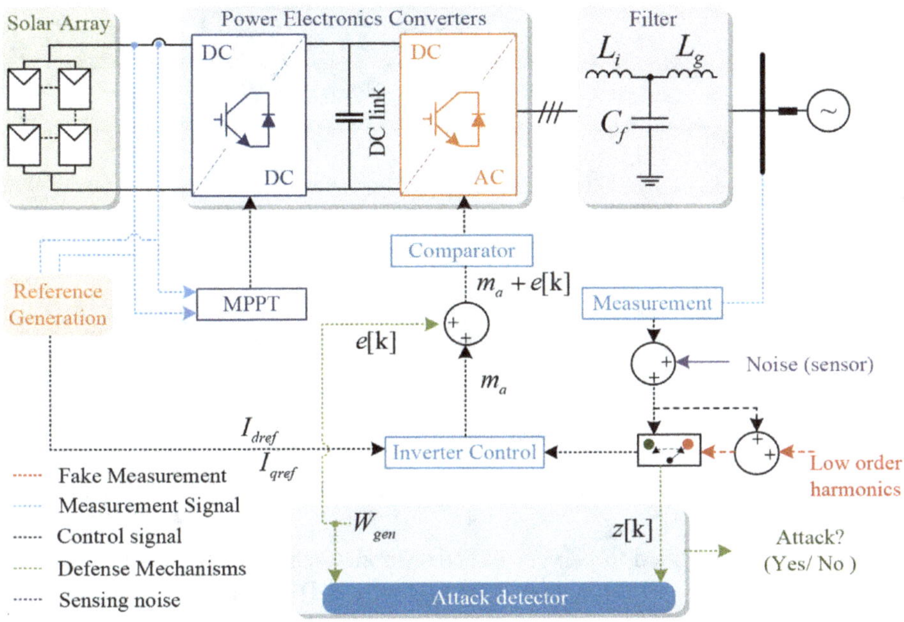

Figure 9.6 Implementation of harmonic injection attack in a GCPVS

noise is small enough that they do not raise any red flags and make the fault look as close as possible.

9.2.5 Harmonic injection attack

The false harmonic measurement is superimposed by the attacker over the recorded sensor measurement, as depicted in Figure 9.6. Lower-order harmonics are mostly added to the actual current measurement.

9.2.6 Other attacks

A certain time delay is introduced in the communication of the sensor data to the command center during the time delay attack, whereas a constant bias is added by the attacker to the measured data during the constant noise bias attack. Furthermore, a random bias can also be superimposed by the attacker during a random noise-biased attack. The attacker may also flip the measured signal from the sensor causing a serious system operation issue as a flip signal attack.

9.3 Physics-based approach

As opposed to the traditional approach of deriving the model of a system using empirical data, the physics-based approach (PBA) models any given system and its operation by utilizing the principles of physics [13]. In the literature, these

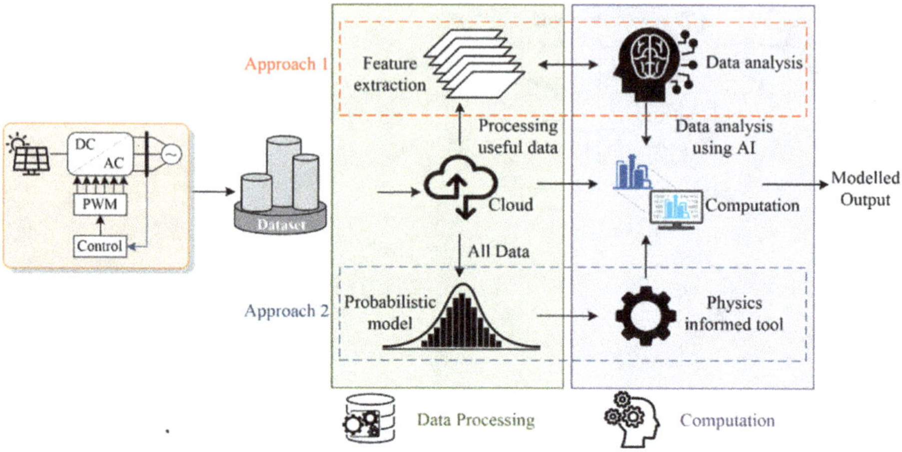

Figure 9.7 Steps for developing a physics-based model of a GCPVS

approaches are identified to be very effective in analyzing the failures in a system and its components and estimating their lifetimes [14]. As it is imperative that a failure or degradation for a specific part of a given system can be predicted under a specific set of conditions and is not an inherent quality of the component [15, 16], the same information can also be accessed to create an FDIA in the system. Therefore, while adapting the PBA for modeling a system, it is necessary to consider the operational conditions and also the environmental variables for better injection of the fault data. A brief overview of how a physical model of a system can be identified is shown in Figure 9.7.

In addition to the above, the control units involved in the system also play a major role in the modeling through PBA. For a GCPVS, the control of the voltage source converter and the boost DC–DC converter are the major components that involve a control unit. These control units are developed based on the data measured from the system and involve various complex operations. Besides, with the advancements in the field of intelligent control and machine learning algorithms, these control units are accommodated with prognosis methods to identify defect initiation and propagation attacks. Hence, it is necessary to consider the aspects of the control unit while developing the PBA to derive dynamic models of FDIA using principles of physics. Furthermore, with the availability of information related to the materials and their microstructural characterization of the power semiconductor modules used in the PECs, the PBA for the FDIA framework can be propagated as a function of time to interfere with the robustness and reliability of the components [17]. These models are considered to involve an immense computational complexity and are not suitable for real-time FDIA implementation as they address the materialistic properties of the device from a single particle level. To ease the development of PBA for such microstructural characterizations, combined and lumped models are developed in the works of Andersson *et al.* and Li *et al.* [18, 19]. These models can be used for creating power module failure as a cyberattack in real time based on the information

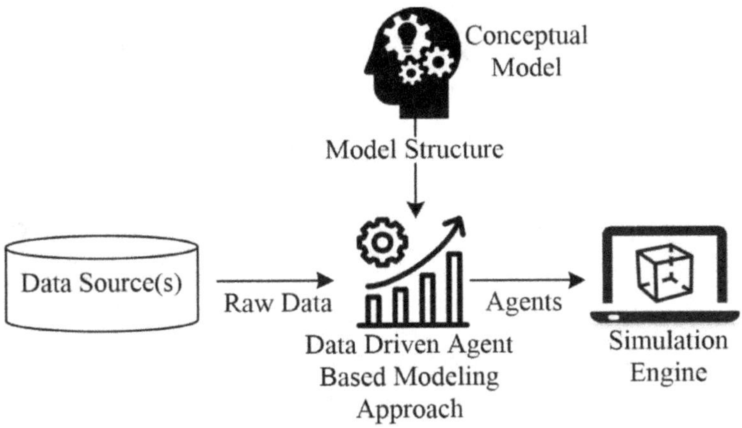

Figure 9.8 Conceptualization of data flow process in a data-oriented model

related to temperature variations, stress ratio, and damage mechanism interactions (fatigue, creep, etc.). The PBA for modeling for FDIA also establishes a scientific basis related to structure, and design modifications in the system that can interrupt planned monitoring, and scheduled maintenance [14]. Besides, they also model the relation between stress failure to emulate the dominant environmentally induced failure mechanisms. Such failure emulations can also lead to miscalculations during the remaining useful life estimation of the system/subsystem.

9.4 Data-oriented approach

Contextually, the external elements or systems are associated with the data-driven approach and have been shown in Figure 9.8. The diagram or text in [20] is the detailed version of the model that is interpreted by the conceptual model. The agents are initiated by generating the data from the data source. These data-driven agents are executed and scheduled using the simulation engine-based software system. Data-based modeling (DBM) is utilized to design these three external elements where conceptual model and data are defined as the input for these external elements and responsible for the creation of data-driven agents. The modeler is assumed to have information about the conceptual model for system analysis/study. Precisely, the modeler is responsible for creating the implicit concepts for the explicit model along with conceptual modeling [21]. The method of conceptual modeling is to create a conceptual model that functions as a commonly discussable medium while communicating with a grid-connected system attacker [22]. The descriptions of model elements, assumptions, and their detailed inter-relationships are necessary to include in the conceptual model [23]. The purpose of conceptual modeling is very imperative in data-driven approaches and accountable for data source guidance, identification, and selection to generate the model. Initially, only high-level model relationships and descriptions are supposed to be included in the conceptual model

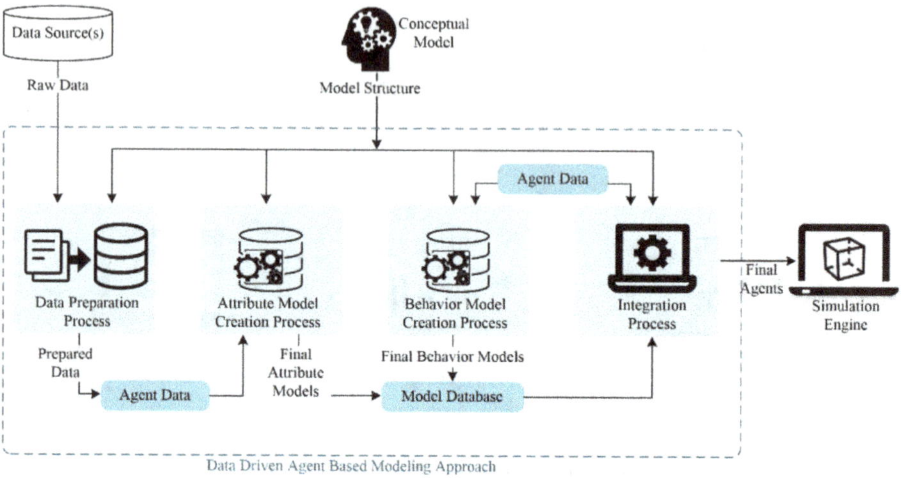

Figure 9.9 Relation between the conceptual model and a data-driven model

for the data-driven case. The conceptual model is required to have a model structure explaining the type of agent with their attributes, a statement of purpose depicting the model goal and signature behavior, and lastly different types of environments and variables.

The raise in agent attribute values and variation in model behavior is attained by the DBM method that uses the conceptual model by identifying and selecting the data sources. The on-demand streams, online repositories, and physical device-based data are available on example data sources. The data source generates indispensable information at the granularity of an independent agent for both attributes and behavior. The behavioral action of the real-world entity is stored in historical data to generate the behavior. These behavioral actions typically are the interactions with other entities over time. In the research by Kavak *et al.* [24], the values of attributes are directed by attribute data or another method of providing the information to vary the values through a process is depicted in []. The agent creation for any given data source and a conceptual model is to be executed by the preferred simulation engine using the data-driven method. The relation between the conceptual model and data sources is elaborated in Figure 9.9 while generating DBM for any given system.

The various processes are named attribute model creation process, data preparation process, integration process, and behavior model creation process. The model database and agent data are the main data stores. The data stored in agent data stores are generated after transforming the raw data into a new form of data and known as the data preparation process. The data issues are solved by passing through various iterative data checks. As the data check process is completed, the data are prepared and stored in the agent data repository. Later, these data are used in the remaining processes such as the behavior model creation process, attribute model creation process, and integration process. Furthermore, these processes are accountable for behavior generation from data, designing the attribute models as per the requirement,

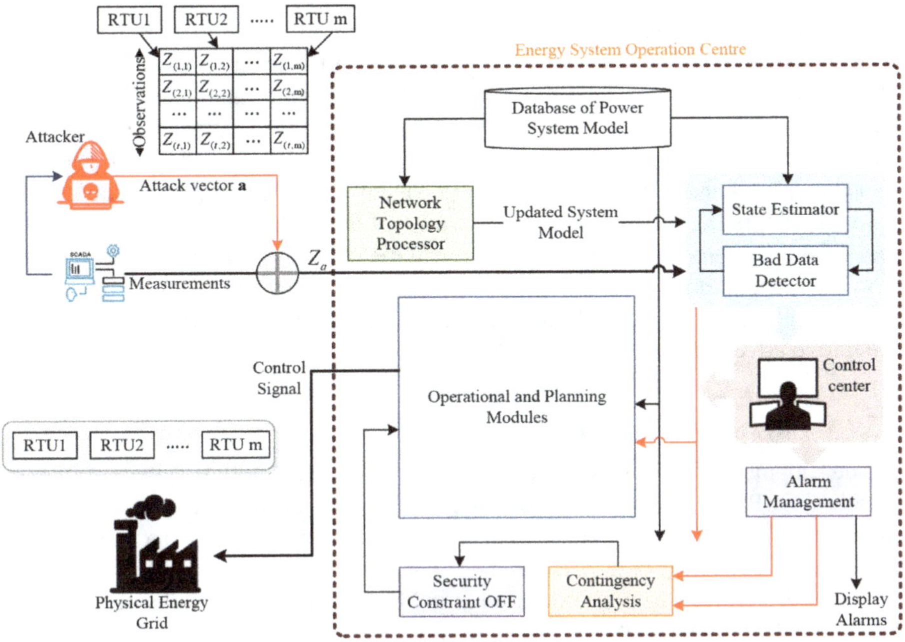

Figure 9.10 Database modeling framework for attack profile generation in a GCPVS

and lastly developing the new agents by integrating the behavior model, attribute model, and agent data based on the structure of the model given in the conceptual model. A generalized framework for developing a DBM approach for a GCPVS is shown in Figure 9.10.

9.5 FDI attack generation

For a GCPVS with m measurements and n state variables, the non-linear relation between both m and n is given as

$$z(t) = h(x(t)) + e(t) \tag{9.5}$$

where $z(t) \in R^{m\times1}$ is a measurement vector, h is a non-linear function between the measurement vector and the state vector, $x(t) \in R^{n\times1}$ is the state vector, and $e \in R^{m\times1}$ is a vector of measurement errors. To launch a perfect FDIA, the attacker must possess the ability to gather information about the system parameters, control architecture, and real-time state estimates of the converter, such that the attack vector $a(t)$ can be designed as

$$a(t) = h(\hat{x}(t) + c(t)) - h(\hat{x}(t)) \tag{9.6}$$

where $x(t)$ indicates the state estimation vector without FDIAs and $c(t)$ is the bias vector injected into the system states. Furthermore, it has been readily demonstrated in the literature that under such perfect availability of information, the attack vector created produces the same measurement residual vector as the attack-free condition. But in reality, achieving perfect attack vectors is a difficult task, and so, the attackers create vectors that are closely related to (9.6) and have a small perturbation to the residual vectors. In one of the previous works of the authors, various physics- and data-oriented approaches are designed to achieve efficient FDIA in PECs [25]. The physics-oriented tools are based on the graph model of the PEC control architecture, whereas the data-oriented tools are based on the game-theoretic (GT) approach and generative adversarial networks (GANs).

9.5.1 Physics-oriented tool

The development of a physics-oriented tool for FDIA is achieved using a graph theory model of GCPVS. The graph network is considered to identify the major and minor components in a system as nodes, and the relation between each component of the system is identified as an edge [26]. Depending on the operational flow of tasks between the nodes, these edges can be both bidirectional and unidirectional. In this chapter, the main focus is on the development of PBM of the grid-connected system control unit such that the attacker can achieve FDIA through the vulnerable points. The graph network for the control unit of the grid-connected inverter is shown in Figure 9.11.

Initially, the voltage and current measuring unit at the point of common coupling are considered as the source node for designing the graph model. Furthermore, the phase locked loop (PLL) and the transformation aspects for both the measurements are considered as the connecting nodes. These nodes are further mapped with the DC link controller, Q controller, and PI control units that are driven to generate pulses for the PEC. At the output of the graph model, the PWM unit sums the connection between all the nodes. As there are no bidirectional flows in the control unit, all the edges in the graph network indicate a unidirectional flow. This makes the model a unidirectional graph model. As illustrated in Figure 9.11, the vulnerable points V_{abc}, I_{abc}, V_d, V_q, I_d, and I_q are easily accessible through the graph model. To emulate a cyberattack with this developed graph model, the attacker can either inject false data at any of the vulnerable points, which will result in a false calculation of reference for the PWM unit, which in turn generates incorrect switching behavior for the power modules in the PEC. In addition to injecting the false data, the attacker can also deceive the controller into a disabled link between any two given nodes or emulate a data disconnection between the measurement unit and the transformation process. Here, the controller can either continue to operate in a failed state or run into a delayed control action (loss of synchronization), where the stable operation of the system is disrupted.

9.5.2 Data-oriented tool

To develop a data-oriented tool for imitating the faults in the power system, the GT approach and GANs are used.

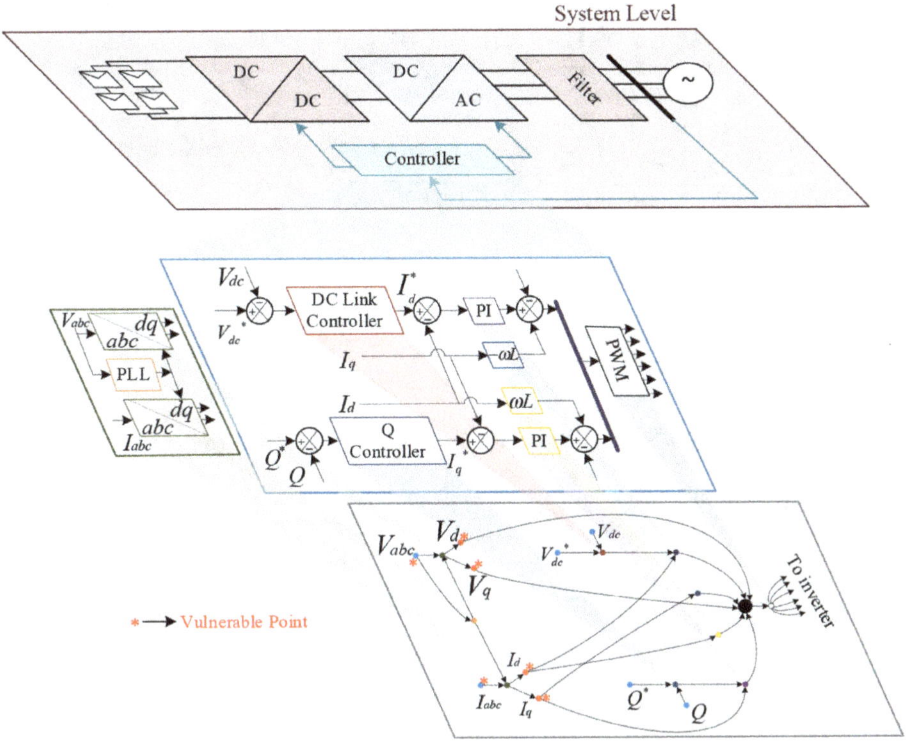

Figure 9.11 Graph network of the control unit in a GCPVS for FDI attack

9.5.3 GT approach

The data-oriented tools, i.e., GANs and GT approaches, have been implemented to imitate faults in the power system. In the GT-based approach, the attack strategy is either pure or mixed operation by idle and active operation players. The attacker aims at gaining control over the convertor control to impact the operation of the power system. Initially, the vulnerable point in the converters is identified and the analysis of the closed-loop defense mechanism is studied. The analysis includes the security features and system response to it before any attack is introduced to the system. The system response analysis is used by the GT to design the attack profile of the model. The GT formulates a relation between the least and best attack technique and optimizes the attack strategy depending on the analysis. An overview of GT-based approach is presented in Figure 9.12.

The development of an attack strategy is based on the understanding that the attacker has about the system. In GT, the attacker man modifies the attack dynamically based on the attack consequences mapping that maximizes the impact of the attack. Furthermore, the attacker can launch multiple attacks at the same location in a short span of time. The information about the vulnerable point in the controller that is exploited by the attacker is presented in Figure 9.13.

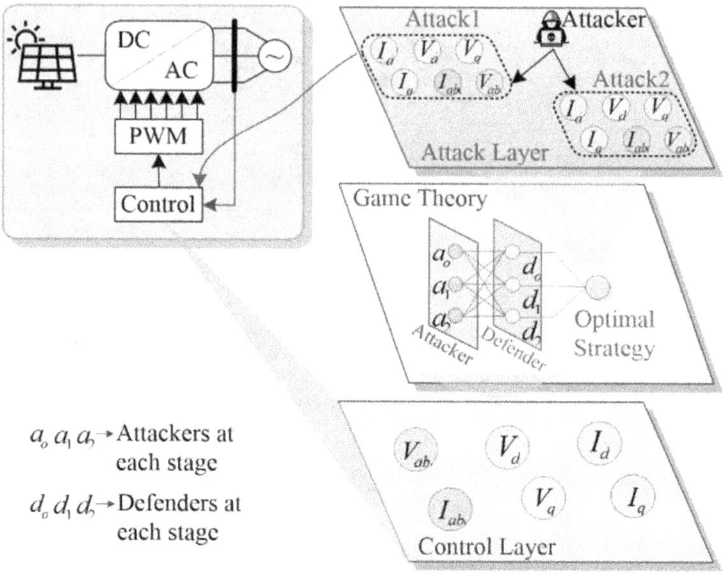

Figure 9.12 GT approach for FDI attack in a GCPVS

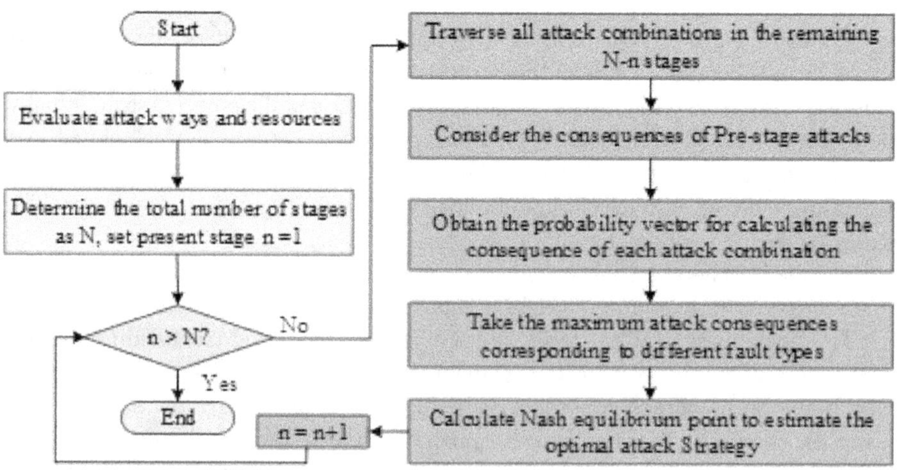

Figure 9.13 Flow diagram for attack generation with the GT approach

Algorithm: Cyber-attack emulation using game theory

Step 1: Let X be the number of attack capabilities of the attacker with R available resources to estimate the possible number of game strategies N.

Step 2: For an initial value of current game strategy $n = 1$, the traverse of all other $N - n$ attack strategies are estimated.

Step 3: A probability vector is obtained by considering the significance of the pre-stage attacks to estimate the attack consequence during FDIA.

Step 4: Based on the outcomes of *Step 2* and *Step 3*, the maximum attack significance in stage n is analysed. This refers to identifying the maximum impact of all the possible FDIAs obtained after considering all the attack combinations in the remaining stages.

Step 5: To identify the maximum impact of all the possible FDIAs, the Nash equilibrium [27] point is established between the attacker and the system. This indicates the optimal outcome of the FDIA at each stage of the game strategy.

Step 6: The above steps are iterated for various measurements from different vulnerable points and by updating the current stage n to n+1 to such that it reaches the pre-set total stage N.

By implementing the above steps, the GT approach can be successfully modeled to generate an FDIA in the GCPVS.

9.5.4 *Generative adversarial networks*

The framework for modeling an FDIA using GANs is shown in Figure 9.14. The GANs develop a generative model of a system using convolutional neural networks [28]. This model is framed as a problem of supervised learning with a generator model and a discriminator model. The generator model trains the historic fault data with a random set of variables to generate an FDIA, whereas the discriminator can be modeled to differentiate between the real data and the generated data. Both the

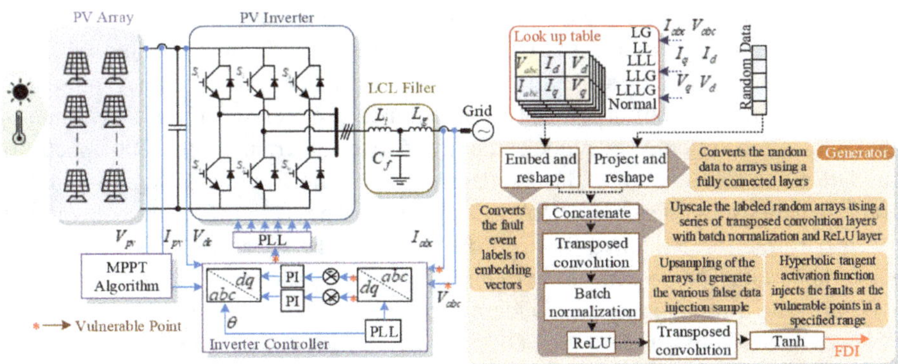

Figure 9.14 Framework for training of GAN for FDI attack in a GCPVS

models are trained in a zero-sum game until the discriminator fails to distinguish between the real data and the generated data for more than half the iterations. To develop an FDIA with GANs for a GCPVS, initially, the information related to faulty event/disturbance is gathered to identify the potential faults that interrupt the operation of a PEC. The identified faults are labeled according to the fault type as line-line (LL), line-ground (LG), line-line-ground (LLG), line-line-line-ground (LLLG), and line-line-line (LLL). The labeled fault types are further embedded into vectors using fully connected layers and reshaped to form an array. Parallelly, a random data generator is used to generate random vectors that can emulate the patterns in the faulty event/disturbance data and is concatenated with the reshaped labeled data to emulate the cyberattack. Before injecting the concatenated data at the vulnerable points in the control unit, the resulting arrays are processed through a series of batch normalization-based transposed convolution layers and rectified linear unit (ReLU) layers. Besides, a hyperbolic tangent activation function is used to inject faults at the vulnerable points within a specified range. The fully connected layer is utilized by the generator network to initiate the reshaping of the random vectors. The reshaping is performed on categorical labels by transforming them into embedded vectors. Furthermore, the results attained by the random layer are concatenated by the generator network, and the data steps are labeled along the channel dimension. Finally, the output of step 3 is up-sampled using a series of $1 - D$ transposed convolution layers with ReLU [29] and batch normalization.

9.6 Result and analysis

To assess the FDIA through the physics- and data-oriented tools developed in section 9.5.2, a three-phase grid-connected system is simulated in MATLAB/Simulink® environment. The simulated system has a 35-kW PV array connected to the grid with a two-stage conversion process. The boost converter is operated with a perturbation and observation of maximum power point tracking, and the design of inverter control is shown in Figure 9.3. All the attack models are trained to emulate the unsymmetrical fault behavior with the grid-connected system.

To begin with, initially, the graph theory approach is implemented with the inverter control unit of the system. The graph network of the control unit is modeled, and the vulnerable points are fed with random data to generate various unsymmetrical faults in the system. The modeled control network is trained for 1 500 epochs to emulate a line-to-line fault in the simulated system. The total time taken to train the graph network to achieve the FDIA is 9 min 16 s. The training accuracy is plotted based on the error estimated between the impact of FDIA at each epoch with the actual fault data. The results of the training process are shown in Figure 9.15.

From the training results, it is identified that the GT approach has an accuracy of approximately 75% before the FDIA converges with a similar set of variations to generate the attack. The impact of the generated data to create a line-to-line fault in the system is shown in Figure 9.16.

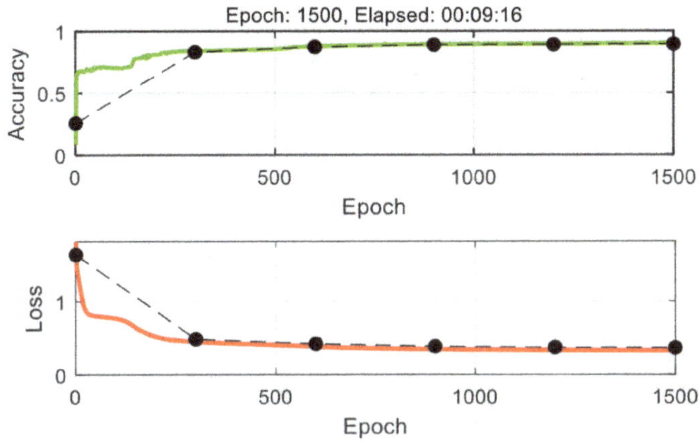

Figure 9.15 Training results of FDI attack with graph network

Furthermore, the data-oriented tools for FDIA are designed through the GT and GAN approaches as discussed in previous sections. The GT approach is implemented by considering the five different faults and six possible fault locations in a grid-connected inverter controller. Initially, the probability of generating a fault in the system by disturbing the data at a particular vulnerable point is identified. Furthermore, based on the minimum level of disturbance injection, the variation is carried out as a step change at all the vulnerable points. In this FDIA approach, the GT is accompanied by the Nash equilibrium, which identifies that with more than 15% variation at the vulnerable points a minimum attack can be launched at game stage 1. Furthermore, this variation is increased in a range of 15% for six steps to disturb the data at the vulnerable points by 90%. Considering the significance of each variation in FDIA, the optimal attack variation is identified. The training results of the GT approach to identify the optimal variation required at each location to generate different faults are shown in Figure 9.17.

Furthermore, the effectiveness of an optimal FDIA to emulate a line-to-line fault in the simulated system is estimated by plotting the game graph between the column graphs and the row strategies, as shown in Figure 9.18. The results signify that

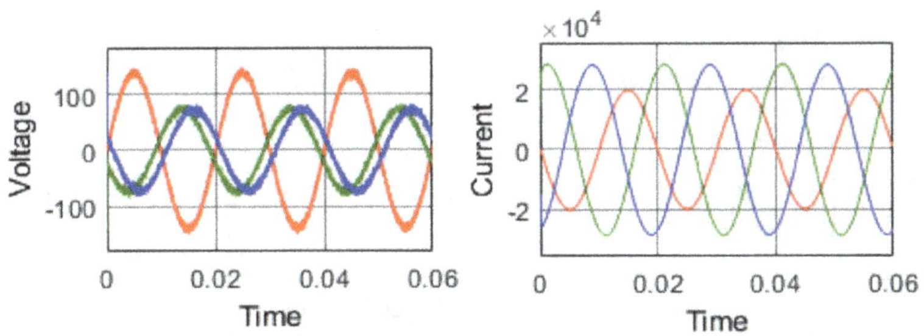

Figure 9.16 Cyberattack emulation of line-to-line fault using graph network

	Attack Location 1	Attack Location 2	Attack Location 3	Attack Location 4	Atack Location 5	Attack Location 6	Row Mix
LG Fault	10	20	30	40	50	60	25
LL Fault	20	10	10	10	40	40	0
LLG Fault	30	30	20	20	30	50	25
LLL Fault	40	40	40	30	20	30	.5
LLLG Fault	50	50	50	50	10	10	0
Column Mix--->	.5	0	0	0	.5	0	
Value of game (to row)	30						

(a) Game theory result

	Col mix 1 * cell payoff	Col mix 2 * cell payoff	Col mix 3 * cell payoff	Col mix 4 * cell payoff	Col mix 5 * cell payoff	Col mix 6 * cell payoff	Expected Value (row sum)
Column's Optimal Mix	.33	0	.17	.17	.17	.17	
LG Fault	5	0	7.5	10	12.5	15	50
LL Fault	10	0	10	12.5	15	2.5	50
LLG Fault	15	0	12.5	15	2.5	5	50
LLL Fault	20	0	15	2.5	5	7.5	50
LLLG Fault	25	0	2.5	5	7.5	10	50
Value of game (to row)							50

(b) Rows expected values

	Optimal Row Mix	Attack Location 1	Attack Location 2	Attack Location 3	Attack Location 4	Attack Location 5	Attack Location 6
Row 1 mix * ...	.17	2.5	5	7.5	10	12.5	15
Row 2 mix * ...	.17	5	7.5	10	12.5	15	2.5
Row 3 mix * ...	.17	7.5	10	12.5	15	2.5	5
Row 4 mix * ...	.17	10	12.5	15	2.5	5	7.5
Row 5 mix * ...	.33	25	30	5	10	15	20
Expected V...		50	65	50	50	50	50
Value of ga...	50						

(c) columns expected values

	Attack Location 1	Attack Location 2	Attack Location 3	Attack Location 4	Attack Location 5	Attack Location 6	Row Minimum	Maximin
LG Fault	15	30	45	60	75	90	15	15
LL Fault	30	45	60	75	90	15	15	15
LLG Fault	45	60	75	90	15	30	15	15
LLL Fault	60	75	90	15	30	45	15	15
LLLG Fault	75	90	15	30	45	60	15	15
Column Maximum	75	90	90	90	90	90		
Minimax	75							
15 <= value <=75								

(d) Maximin/Minimax

Figure 9.17 Mapping of variation (in %) between different fault types and vulnerable points

the developed GT approach has 66.67% accuracy for the estimated optimal FDIA. The FDIA with the trained GT approach to emulate a line-to-line fault is shown in Figure 9.19.

Finally, the FDIA is implemented with the GAN-based data-oriented tool, where the data gathering process is similar to that of the GT approach. For the development of the GAN approach, the historical fault data related to unsymmetrical faults

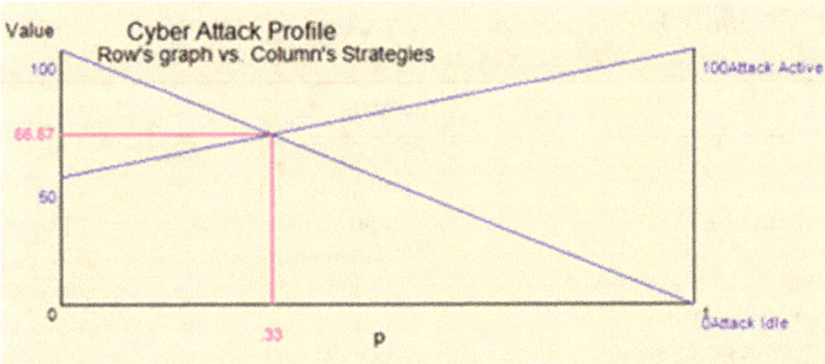

Figure 9.18 Accuracy of GT approach for active attack condition

of a three-phase GCPVS are obtained. Initially, the discriminator model is developed to generate the attack data using the procedure discussed in the above sections. Furthermore, the discriminator tries to distinguish between the actual data and the generated data until it fails to distinguish for more than 50% of the training times. Once this process is achieved, the finally generated attack data are used to emulate a certain fault as a cyberattack in the system. The parameters for training the GAN are shown in Table 9.1.

During the development of the generator model and the discriminator model, the actual and generated data are visualized by identifying the statistical features of the data sets using the principal component analysis (PCA). The data visualization between the different principal components in a PCA subspace is shown in Figure 9.20a.

Furthermore, to illustrate the performance of the trained GAN, the confusion matrix is plotted between the true data and false data from the actual and generated data sets. The results shown through the confusion matrix in Figure 9.20b indicate that the classifier training based on the generated signals achieves higher accuracy. In addition to the truly and falsely classified samples, the spectral characteristics of the real and generated signals for both healthy and faulty data are identified at multiple frequency levels in Figure 9.20c,d respectively. The results identify the deviation of the generated data from the actual data for

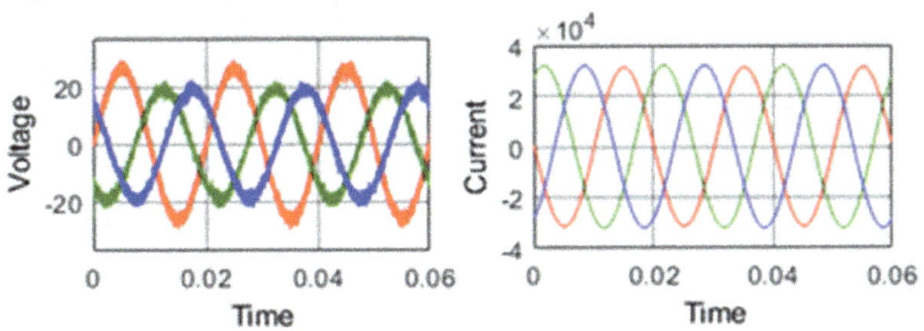

Figure 9.19 Cyberattack emulation of line-to-line fault using GT approach

Table 9.1 Training parameters for FDI attack with GANs

Parameter	Value
Epochs	1 000
Iterations	7 000
Minimum batch size	100
Optimizer	Adam optimizer
Learning rate	0.005
Gradient decay factor	0.5
Squared gradient decay factor	0.999

both conditions. From all the above results, the training accuracy with GANs is identified to be very high. With such high accuracy, the line-to-line fault emulated as a cyberattack is shown in Figure 9.21. Furthermore, a brief comparison of all the three approaches, i.e., graph theory approach, GT approach, and GANs approach are given in Table 9.2.

A brief comparison between the three techniques is presented in Table 9.2.

All the above results and comparison indicate that the GAN-based FDIA effectively emulates a line-to-line fault in the system with the highest accuracy of 99.4% and within 1 min 36 s. These statistics are very much better when compared with the other approaches developed for generating similar fault types on the same three-phase grid-connected system.

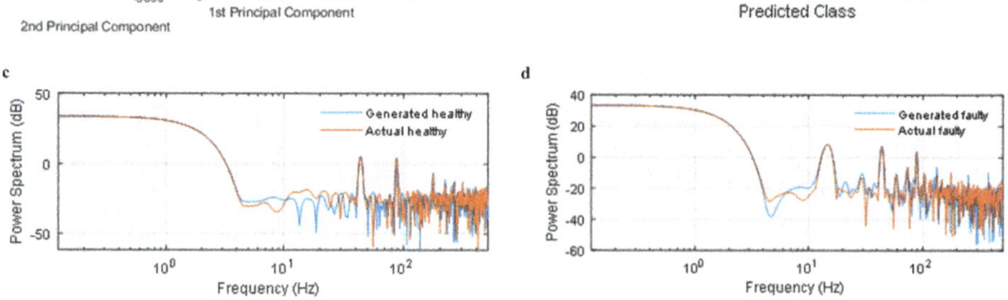

Figure 9.20 Results for GANs

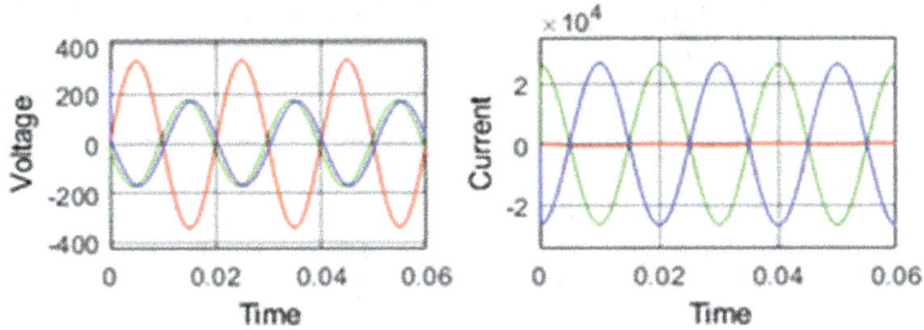

Figure 9.21 Cyberattack emulation of line-to-line fault using GANs

Table 9.2 Comparison between physics- and data-oriented approaches for fault data injection attack in a GCPVS

	Technique		
	Physics-oriented tool	Data-oriented tool	
Parameter	**Graph theory**	**Game theory**	**GANs**
Iterations	1 500	–	7 000
Efficiency	75%	66.67%	99.4%
Generation time	12 min 35 s	5 s	1 min 36 s

9.7 Conclusion

This chapter provides a detailed understanding of physics- and data-oriented approaches for emulating cyberattack in a grid-connected system through FDI attacks. The physics-oriented tools are developed based on the graph model of the systems control architecture, whereas the data-oriented tools are based on the GT approach and GAN. The design process for all three approaches is carried out on the same system. The physics-oriented tools are identified to have an average FDIA accuracy of 75 to emulate the fault with long training times. Furthermore, the data-oriented approach with game theory has less training accuracy with less training time, and the GANs have the highest training accuracy of 99.4% with a training time of 1 min 36 s for 7 000 iterations. Besides, the GAN-based FDIA is also identified to have a maximum-likelihood ratio while emulating the cyberattack profile.

References

[1] Misbrener K. 'Global solar installations expected to surpass 200 GW for first time'. *Solar Power World*. 2022.

[2] 'Ukraine power cut "was cyber-attack"'. *BBC*. 2017.

[3] Staff R. 'Cyber attack on U.S. power grid could cost economy $1 trillion: report'. *Reuters*. 2015.

[4] Ding D., Han Q.-L., Xiang Y., Ge X., Zhang X.-M. 'A survey on security control and attack detection for industrial cyber-physical systems'. *Neurocomputing*. 2018;275(1):1674–83.

[5] Teixeira A., Shames I., Sandberg H., Johansson K.H. 'A secure control framework for resource-limited adversaries'. *Automatica*. 2015;51(6):135–48.

[6] Wang Q., Tai W., Tang Y., Zhu H., Zhang M., Zhou D. 'Coordinated defense of distributed denial of service attacks against the multi-area load frequency control services'. *Energies*. 2019;12(13):2493.

[7] Zhong X., Jayawardene I., Venayagamoorthy G.K., Brooks R. 'Denial of service attack on tie-line bias control in a power system with pv plant'. *IEEE Transactions on Emerging Topics in Computational Intelligence*. 2017;1(5):375–90.

[8] Mahmoud M.S., Hamdan M.M., Baroudi U.A. 'Modeling and control of cyber-physical systems subject to cyber attacks: a survey of recent advances and challenges'. *Neurocomputing*. 2019;338(3):101–15.

[9] Sridhar S., Manimaran G. 'Data integrity attacks and their impacts on SCADA control system'. *IEEE PES General Meeting*; IEEE, 2010. pp. 1–6.

[10] Pasqualetti F., Dorfler F., Bullo F. 'Attack detection and identification in cyber-physical systems'. *IEEE Transactions on Automatic Control*. 2013;58(11):2715–29.

[11] Fritz R., Zhang P. 'Modeling and detection of cyber attacks on discrete event systems'. *IFAC-PapersOnLine*. 2018;51(7):285–90.

[12] Physical Authentication of Control Systems. 'Designing watermarked control inputs to detect counterfeit sensor outputs'. *IEEE control systems*. 2015;35(1):93–109.

[13] Halliday D., Resnick R., Walker J. *Principles of physics, 10th edition international student version*. 10th ed. Wiley; 2014.

[14] Hansen P., McCluskey P. 'Failure models in power device interconnects'. *European Conference on Power Electronics and Applications*; IEEE, 2007. pp. 1–9.

[15] Snook I., Marshall J.M., Newman R.M. 'Physics of failure as an integrated part of design for reliability'. *Annual Reliability and Maintainability Symposium*; 2003. pp. 46–54.

[16] Pecht M., Dasgupta A. 'Physics-of-failure: an approach to reliable product development, final report'. *IEEE 1995 International Integrated Reliability Workshop*; IEEE, 1995. pp. 1–4.

[17] Xiong J.J., Shenoi R.A. 'A two-stage theory on fatigue damage and life prediction of composites'. *Composites Science and Technology*. 2004;64(9):1331–43.

[18] Andersson M., Streb M., Ko J.Y., *et al.* 'Parametrization of physics-based battery models from input–output data: a review of methodology and current research'. *Journal of Power Sources*. 2022;521(1):230859.

[19] Li J., Wang Q., Reutzel E.W., Reutzel E.W., Nassar A.R. 'An extended lumped-parameter model of melt–pool geometry to predict part height for directed energy deposition'. *Journal of Manufacturing Science and Engineering*. 2017;139(9).

[20] McKenzie F.D. *Systems modeling: analysis and operations research modeling and simulation fundamentals*. Hoboken, NJ: John Wiley & Sons, Inc; 2010. pp. 147–80.

[21] Robinson S., Arbez G., Birta L.G., Tolk A., Wagner G. 'Conceptual modeling: definition, purpose and benefits'. *Winter Simulation Conference (WSC)*; 2015. pp. 2812–26.

[22] Balci O., Ormsby W.F. 'Conceptual modelling for designing large-scale simulations'. *Journal of Simulation*. 2007;1(3):175–86.

[23] Robinson S. 'Conceptual modeling for simulation: issues and research requirements'. *Proceedings of the 2006 Winter Simulation Conference*; 2006. pp. 792–800.

[24] Kavak H., Vernon-Bido D., Padilla J.J., Diallo S.Y., Gore R.J. 'The spread of wi-fi router malware revisited'. *Simulation Series*. 2017;49(2):80–9.

[25] Khan M.A., Bharath Kurukuru V.S., Sahoo S., Blaabjerg F. 'From physics to data oriented cyber attack profile emulation in grid connected PV systems'. *IEEE 22nd Workshop on Control and Modelling of Power Electronics (COMMPEL)*; 2021. pp. 1–8.

[26] Jorjani M., Seifi H., Varjani A.Y. 'A graph theory-based approach to detect false data injection attacks in power system AC state estimation'. *IEEE Transactions on Industrial Informatics*. 2021;17(4):2465–75.

[27] Holt C.A., Roth A.E. 'The NASH equilibrium: a perspective'. *Proceedings of the National Academy of Sciences*. 2004;101(12):3999–4002.

[28] Goodfellow I., Pouget-Abadie J., Mirza M., *et al*. 'Generative adversarial networks'. *Communications of the ACM*. 2020;63(11):139–44.

[29] Dittmer S., King E.J., Maass P. 'Singular values for ReLU layers'. *IEEE Transactions on Neural Networks and Learning Systems*. 2020;31(9):3594–605.

Chapter 10

Stealth cyber attacks in microgrids: detectability and observability

Subham Sahoo[1], Tomislav Dragicevic[2], and Frede Blaabjerg[1]

This chapter discusses a cooperative mechanism for detecting potentially deceptive cyber attacks that attempt to disregard average voltage regulation and current sharing in cyber-physical microgrids. Considering a set of conventional cyber attacks, the detection becomes fairly easy for distributed observer-based techniques. However, a well-planned set of balanced attacks, termed as the *stealth attack*, can bypass the conventional observer-based detection theory as the control objectives are met without any physical errors involved. In this chapter, we discuss the formulation and associated scope of instability from stealth attacks to deceive distributed observers realizing the necessary and sufficient conditions to model such attacks. To address this issue, two disagreement indices (DIs) for each agent are introduced to detect potential threats to voltage and current, which accurately identify the attacked agent(s) under various scenarios. To facilitate detection under worst cases, the DIs from the secondary voltage control sublayer are strategically cross-coupled to the current sublayer, which ultimately disorient the control objectives in the presence of stealth attacks and provide a clear norm for triggering defense mechanisms. Finally, its performance is simulated under many potential threats on sensors and communication links.

10.1 Introduction

DC microgrids are an effective means of integrating renewable energy sources, storage devices, and modern electronic loads, capable of operating independently of the utility grid [1, 2]. Moreover, the operating nature of these units in the DC paradigm makes it a vivid option to enhance the efficiency [3]. For enhancing the scalability and robustness, distributed controllers are desirable in microgrids [4, 5] to avoid

[1]Department of Energy, Aalborg University, 9220 Aalborg, Denmark
[2]Centre of Electric Power and Energy, Technical University of Denmark, Lyngby, Denmark

single point of failure as compared to the centralized communication, owing to their highly reliable operation during link failures. Moreover, distributed control philosophy is an economic option since it can be easily accommodated by transmitting lesser volume of data without entailing much traffic in contrast to the centralized communication [6]. In DC microgrids, cooperative secondary controllers have been deliberately used for various objectives such as average voltage regulation [7], proportional load sharing [8], and energy balancing [9].

To enhance the scope of reliability, system security plays an increasingly important role to maintain *unbiased* coordination among the sources since it directly affects the technological aspects based on penalties, specifically allocated for poor performance metrics [10]. Few potential ways to violate security measures are cyber attacks, which typically include false data injection attacks (FDIAs) [11], denial of service (DoS) [12], replay attacks [13], and others. Such attacks are adept at disrupting the network stability as well as control structures. Several instances have been reported in the past, which became a critical concern for the control centers [14]. FDIAs alter the system state by injecting a false data into any of the compromised sensors/actuators. An example of implementation of such attacks is given in [11]. To analyze the impact of such attacks, further investigation is done in [15] to assess its impacts on the economic load dispatch that is realized in a cooperative manner. In this respect, the system under attack reaches a consensus stage, which is not optimal. Broadly, detection and mitigation of conventional attacks are already well classified in the literature since such attacks disrupt the operation of observers, which become simple criteria for detection. However, it is reported that generalized FDIAs, commonly known as *stealth* attacks [16], can easily penetrate into networked systems without altering the system observability. These attacks can be specifically classified as *coordinated intelligent attacks* [17], which involve coordinated attack vectors in multiple nodes to nullify system dynamics. As a result, the system/agent operator would be unaware of any online attack vectors present in the system. Prior to this, the attacker could cause an unfair increase in the magnitude of attack vectors, which may cause system shutdown depending upon the severity of the attack. Additionally, implementation of such attacks gets easier when the attacker has obtained *apriori* knowledge about the system using adequate system monitoring [18]. More instances of coordinated attacks on large power systems and its vulnerability assessment are provided in [19, 20]. In this regard, risk assessment, along with control vulnerabilities, is crucial since the modeling of coordinated attacks for microgrids can be easier owing to their small system size without significant security measures [21, 22].

In [23], the authors have identified aberrant operation of a microgrid when a false data is injected into the voltage controller of the substation. Apart from stability, it is also crucial to analyze if the proposed strategies can attain economic vulnerabilities in a microgrid. In fact, this attribute is well addressed [24], where the FDIAs are categorized by their utilization levels having monitored the stability of microgrids under different conditions. On the other hand, Beg *et al.* [25] have stressed on the identification of the varying of candidate invariants to detect the presence of FDIAs. Moreover, it has been demonstrated that stealth attacks in DC microgrids can deceive the control system without creating any negative impact/

disturbance. However, it is crucial to understand that such undetectable attacks, which are able to penetrate while maintaining discretion, can cause network instability in unforeseeable ways.

Since distributed observer-based strategy [26] is more prone to cyber attacks for a well-spanned distributed graph as the injected false data propagates in the entire network, proper analysis has to be carried out toward the detection of the attacked agent in a microgrid to establish corrective action. False data propagation in DC microgrids may lead to loss of generality from an economic point of view, cause current sharing errors, which lead to circulating currents between each converter. Using distributed computation, the estimated states will converge to a non-zero steady value under FDIAs, which makes them simple to detect. In [27], the compromised agent with false data is detected using cooperative-based trust and confidence factors to realize mitigation of the propagation of false data in the cyber network. However, considering the worst case for such attacks, the abovementioned factors can also be manipulated by adding/subtracting a large constant value while the controller is attacked, which may lead to false values corresponding to the attacked node. Consequently, it will result in maloperation of the mitigation strategy, since it operates on non-attacked agent(s). In [28], Fawzi *et al.* have determined a theoretical limit on the number of compromised sensors in a system beyond which it is impossible to characterize the detection of such attacks. Considering this view point, theoretical analysis for stealth attacks at multiple sensors/actuators in a cooperative network to create instability and the corresponding detection methodologies in DC microgrids have not gained significant attention yet. On the other hand, Zhao *et al.* [29] have addressed this issue for an economic dispatch problem as it decreases the overall efficiency with an increase in the generation cost by dislocating toward a non-optimal point. However, it does not administer a mechanism for detection of the compromised agent during a stealth attack, which is crucial to cease its propagation into the network and may consequently lead to instability.

The idea behind stealth attack detection in this work is identification of the merits of a well-spanned network in cooperative control mechanism. In particular, the difference between the secondary output of voltage sublayers, termed as cooperative vulnerability factor (CVF) for stealth attack on voltage, converges to zero if the system is not under attack. Furthermore, the necessary and sufficient conditions for modeling of worst-case stealth attack involving multiple sensors/communication links are studied extensively. Moreover, the impact of FDI and stealth attacks on sensors and communication links is studied for intrusion in voltage and current information separately to preserve system security and energy efficiency simultaneously. In addition, a similar detection strategy to detect prospective attacks on currents is also presented exploiting the same philosophy. On the other hand, the agent(s) representing positive value of CVF is resolved as attacked, who suggests that their respective measurements are *untrue*. This can be easily extended to trigger the likely defense mechanisms to prevent further instability.

To sum up, the research contributions of this chapter are as follows:

- To ascertain the possibility of FDI and stealth attacks in DC microgrids, a new methodology based on CVF is proposed using the outputs from secondary sublayers used for global average voltage regulation in DC microgrids. Generalization of distributed observers is done to detect such attacks and how it can be circumvented for a multiple sensor/link-based stealth attacks. For detection of the compromised/attacked agent, CVF of each agent is locally monitored to isolate the agent corresponding to a positive valued CVF in the network, which represents the attacked agent(s). This technique is used as an apparent method of detecting attacks locally such that corrective actions can take place. To the best of authors' knowledge, CVF has never been proposed in the realm of cyber attack detection in microgrids.
- A new cross-coupling methodology of CVF output of each agent from the secondary voltage sublayer is proposed to strategically disorient the control operation for the worst case of consecutive attacks when the attacker can attempt to reduce CVF into a negative value so as to deceive the above-mentioned detection philosophy. Hence, the cross-coupling approach ensures accurate detection of the attacked agent(s) by preventing against further attacks into the proposed detection metric, i.e., CVF.

10.2 Cyber-physical preliminaries of microgrids

This chapter considers an autonomously operating DC microgrid system with K sources. The architecture of the microgrid is shown in Figure 10.1. Each of the sources (interfaced using DC/DC buck converters for regulated power conversion) is connected to one another via tie-lines. These elements collectively represent the microgrid physical layer. Operation of the power electronic converters occurs in voltage-controlled mode. Proper voltage regulation and current sharing are achieved using a cooperative secondary control framework where a local controller is associated with each of the DGs [7]. All the local controllers are connected through a distributed communication network, which requires each controller to share information only with its neighboring controller(s). The cyber layer can be considered as a graph (consisting of multiple nodes and edges), where each node represents an agent and each edge represents a communication link that connects two agents. Elements of the network compose an adjacency matrix, $\mathbf{A} = [a_{kj}] \in \mathbb{R}^{N\times N}$, where the communication weights may be expressed as $a_{kj} > 0$, if $(\psi_k, \psi_j) \in \mathbf{E}$ ($\mathbf{E}$ denotes an edge that connects ψ_k, i.e., the local node and ψ_j, i.e., the neighboring node). Else, $a_{kj} = 0$. The matrix for inbound cyber information can be represented as $Z_{in} = \mathtt{diag}\left\{\sum_{k\in k} a_{kj}\right\}$. The Laplacian matrix $\mathbf{L}$ is said to be *balanced*, if $\mathbf{A}$ and $\mathbf{Z}_{in}$ are equal (since, $\mathbf{L} = \mathbf{Z}_{in} - \mathbf{A}$).

Each of the controller units can be represented as an *agent* in the cyber layer, sending and receiving a group of measurements:

$$\mathbf{x} = \{\bar{\mathbf{V}}, \mathbf{I}^{pu}\} \tag{10.1}$$

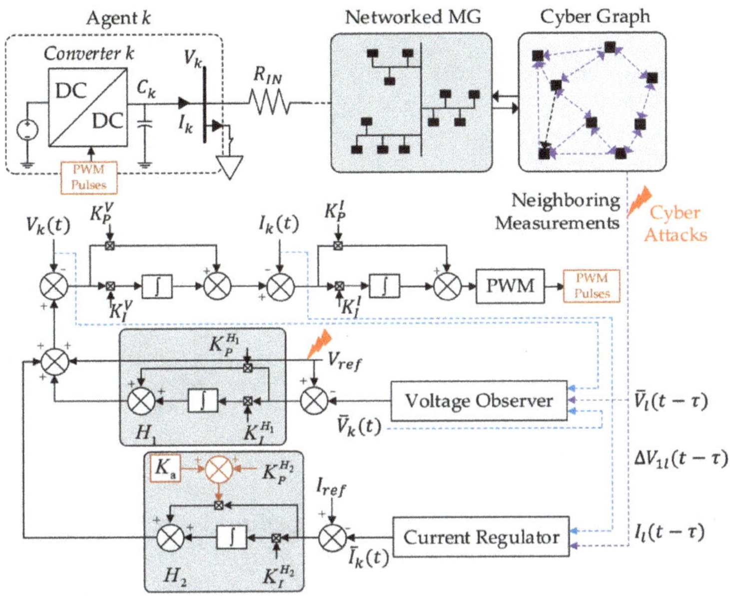

Figure 10.1 *Control structure of a networked DC microgrid with many agents operating with a distributed cyber graph under the presence of cyberattacks*

with their respective neighboring agents to attain average voltage regulation and proportionate current sharing.

Considering preliminaries of the communication graph, control input of the local secondary controller (associated with each DG) can be stated as

$$u_k(t) = \sum_{j \in M_k} \underbrace{a_{kj}(x_j(t) - x_k(t))}_{e_k(t)} \tag{10.2}$$

where $u_k = \{u_k^V, u_k^I\}$, $e_k = \{e_k^V, e_k^I\}$ (according to the elements present in x). Additionally, M_k is the set of neighbors of agent k. To clarify the error formulation in (10.2), we can simplify it using:

$$\begin{aligned} e_k^V(t) &= a_{kj}(\bar{V}_j(t) - \bar{V}_k(t)) \\ e_k^I(t) &= a_{kj}(I_j^{pu}(t) - I_k^{pu}(t)) \end{aligned} \tag{10.3}$$

boxes A similar extrapolation can be done to represent u_k.

Remark I: According to the cooperative synchronization law, consensus will be achieved by all agents (who participate in distributed control) using $\dot{\mathbf{x}}(t) = -\mathbf{L}\mathbf{x}(t)$ to finally converge to $\lim_{t \to \infty} x_k(t) = c, \ \forall\, k \in K$.

Using (10.3), the local control inputs necessary to achieve the control targets (average voltage regulation and proportionate sharing of load current) can be acquired from the secondary controller by using the voltage correction terms as mentioned below (for k^{th} agent) [30]:

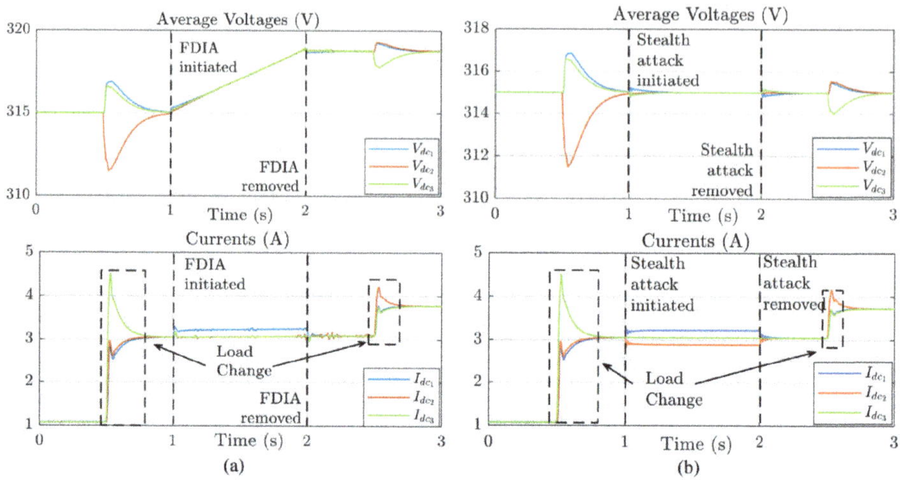

Figure 10.2 *Performance of DC microgrid consisting of* M = 3 *agents for (a) FDI attack on current sensor of agent I and (b) stealth attack on current sensors of agents I and II: deteriorates current sharing profile*

Average voltage regulation:

$$\Delta V_{1_k} = H_1(s)(V_{ref} - \bar{V}_k) \tag{10.4}$$

Proportionate current sharing:

$$\Delta V_{2_k} = H_2(s)(I_{ref} - u_k^I) \tag{10.5}$$

where, $\bar{V}_k = V_k + \int_0^{\tau} \sum_{j \in M_k} u_k^V d\tau$. For proportionate current sharing, $I_{ref} = 0$. Correction terms acquired in (10.4) and (10.5) can be added to the global reference voltage for the achievement of local voltage references (for the kth agent) using:

$$V_{ref}^k = V_{ref} + \Delta V_{1_k} + \Delta V_{2_k} \tag{10.6}$$

The target objectives mentioned in (10.4) and (10.5) are achieved by using (10.6) as the local reference voltage (for the kth agent).

As per the distributed consensus algorithm for a heavily connected digraph (in the DC microgrid) [31], the system objectives [using (10.1)–(10.6)] shall converge to

$$\lim_{t \to \infty} \bar{V}_k(t) = V_{ref}, \ \lim_{t \to \infty} u_k^I(t) = 0 \ \forall \ k \in K \tag{10.7}$$

As shown by the red symbols in Figure 10.1, malicious attackers may try to corrupt the cyber layer in several ways (e.g., false data injection, DoS) to disturb the achievement of the objectives mentioned in (10.7). In case of a stealth attack, the attack vector penetrates deep in the control layer by deceitfully hiding from the system operator. The ability to access multiple nodes allows such vectors to create disturbances that can be continued over an elongated stretch of time and enables

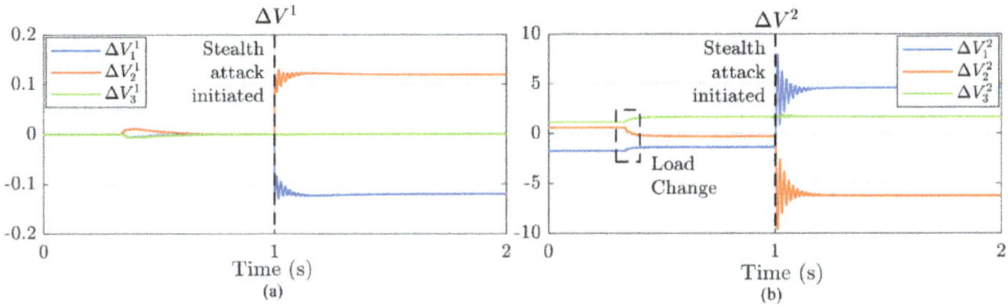

Figure 10.3 Case study II: Performance of (a) average voltage regulation and (b) current sharing for a strong case of stealth attack on voltage sensors of agents II and III

them to forcefully cause generation outages. This may ultimately result in system shutdown. Hence, identifying the compromised node(s) is essential to prevent malware propagation (reducing chances of further destabilization).

Such attacks can perform coordinated manipulation to fool the system observer via the following additions in (10.2):

$$\mathbf{u}^a(t) = \mathbf{L}\mathbf{x}(t) + \mathbf{W}\mathbf{x}_{attack} \tag{10.8}$$

where $\mathbf{u}^a$,$\mathbf{x}$ and $\mathbf{x}_{attack}$ denote the vector representation of the attacked control input $u_k^a = \{u_k^{Va}, u_k^{Ia}\}$, the states $x_k = \{\bar{V}_k, I_k^{pu}\}$, and the attack elements $x_{attack_k} = \left[x_{attack_k}^V, x_{attack_k}^I\right]^T$, respectively. It should be noted that $\mathbf{x}_{attack}$ could be a step, sawtooth, sinusoidal, or an unbounded signal. Furthermore, $\mathbf{W} = [w_{kj}]$ depicts a row-stochastic matrix with its elements expressed by

$$w_{kj} = \begin{cases} -\frac{1}{M_k+1}, \; j \in M_k \\ 1 + \sum_{j\varepsilon M_k} w_{kj}, \; j = k \\ 0, \; j \notin M_k, \; j \neq k \end{cases} \tag{10.9}$$

The diagonal entries denote the placement of attack elements in locally measured $\mathbf{x}$. Moreover, the non-zero entries in off-diagonal elements in $\mathbf{W}$ represent the communicated measurements. Using (10.9), we formalize that an undetectable attack can be maintained if and only if the sum of the change in state produced by the attack and the zero input evolution of the state induced by the attack belong to the system's weakly unobservable

Table 10.1 Stealth attacks in DC microgrids in [30] and [33]

Affected counterparts	**Modeling**
Voltage [30]	$\mathbf{W}x_{attack}^V = 0$
Current [33]	$\mathbf{W}x_{attack}^I = 0$

subspace. Although $\mathbf{Wx}_{attack}$ will always be equal to zero from a system-level perspective, the change identified across an agent is suppressed by the opposite shift in the remaining agents, without contributing any significant dynamics into the system.

10.3 Proposed stealth attack detection strategies

10.3.1 Stealth attack on voltages

This section discusses about the detection of the attacked nodes in a cooperative network-based DC microgrid. Considering the attacker injecting false data into multiple sensors/communication links to formulate a stealth attack, an analysis of how the convergence in (10.7) can be guaranteed is provided in this section. Furthermore, the necessary and sufficient conditions to formulate a stealth attack on multiple sensors in a cooperative network are given in detail.

For each agent, the local power balancing equation can be expressed in terms of

$$\chi_i(k) = I_{dc_i}(k) - I_{o_i}(k) \tag{10.10}$$

where I_{o_i} denotes the total output current from i^{th} agent. Using (10.10), the consensus algorithm in (10.4)–(10.5) under attacks can be re-written as

$$\begin{cases} \bar{V}_{dc_i}(k+1) = \sum_{i\in M} l_{ij}\bar{V}_{dc_j}(k) + \sigma\chi_i(k) + u^a_{V_i}(k) \\ I_{dc_i}(k+1) = \sum_{j\in M} l_{ij}I_{dc_j}(k) + u^a_{I_i}(k) \\ \chi_i(k+1) = \sum_{j\in M} l_{ij}\chi_j(k) + (I_{dc_i}(k+1) - I_{dc_i}(k)) \end{cases} \tag{10.11}$$

where $u^a_{V_i}(k)$ and $u^a_{I_i}(k)$ denote the attack vectors imposed into voltage and current secondary sublayer in i^{th} agent at k^{th} instant, respectively. It should be noted that since χ is not a physical measurement entity, the possibility of attack in χ_i will be entirely due to $u^a_{I_i}$. Here, we obtain the necessary and sufficient conditions for the convergence of system under such attacks in (10.11).

Problem Statement: If there exist a constant R such that

$$\sum_{k=0}^{\infty} |u^a_V(k)| \le R, \sum_{k=0}^{\infty} |u^a_I(k)| \le R \;\forall i \in M \tag{10.12}$$

then (10.11) in the presence of stealth attack shall converge as per (10.7) with $\lim_{k\to\infty} \chi_i = 0$.

Proof: Representing (10.11) in the form of $x(k+1) = Ax(k) + Bu(k)$, we have

$$x(k+1) = Ax(k) + Bu(k) = A^{k+1}x(0) + \sum_{t=0}^{k} A^{k-t}u(t) \tag{10.13}$$

As A is primarily composed of Laplacian matrices in (10.11), its eigenvalues lie around zero and unit plane [32]. Since $\lim_{k\to 0} \sum_{t=0}^{k} A^{k-t}u(t)$ will converge to zero for a well-connected graph, as per (10.7), $\lim_{k\to 0} A^{k+1}x(0)$ should converge to $V_{dc_{ref}}d$, where

$d = \left[1, \ldots, 1, 0, \ldots, 0, 0, \ldots, 0\right]^T \in R^{3MX1}$ with N elements equal to 1 and $2N$ elements equal to 0. Hence, this proves the convergence of a stealth attacked system to the global reference set-points in (10.7).

Additionally, the above-mentioned proof can be extrapolated to justify

$$\sum_{i \in M} I_{dc_i}(k) = \sum_{i \in M} I_{l_i}(k) \tag{10.14}$$

under a stealth attack where I_{l_i} is the local load at i^{th} agent. Due to (10.14), convergence of (10.10) is guaranteed. By the iterative rule, subtracting $I_{dc_i}(k+1)$ from $\chi_i(k+1)$, we get

$$\sum_{i \in M} \chi_i(k+1) - \sum_{i \in M} I_{dc_i}(k+1) = \sum_{i \in M} \chi_i(k) - \sum_{i \in M} I_{dc_i}(k) - u^a_{I_i}(k) \tag{10.15}$$

$$= \sum_{i \in M} \chi_i(k-1) - \sum_{i \in M} I_{dc_i}(k-1) - (u^a_{I_i}(k-1) + u^a_{I_i}(k)) \tag{10.16}$$

$$= \sum_{i \in M} \chi_i(0) - \sum_{i \in M} I_{dc_i}(0) - \sum_{z=0}^{k} \sum_{i \in M} u^a_{I_i}(z) \tag{10.17}$$

Substituting for $I_{dc_i}(k+1)$ from (10.11) in (10.17) and taking limitation on both sides considering (10.10) as $k \to \infty, \forall i \in M$, we get

$$\sum_{z=0}^{k} \sum_{i \in M} u^a_{I_i}(z) = 0 \tag{10.18}$$

A similar analysis can be carried out to determine the effect of u^a_V in the convergence of the algorithm to get

$$\sum_{z=0}^{k} \sum_{i \in M} u^a_{V_i}(z) = 0 \tag{10.19}$$

using $\chi_i(k+1)$ and $\bar{V}_{dc_i}(k+1)$ in (10.15).

Remark II: Following the concept of cooperative synchronization [7], $\bar{\mathbf{V}}_{dc}$ tends to achieve consensus for all its elements for a spanning cyber graph such that $\mathbf{L}\bar{\mathbf{V}}_{dc} = 0$ during steady state to reach a steady-state value of $V_{dc_{ref}}$. Alternatively, a similar representation can be given using $e^i_1(k)$ in (10.4) such that $\mathbf{LE}_1 = 0$ reaches a steady-state solution of zero, where $\mathbf{E}_1$ denotes the vector notation of $e^i_1(k)$. Using (10.19) as an attack vector for the above-mentioned consensus theory, it can be concluded that the steady solution is not affected for $\mathbf{E}_1$ owing to the consensus properties of a Laplacian graph [7]. A similar relation can be derived for $e^i_2(k)$ in (10.5) using (10.36). Hence, it can be concluded that the final state convergence as per (10.7) is not affected even under stealth attacks since it gets nullified by the sum of false data injection in multiple sensors/links for a cooperative network as established in (10.36) and (10.19).

As opposed to the centralized systems where the global information is present at a single node, it is a complicated task to apprehend the attacked node in cooperative

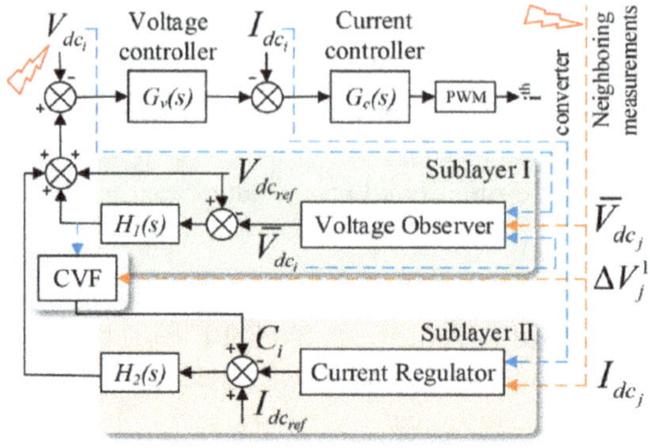

Figure 10.4 Proposed attack resilient controller for i[th] *agent to detect attack on sensors and communication links in DC microgrids*

systems as intrusion in any agent affects the entire system for a strongly connected graph. To address the issue, this chapter utilizes the concept of control output synchronization to detect the attacked node in a cooperative network where the input signals with attack vectors are deemed to achieve consensus. Following the convergence of the inputs, it is shown how the difference in their respective PI controller outputs achieves consensus for the same global reference voltage.

Remark V: Since output current from an agent, as shown in Figure 10.1, is based on the voltage levels between two different points, a stealth intrusion in the agents' current values for operation at a particular load leads to change in voltage levels across the network thereby disproving (10.7). In simple terms, it can be stated that the agent can recognize such attacks as it would result in the current sharing error. Such error may in turn cause undesirable effects such as overloading of individual converters or reduced energy efficiency. This has been justified by a case study in Figure 10.2 for FDI and stealth attack on current sensors in a DC microgrid shown in Figure 10.1 of $M = 3$ agents. In Figure 10.2a, a false data of −0.5 A is injected into the current sensor in agent I at $t = 1$ s, which immediately results into improper sharing, thereby reducing energy efficiency. Similarly, in Figure 10.2b, a stealth attack is attempted at $t = 1$ s by injecting a balanced set of zero sum attacks of ±0.5 A into the current sensors in agents I and II simultaneously, which deteriorates the current sharing profile. However, the average voltage is still maintained in Figure 10.2b in contrast to the case for FDI attack. With the basic assumption that each agent operator bears knowledge that the system is equipped with proportionate current sharing controller, the sharing error shown in Figure 10.2a b should be sufficient criteria to identify attacks on current sensors such that corrective action can take place. Hence, it becomes an easier task to determine such attacks in a cooperative network. However, stealth attacks on voltage sensors in case of multiple sensors/communication links can be inconspicuous to identify. In other words, the agent voltages are maneuvered in such a way that the control operation in (10.7) still holds true even in the presence of such attacks.

Using Remark V, the control input for voltage regulation is particularly used to present a strong case for stealth attack. Hence, the control input for average voltage regulation [31] at i^{th} agent is given by

$$u_i(k) = \underbrace{\sum_{j\in N_i} a_{ij}(\bar{V}_{dc_j}(k) - \bar{V}_{dc_i}(k))}_{u_{ij}(k)} + b_i e_1^i(k) \tag{10.20}$$

For various attacks in i^{th} controller, the attacked control input can be modeled as

Sensor attack:

$$u_i^f = u_i + \kappa u_i^a \tag{10.21}$$

Cyber link attack:

$$u_{ij}^f = u_{ij} + \kappa u_i^a \tag{10.22}$$

where $\kappa = 1$ denotes the presence of attack vector, or 0 otherwise and u_i^a denote the attack vector in i^{th} agent. By local investigation of $u_i^f(k)$ in each agent, non-zero synchronization error can be detected with residual output; however, it is not a sufficient criterion for detection of the attacked node(s) in a cooperative network since comparison of each residue requires global information that contradicts our case. To verify this case, the effort of the controller to synchronize the output for a given reference voltage is strategically used to indicate the occurrence of attack. It can be ensured using (10.4) and (10.5) in sublayer I to give

$$\xi_i(k) = u_i^f(k) - u_i(k-1) \tag{10.23}$$

for an attack within $[k-1, k]$ instant, which changes due to the momentary increase/decrease in (10.23) as input for the attacked agents and its neighbors at the instant of attack vector injection in multiple sensors/cyber links in a microgrid. As a result, the change in PI output in sublayer I can be written as

$$\delta\Delta V_i^1(k) = K_p^{H_1}\xi_i(k) + K_i^{H_1} u_i^f(k) \tag{10.24}$$

where $\delta\Delta V_i^1(k) = \Delta V_i^1(k) - \Delta V_i^1(k-1)$. Using the change in outputs obtained in (10.24), CVF, C_i is calculated using the PI controller outputs for each agent, which has been used in this chapter to determine the attacked nodes accurately. Mathematically, it can be represented as

$$C_i(k) = h_i \underbrace{[\sum_{j\in N_i} a_{ij}(\Delta V_j^1(k) - \Delta V_i^1(k))]}_{y_1^i(k)} \underbrace{[\sum_{j\in N_i} a_{ij}(\Delta V_j^1(k) + \Delta V_i^1(k))]}_{y_2^i(k)} \tag{10.25}$$

for i^{th} agent, where h_i is a positive constant. Moreover, using (10.5) and (10.30), we get

$$\dot{\mathbf{Y}}_1 = \mathbf{L}_1 \bar{\mathbf{V}}_{dC} = \mathbf{L}\mathbf{Y}_1 = 0 \tag{10.26}$$

Using (10.31), it can be concluded that cooperative synchronization law [7] holds true in the absence of attacks where $\mathbf{L}_1 = diag\{\mathbf{H}_1\}\mathbf{L}$ and $\mathbf{Y}_1$ denote the vector notation of y_1^i in (10.30). However, in the presence of attacks, equation (10.31)

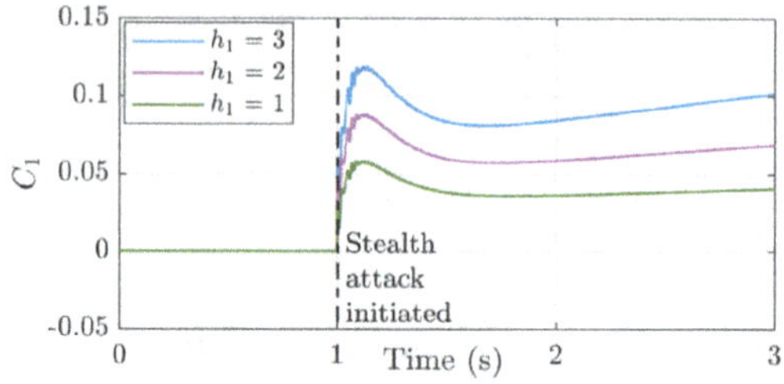

Figure 10.5 Variation of C_l *for different values of the design parameter* h_l

synchronizes to a non-zero value, which varies on the magnitude of injected attack vector. The above action can be justified by observing each secondary sublayer output in Figure 10.3 for a stealth attack on multiple voltage sensors on agents II and III in a DC microgrid of different line resistances. It can be seen that the voltage correction terms from average voltage sublayer in Figure 10.3(a) change symmetrically as compared to current sharing sublayer in Figure 10.3(b) following a stealth attack at $t = 1$ s. This attribute can be explained using Remark II. Considering the system operating at steady state, a step change of balanced zero sum attack $u_i^a(k)$ is injected into two agents during $(k-1)^{th}$ instant, equation (10.24) can be represented as

$$\Delta V_i^1(k) = K_p^{H_1} u_i^a(k) + \sum_{z=0}^{k} K_i^{H_1}(u_i(z)) + \underbrace{\sum_{z=(k-1)}^{k} K_i^{H_1}(u_i^a(z))}_{\Gamma(z)} \tag{10.27}$$

Eliminating the first two terms in RHS of (10.27), using Remark IV and substituting (10.27) in (10.30), it can be concluded that $y_1^i(k)$ and $y_2^i(k)$ will always lead to positive/negative values due to $\Gamma(k)$ for a balanced sum zero attack only on the attacked nodes. As a result, C_i of the attacked nodes will always reflect a positive value. This provides sufficient criteria for the detection of the attacked nodes in case of multiple sensor/link-based stealth attacks in DC microgrids. Concluding the above discussion, the CVF algorithm for each agent will result into

$$C_i = \begin{cases} 0 & , \text{if } \kappa = 0 \\ > 0 & , \text{else} \end{cases} \tag{10.28}$$

However, under worst cases, C_i can also be manipulated by the attacker using subtraction to make it negative, which displeases our sufficient criteria. To handle these discrepancies, C_i is tactically added to (10.6), which can now be rewritten as

$$\Delta V_i^2 = k_p^{H_2} \underbrace{(I_{dc_{ref}} + C_i(k) - \bar{I}_{dc_i}(k))}_{\bar{e}_2^i} + K_i^{H_2} \sum_{z=0}^{k}(I_{dc_{ref}} + C_i(k) - \bar{I}_{dc_i}(z)) \tag{10.29}$$

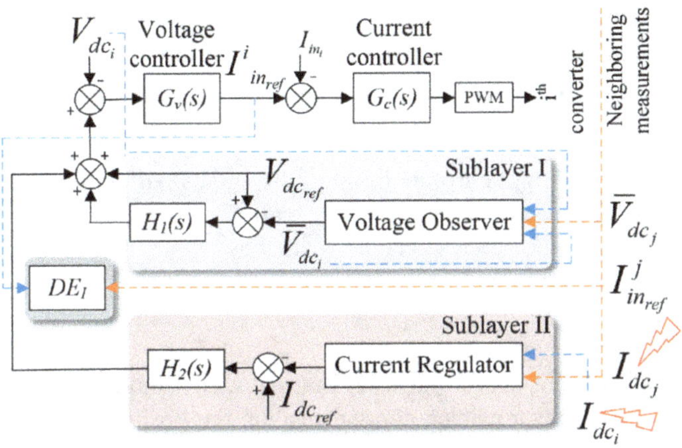

Figure 10.6 Proposed discordant element-based detection controller using local and neighboring measurements for i[th] *agent in DC microgrids – DE_I to detect anomalies in the current counterparts*

such that the control operation will be disoriented locally, thereby allowing the agents to reliably detect the attacks. Since $I_{dc_{ref}} = 0$, the cross-coupling of the CVF suggested in (10.32) will supplement for an accurate detection and facilitate protection against attacks on CVF since C_i now forms the forward path between both secondary control sublayers. By doing so, further attacks on C_i will disorient the objectives laid down for the outer voltage controller in sublayer I since it disregards (10.7). As shown in Figure 10.4, the CVF output C_i when cross-coupled into sublayer II introduces ramp up of ΔV_i^2 for proportionate current sharing. The ramp up/down of C_i can be explained using the addition of the term $\Gamma(z)$ in (10.27), which ramps up/down indefinitely for $z \to \infty$ unless the positive/negative attack vector is removed from i^{th} agent. Hence, the ramp up/down of C_i in the positive region qualifies as a sufficient criterion for the corresponding node to be declared as *attacked* in the cooperative realm for DC microgrids.

Moreover, in Figure 10.5, it can be seen that the slope of C_1 increases with increase in h_1 for a particular stealth attack in two sensors. Since the main focus of the chapter is to detect the attacked unit accurately alongside prevention of further coordinated attacks, it is a seemingly fair approach to include the cross-coupling strategy such that the defense mechanism can take place immediately without disrupting stability for lower values of h_i.

10.3.2 Stealth attack on currents

This section involves discussion of the proposed discordant element-based detection approach for both categories of FDI attacks in cooperative DC microgrids. The basic challenge lies in determining the attacked agent in cooperative network, which becomes more appealing as the transmitted false data in any given agent also propagates to its neighbors.

10.3.2.1 Destabilization attacks

Considering a constant signal attack on current sensor in i^{th} agent using Remarks II and III, it can be written as

$$\mathbf{L}\mathbf{I}^{a}_{dc} \neq 0 \tag{10.30}$$

Under such case, the output currents from all the agents will not be shared equally. Consequently, the solution to cooperative synchronization error for output currents using Remark I under such attacks can be written as

$$\mathbf{e}^{I}(t) = \mathrm{e}^{-h_V \mathbf{L}_t}\mathbf{e}^{I}(0) + \int_0^t \mathrm{e}^{-h_V \mathbf{L}(t-\tau)}\mathbf{k}\,d\tau \tag{10.31}$$

where $\mathbf{k} = \eta \mathbf{L}\mathbf{e}^{Ia}$, where η is a diagonal matrix, which indicates the presence of attack in the current sensor using a non-zero value. For positive definite values of L and h_V, the first term in (10.31) goes to zero. Using $\mathrm{e}^{At} = \sum_{i=1}^{\infty}(At)^i$, the final steady-state value of (10.31) is given by

$$e^{I}(t) \rightarrow \sum_{i=1}^{\infty} \int_0^t (-h_v(L(t-\tau))^i d\tau \tag{10.32}$$

Hence, for non-zero elements in the Laplacian graph, $l_{ij} \neq 0$, the synchronization error in (10.32) converges to a non-zero value. As a result, the synchronization error leads to a ramped up/down quantity for ΔV_{2_i} corresponding to the output of PI controller in (10.4). Using this condition as a sufficient criterion, it can be alternatively termed as output currents from each bus are in *discord* with each other for a non-zero synchronization error. Hence, the operational dynamics of secondary sublayer I will not obey consensus theory as per [30].

10.3.2.2 Deception attacks

However, in case of deception attacks, the cooperative synchronization law holds true as per

$$\dot{\bar{\mathbf{V}}}_{dc} + \mathbf{L}\bar{\mathbf{V}}_{dc} = \dot{\mathbf{V}}_{dc} = 0 \tag{10.33}$$

Moreover, the voltage dynamics at each bus in vector form can be written as

$$\dot{\mathbf{V}}_{dc} = \mathbf{C}^{-1}(\mathbf{D}_k\mathbf{I}_{in} - \mathbf{I}^{f}_{dc}) \tag{10.34}$$

where $\mathbf{D}_k = 1 - \mathbf{D}$. Furthermore, $\mathbf{I}_{in}$, $\mathbf{D}$, $\mathbf{C}$ and $\mathbf{I}^{f}_{dc}$ denote the diagonal matrices of the input current I_{in_i}, duty ratio D_i, DC link capacitance C_i and the attacked output current measurement I_{dc_i}, respectively, for N agents. Since the average voltage estimates are not compromised, they adhere to the global reference.

Considering the error dynamics into the voltage controller under steady-state conditions, we get

$$\mathbf{L}^{T}\Delta\mathbf{V}_1 + \mathbf{L}^{T}\mathbf{H}_2\mathbf{e}^{Ia} + \mathbf{V}_{dc_{ref}} = \mathbf{L}^{T}\mathbf{V}_{dc} \tag{10.35}$$

where $\mathbf{e}^{Ia}$ denotes a diagonal matrix of the error quantity in (LABEL:1) with attacked current signals.

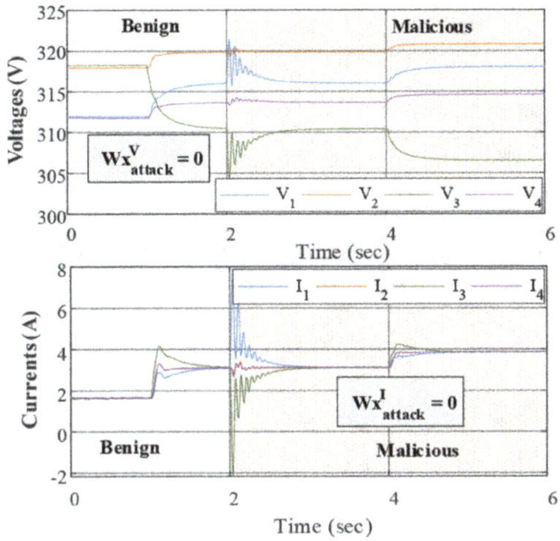

Figure 10.7 Local voltage and current for each DG. Attack is initiated at

Remark VI: Since the system objectives in (10.7) are met for a deception attack, $\mathbf{L}^T\Delta\mathbf{V}_1 = 0$ holds true [30].

Using Remark VI and differentiating (10.35), we get

$$\mathbf{L}^T\mathbf{K}_p^{H_2}\dot{\mathbf{e}}^{Ia} + \mathbf{L}^T\mathbf{K}_I^{H_2}\mathbf{e}^{Ia} - \mathbf{L}^T\dot{\mathbf{V}}_{dc} = 0 \tag{10.36}$$

In steady state, $\mathbf{e}^{Ia} = 0$. Further using Remark III, $\mathbf{L}^T\mathbf{C}^{-1}\mathbf{I}^a_{dc} = 0$. Using these equalities after substituting (10.33) in (10.34), we get

$$\mathbf{L}^T\mathbf{K}_p^{H_2}\dot{\mathbf{e}}^{Ia} - \mathbf{L}^T\mathbf{C}^{-1}\mathbf{D}_k\mathbf{I}_{in} = 0 \tag{10.37}$$

Remark VII: Due to the injected attack signal, the first term of (10.37) will be asymmetric, as explained in (10.32). For (10.37) to hold true, this property will be reflected in the second term of (10.37), which becomes the basis of detection for FDIAs in cooperative DC microgrids.

Using Remark VII, it is intuitional that $\mathbf{L}^T\mathbf{I}_{in}$ will always converge to zero under no attacks for normalized duty ratios D_i across the microgrid. Hence, it has been proved that the normalized input current reference quantities also achieve consensus among themselves for a constant global reference voltage at the output of their respective DC/DC converters in DC microgrids, as shown in Figure 10.6.

The summary of stealth attacks on voltages and currents can be found in Table 10.1. Its impact on DC microgrids has been shown in Figure 10.7.

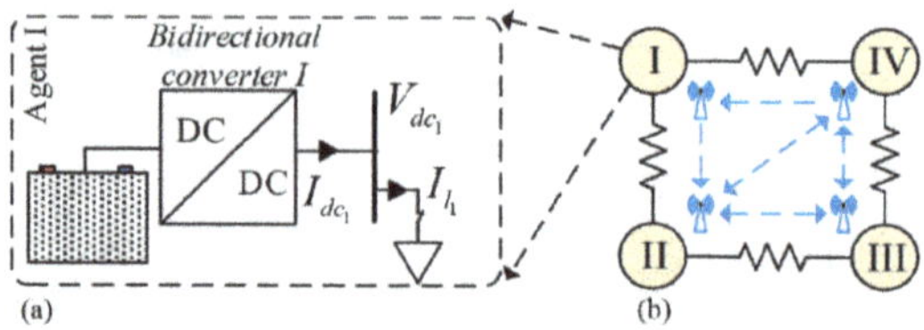

Figure 10.8 *Considered system: (a) Agent modeling and (b) Cyber-physical DC microgrid with four sources*

10.4 Results

10.4.1 Stealth voltage attacks

The proposed attack-detection strategy is tested on a cyber-physical DC microgrid, as shown in Figure 10.8b, with $V_{dc_{ref}} = 315$ V consisting of four agents of equal capacities interconnected to each other via resistive lines. It should be noted that each agent consists of a battery accompanied via DC/DC bidirectional converters, respectively, as shown in Figure 10.8a. To test the performance of the proposed attack-resilient strategy for cooperative DC microgrid, it has been tested against several disturbances such as FDIA, stealth attack in multiple sensors, which usually goes undetected by distributed observers, communication links to detect the affected node such that necessary action can be taken to maintain security. The system and control parameters are provided in Appendix. It should be noted that each event in the above-mentioned scenarios is separated by a certain time gap to provide clear understanding.

In scenario I, the outgoing cyber links from agent III are attacked with a set of attack vectors of ±3 V at $t = 1$ s such that the cumulative effect seen in a cooperative network is zero. Prior to initiating the attack, it is difficult to denote the attacked node from the average voltage estimate as both estimates diverge symmetrically. Considering this problem using a distributed observer-based approach, norm of these errors would mistranslate into two attacked nodes, i.e., agents III and IV. This issue is well addressed using the proposed approach since C_3, as shown in Figure 10.9, shoots up to 0.18, thereby suggesting that agent III is attacked. As a protective measure of security, the outgoing links from agent III are deactivated, which brings the average voltage estimate into synchronism by tracking the desired reference value of 315 V. For load changes highlighted as *A* and *B*, the system performs satisfactorily. To test the robustness of the proposed approach under worst-case scenarios, another consecutive attack at $t = 2$ s is preempted by the attacker to manipulate C_3 by reducing it to a negative value. However, due to cross-coupling of C_i into the current secondary sublayer in (10.32), it prevents further exploitation as it cannot disorient the nested control output for a particular operating point.

In scenario II, a balanced attack of ±10 V in sensors of agents I and IV, respectively, at $t = 1$ s is practiced as in Figure 10.10 to test the fidelity of the proposed approach. As C_1 and C_4 shoot up in the positive region, agents I and IV are plugged

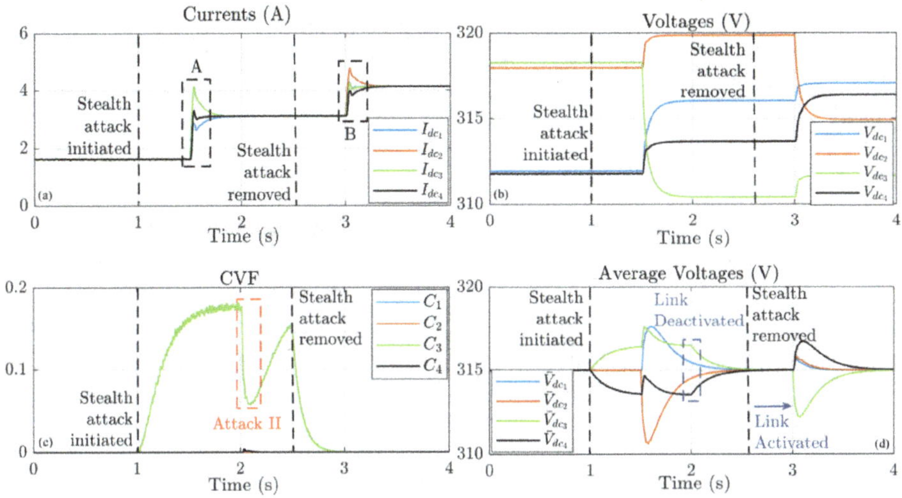

Figure 10.9 *(a) Currents, (b) Voltages, (c) CVF, (d) Average voltages: Stealth attack on two outgoing cyber links at* t = 1 *s from agent III. It is shown that the average voltage estimates diverge symmetrically on initiating the attack. The proposed strategy accurately detects the* attacked *agent.*

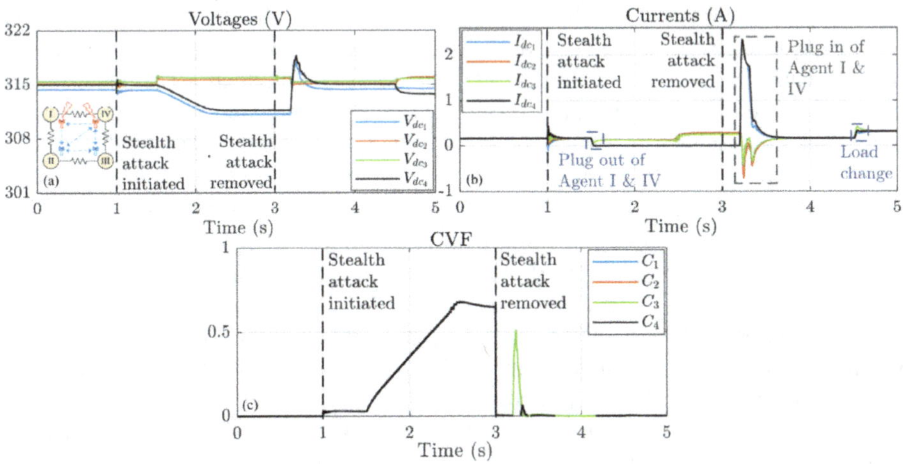

Figure 10.10 *(a) Currents, (b) Voltages, (c) CVF: Consecutive stealth attack on voltage sensors at* t = *1 s in agents I and IV. Upon initiating the attack, the average voltages and current sharing remain intact. As seen, the proposed strategy identifies the* attacked *agents instantly with the CVF for agents I and IV in the positive region.*

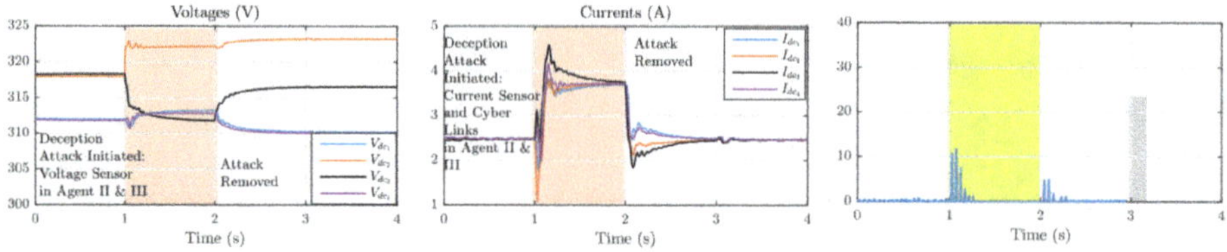

Figure 10.11 Performance of the proposed detection strategy under deception attack on current sensors and cyber links in agents II and III – DE_I for agents II and III becomes positive as soon as the attack is initiated. Attack II carried out in the input current sensor of agent II cannot be further exploited as $I_{in_{ref}}$ *is a control output in the closed loop voltage regulation.*

out of the system at $t = 1.5$ s. Based on *Assumption 2* in [27], the network connectivity is affected due to plugging out of $M/2$ agents, which leads to change in system dynamics. On clearing out of the attack at $t = 3$ s, indicated by C_1 and C_4 dropping to zero, the converters are plugged back in around $t = 3.2$ s, resulting into restoration of the average voltage estimates to 315 V.

10.4.2 Stealth current attacks

A deception attack is injected in Figure 10.11 using the attack elements $I^a_{dc} = \{0, 1.6, 1.8, 0\}$ A into the current sensors and cyber link, respectively, into agents II and III at $t = 1$ s. When the attack is initiated, it can be seen in Figure 10.11 that the voltages at each bus are still distributed around the global reference of $V_{dc_{ref}} = 315$ V. Moreover, as a result of the false data injection into the current sensors and cyber link, a visual imprint of load change is created without any actual physical disturbance. This behavior deceives the system operator, thereby adhering to the control objectives, otherwise. Such attacks can be critical as this action can be deceitfully used by the attacker to cause destabilization in DC microgrids later. Identifying such conditions as a severe risk to infiltrate large attack vectors later, it can cause critical damage to the system by shutting down both converters (MO > 2). Hence, the *RA* index lies in the range of [4, 6]. As per the proposed attack-detection theory for deception attacks, DE_I^2 and DE_I^3 go positive at $t = 2$ s to indicate the presence of false data elements for the current counterparts in agents II and III. When the attack is removed, the proportionate current sharing and voltage regulation inputs operate normally with *unbiased* measurements.

To differentiate between faults and cyber attacks, a case study is presented in Figure 10.12. for a short timescale ($\approx$100 ms) illustrating the response of a DC/DC converter in case of FDI attacks and faults. It should be noted that the origin in Figure 10.12b is ($V_{dc_{ref}}$, I_{dc}), where I_{dc} will vary between the minimum and maximum current limit based on the loading level. A *boundary of operation* region is defined in Figure 10.12a, which varies within $\{X \in [V_{in}, 1.3V_{dc_{ref}}], Y \in [I_{min}, I_{max}]\}$. It can be observed that positive and negative destabilization attacks cause the

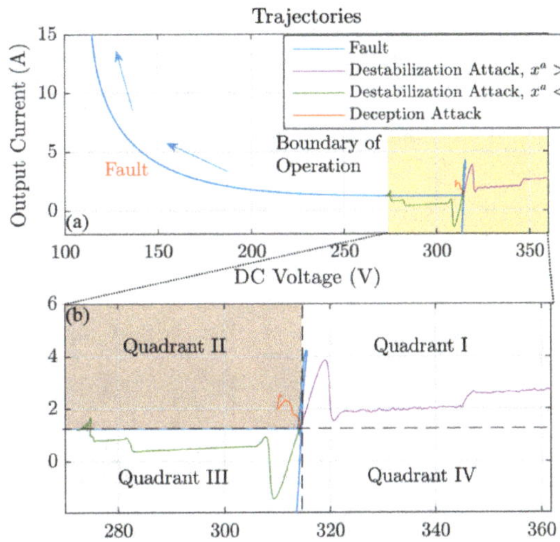

Figure 10.12 *(a) Response of DC/DC converter in a distributed control-based DC microgrid to attacks and faults for 100 ms – Quadrant II shows the faulted area outside the boundary of operation, (b) Zoomed picture of the boundary of operation*

trajectories to move into Quadrant I and III, respectively. Furthermore, deception attacks with a feasible solution operate either in Quadrant II/IV corresponding to an increase in load in the same/different bus. This behavior can be attributed to the response of the distributed secondary controller in (10.4)–(10.6). However, in case of DC line-to-line faults, the response of the primary control layer results in a large increase in the output current along with a decrease in output voltage. This behavior can be clearly seen in Figure 10.12a, where the fault trajectory goes out of the boundary of operation in less than 100 ms. Since the timescale separation between the secondary and primary layers is considerably large, this evaluation theory (within a certain time frame ≈100 ms) can be used locally as a substantial indicator to assist the proposed detection scheme in differentiating between faults and cyber attacks.

To test the robustness of the proposed detection strategy, a deception attack $I^a_{dc_3}$ = (1 + 0.8sin 0.4πt) A is injected into agent III at t = 1 s in Figure 10.13a. It can be seen that the sinusoidal trace is in the positive DE_I region for agent III. Furthermore, two deception attacks $\{I^a_{dc_1}, I^a_{dc_3}\}$ = 1.5t A are injected into agents I and III simultaneously at t = 1 s in Figure 10.13b. As per the proposed detection criteria, ramp traces are observed for both DE^1_I and DE^3_I. This establishes robustness of the proposed scheme in detecting false data attacks in the realm of DC microgrids.

10.5 Conclusion

This chapter proposes a general cyber-attack resilient framework for cyber-physical DC microgrids. The vulnerability of the conventional cooperative techniques in DC

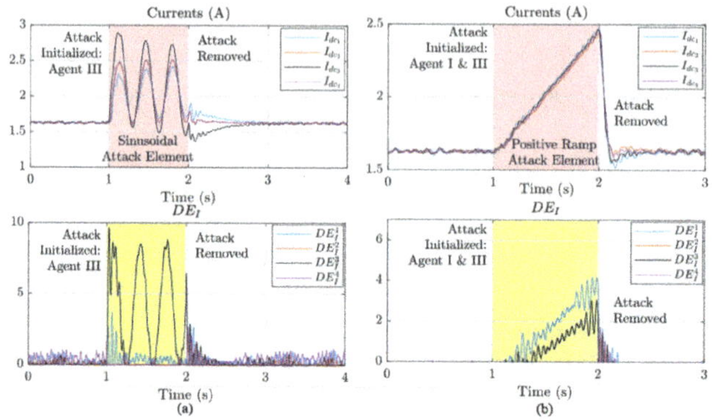

Figure 10.13 Performance of the proposed detection strategy using two different attack models: (a) A deception attack (IA = 2) modeled as $I^a_{dc_3} = (1 + 0.8\sin 0.4\pi \mathrm{t})$ *A on agent III at* t = *1 s, (b) Deception attack (IA = 3) modeled as* $\{I^a_{dc_1}, I^a_{dc_3}\} = 1.5\mathrm{t}$ *A on agents I and III at* t = *1 s*

microgrids under false data injection is investigated in detail. In addition to that, the modeling of stealth attacks, which manage to deceive distributed observers, is carried out using necessary and sufficient conditions. To address this issue, two DIs are presented, which operate on the secondary controller output so as to provide an accurate identification strategy of the attacked agent(s). In this way, the requirement of additional observers is neglected making it computationally efficient against smart attacks. To add fidelity, it is cross-coupled with the secondary current sublayer such that it operates under worst scenarios of attacks. Its robustness is evaluated using both simulation and experimental results for both false data injection and stealth attacks on multiple sensors/links.

Appendix

Simulation parameters

The considered system consists of four sources rated equally for 3 kW. It is to be noted that the line parameter R_{ij} is connected from i^{th} agent to j^{th} agent. Moreover, the controller gains are consistent for each agent.

Plant: $R_{12} = 1.3\ \Omega$, $R_{13}=1.8\ \Omega$, $R_{23}=1.2\ \Omega$, $R_{43}= 1.5\Omega$

Converter: $L_i = 3$ mH, $C_{dc_i} = 250\ \mu$F, $i^{max}_{b_1} = i^{max}_{b_2} = 9.5$ A

Controller: $V_{dc_{ref}}$ =315 V, $I_{dc_{ref}} = 0$, $K_p^{H1} = 0.1$, $K_i^{H1} = 3$, $K_p^{H2} = 0.2$, $K_i^{H1} = 9.2$, $G_{vp} = 2.8$, $G_{vi} = 12.8$, $G_{cp} = 0.56$, $G_{ci} = 21.8$, h =1, $c = 0.4$.

References

[1] Lasseter R.H. 'Smart distribution: coupled microgrids'. *Proceedings of the IEEE*. 2011;99(6):1074–82.

[2] Dragicevic T., Lu X., Vasquez J.C., Guerrero J.M. 'DC microgrids–Part I: a review of control strategies and stabilization techniques'. *IEEE Transactions Power Electronics*. 2016;31(7):4876–91.

[3] Meng L., Dragicevic T., Guerrero J.M., Vasquez J.C. 'Dynamic consensus algorithm based distributed global efficiency optimization of a droop controlled DC microgrid'. *Energy Conference (ENERGYCON), 2014 IEEE International*; Cavtat, Croatia, 13–16 May; 2014. pp. 1276–83.

[4] Yazdanian M., Mehrizi-Sani A. 'Distributed control techniques in microgrids'. *IEEE Transactions on Smart Grid*. 2014;5(6):2901–9.

[5] Sahoo S., Mishra S. 'An adaptive event-triggered communication-based distributed secondary control for DC microgrids'. *IEEE Transactions on Smart Grid*. 2017;9(6):6674–83.

[6] Dragičević T., Lu X., Vasquez J.C., Guerrero J.M. 'DC microgrids—Part II: a review of power architectures, applications, and standardization issues'. *IEEE Transactions on Power Electronics*. 2016;31(5):3528–49.

[7] Nasirian V., Moayedi S., Davoudi A., Lewis F.L. 'Distributed cooperative control of DC microgrids'. *IEEE Transactions on Power Electronics*. 2015;30(4):2288–303.

[8] Anand S., Fernandes B.G., Guerrero J. 'Distributed control to ensure proportional load sharing and improve voltage regulation in low-voltage DC microgrids'. *IEEE Transactions on Power Electronics*. 2013;28(4):1900–13.

[9] Morstyn T., Hredzak B., Agelidis V.G. 'Cooperative multi-agent control of heterogeneous storage devices distributed in a DC microgrid'. *IEEE Transactions on Power Systems*. 2016;31(4):2974–86.

[10] Veitch C.K., Henry J.M., Richardson B.T., Hart D.H. Microgrid cyber security reference architecture. Tech. Rep. SAND2013-5472. Albuquerque, NM: Sandia National Laboratories (Hierarch. SNLNM); 2013.

[11] Liu Y., Ning P., Reiter M.K. 'False data injection attacks against state estimation in electric power grids'. *ACM Transactions on Information and System Security*. 2011;14(1):1–33.

[12] Liang G., Zhao J., Luo F., Weller S.R., Dong Z.Y. 'A review of false data injection attacks against modern power systems'. *IEEE Transactions on Smart Grid*. 2017;8(4):1630–8.

[13] Pasqualetti F., Dorfler F., Bullo F. 'Attack detection and identification in cyber-physical systems'. *IEEE Transactions on Automatic Control*. 2013;58(11):2715–29.

[14] Yan Y., Qian Y., Sharif H., Tipper D. 'A survey on cyber security for smart grid communications'. *IEEE Communications Surveys & Tutorials*. 2010;14(4):998–1010.

[15] Li P., Liu Y., Xin H., Jiang X. 'A robust distributed economic dispatch strategy of virtual power plant under cyber-attacks'. *IEEE Transactions on Industrial Informatics*. 2018;14(10):4343–52.

[16] Zhao J., Mili L., Wang M. 'A generalized false data injection attacks against power system nonlinear state estimator and countermeasures'. *IEEE Transactions on Power Systems*. 2018;33(5):4868–77.

[17] Sridhar S., Govindarasu M., Liu C.C. 'Risk analysis of coordinated cyber attacks on power grid'. *Control Optimization Methods for Electric Smart Grids*. 2012;3:275–94.

[18] Salmeron J., Wood K., Baldick R. 'Analysis of electric grid security under terrorist threat'. *IEEE Transactions on Power Systems*. 2004;19(2):905–12.

[19] Li Z., Shahidehpour M., Alabdulwahab A., Abusorrah A. 'Bilevel model for analyzing coordinated cyber-physical attacks on power systems'. *IEEE Transactions on Smart Grid*. 2016;7(5):2260–72.

[20] Chakhchoukh Y., Ishii H. 'Coordinated cyber-attacks on the measurement function in hybrid state estimation'. *IEEE Transactions on Power Systems*. 2018;30(5):2487–97.

[21] Manshadi S.D., Khodayar M.E. 'Resilient operation of multiple energy carrier microgrids'. *IEEE Transactions on Smart Grid*. 2015;6(5):2283–92.

[22] Rana M.M., Li L., Su S.W, . 'Cyber attack protection and control of microgrids'. *IEEE/CAA Journal of Automatica Sinica*. 2018;5(2):602–9.

[23] Hao J., Kang E., Sun J., *et al.* 'An adaptive Markov strategy for defending smart grid false data injection from malicious attackers'. *IEEE Transactions on Smart Grid*. 2018;9(4):2398–408.

[24] Zhang H., Meng W., Qi J., Wang X., Zheng W.X. 'Distributed load sharing under false data injection attack in an inverter-based microgrid'. *IEEE Transactions on Industrial Electronics*. 2018;66(2):1543–51.

[25] Beg O.A., Johnson T.T., Davoudi A. 'Detection of false-data injection attacks in cyber-physical DC microgrids'. *IEEE Transactions on Industrial Informatics*. 2017;13(5):2693–703.

[26] Derui D., Zidong W., Ho D.W.C., Guoliang W. 'Observer-based event-triggering consensus control for multiagent systems with lossy sensors and cyber-attacks'. *IEEE transactions on cybernetics*. 2017;47(8):1936–47.

[27] Abhinav S., Modares H., Lewis F.L., Ferrese F., Davoudi A. 'Synchrony in networked microgrids under attacks'. *IEEE Transactions on Smart Grid*. 2017;9(6):6731–41.

[28] Fawzi H., Tabuada P., Diggavi S. 'Secure estimation and control for cyber-physical systems under adversarial attacks'. *IEEE Transactions on Automatic Control*. 2014;59(6):1454–67.

[29] Zhao C., He J., Cheng P., Chen J. 'Analysis of consensus-based distributed economic dispatch under stealthy attacks'. *IEEE Transactions on Industrial Electronics*. 2017;64(6):5107–17.

[30] Sahoo S., Mishra S., Peng J.C.-H., Dragičević T. 'A stealth cyber-attack detection strategy for DC microgrids'. *IEEE Transactions on Power Electronics*. 2019;34(8):8162–74.

[31] Sahoo S., Mishra S. 'A distributed finite-time secondary average voltage regulation and current sharing controller for DC microgrids'. *IEEE Transactions on Smart Grid*. 2017;10(1):282–92.
[32] Brualdi R.A., Herbert J.R. *Combinatorial Matrix Theory*. 39. Cambridge: Cambridge University Press; 1991.
[33] Sahoo S., Peng J.C.-H., Devakumar A., Mishra S., Dragičević T. 'On detection of false data in cooperative DC microgrids—a discordant element approach'. *IEEE Transactions on Industrial Electronics*. 2020;67(8):6562–71.

Chapter 11

Resilient distributed control strategies in microgrids against cyber attacks

Mahdieh S. Sadabadi[1], *Subham Sahoo*[2], *and Frede Blaabjerg*[2]

Microgrids rely on cyber networks in their monitoring and control systems. This poses significant challenges in terms of cybersecurity in microgrids. This chapter addresses this problem by the design of a resilient cooperative distributed control system for DC microgrids that are resilient to stealthy false data injection (FDI) cyberattacks, which adversely impact the data integrity of the control systems and the communication networks, while simultaneously remaining undetected by anomaly detection algorithms. The chapter also reviews some of the recent scientific contributions in the resilient control systems for DC microgrids and discusses some of the research challenges in this area.

11.1 Introduction

Microgrids represent a class of cyber-physical systems that are reliant on reliable and low-latency communication networks and distributed control infrastructures. With this high dependency on information exchange and communication links, cybersecurity issue in microgrids has become prevalent. The adversaries might compromise (a) the data availability using denial of service (DoS) attacks and (b) the integrity of data using false data injection (FDI) attacks, in the control systems of microgrids. The existence of cyberattacks in microgrids' control systems might result in large disturbances and cause control failures, which affect the normal operation of microgrids and threaten their stability simultaneously [1].

[1]Department of Automatic Control and Systems Engineering, University of Sheffield, Sheffield, UK
[2]Department of Energy Technology, Aalborg University, Aalborg, Denmark

11.1.1 Related works

The current state-of-the-art cybersecurity strategies in microgrid control systems can be divided into protection and attack detection/mitigation mechanisms [1]. In protection strategies, cryptographic protection is a common approach; however, cryptographic protection for the privacy-preserving of exchanged data increases latency in microgrid control systems. Consequently, it might not be a feasible protection scheme considering the fast dynamics of microgrid control systems aimed at maintaining the stability of microgrids [2, 3].

Due to the importance of cybersecurity in microgrids, several cyberattack detection and mitigation methods in power-electronics-dominated microgrids have recently emerged. Various approaches for the detection and mitigation of FDI attacks in islanded DC microgrids [4–10] and islanded AC microgrids [11–15] have been reported in the literature. An FDI attack detection scheme has been proposed in Reference 4, which relies on identifying changes in the sets of inferred candidate invariants. Several methods for the detection and/or mitigation of FDI attacks on current sensors, communication links in DC microgrids' controllers, sensors and communication channels, control input channels, as well as controller hijacking have respectively been proposed in [5–8, 9]. A data-driven attack detection method has been developed in [10]. Multiple detection algorithms for other types of cyber attacks in islanded AC and DC microgrids have been developed, such as DoS [7, 12, 16, 17], stealthy attacks [18–20], hijacking attacks [21], and man-in-the-middle (MITM) attacks [22]. However, in these attack detection/mitigation approaches, potential attack detection and mitigation might not be instantaneous; as a result, detection and mitigation delays might cause microgrids exposed to high cybersecurity risks, which might deteriorate the performance and/or stability of microgrids [23]. Furthermore, the proposed attack detection and mitigation algorithms typically rely on some key yet restrictive assumptions on the number of attacked units in microgrids and/or required computational burden for attack detection and mitigation. Considering these restrictions and the limitations of the disruptive resources, it is essential to build a cyberattack-resilience layer in microgrids' control infrastructures such that the stability and normal operation of microgrids can be guaranteed despite the presence of cyberattacks. In this case, microgrids' control systems are referred to as resilient controllers. Resilient control systems contribute to a more effective control design by considering an accepted level of operational normality in response to disturbances, including cyberattacks [24]. More specifically, in a resilient control setting, the resilient controller effectively mitigates the adverse impact of cyberattacks without relying on (a) any information about the number of attacked units and (b) the prompt detection of cyber attacks in microgrids.

For resilient distributed control in DC microgrids, a few approaches have recently been developed. In [8], a resilient adaptive cooperative secondary control approach for DC microgrids exposed to unknown FDI cyberattacks in control input channels has been developed. The modeled FDIs can be a constant or time-varying signal, either bounded or unbounded. In [25], a novel resilient cooperative control mechanism, which guarantees average voltage regulation and proportional load sharing in DC microgrids under cyberattacks, has been proposed. The cyberattack considered in [25] is modeled as uniformly bounded FDI and MITM attacks on communication links. Both distributed control methods in [8, 25] are fully distributed and do not require any global-level information, hence facilitating the plug-and-play capability in microgrids. However, the proposed resilient control approach in [8] relies on exchanging both current and voltage or estimates of voltages of distributed generation (DG) units among their neighbors, which increases the vulnerability of DC microgrids to cyberattacks on communication links.

11.1.2 Chapter focus

In this chapter, the proposed resilient control strategy of DC microgrids in Reference 25 is extended to the case of strategic stealthy FDI cyberattacks. In such an attack scenario, adversaries with detailed knowledge of microgrids' control systems aim at corrupting the integrity of exchanged data in controllers to affect the normal operation of microgrids without being detected by anomaly detection mechanisms. Questions that are of central relevance in the context of resilient control in microgrids and that will be discussed in detail in this chapter are as follows:

- How to design a resilient distributed control algorithm that simultaneously stabilizes DC microgrids and guarantees desired performance while under cyberattacks?
- What information is required for the design of resilient controllers for microgrids? Is the control design based on global-level or local-level information?

11.1.3 Chapter structure

Section 11.2 focuses on physical and cyber networks in power-electronics-interfaced DC microgrids considered in this chapter. In section 11.3, a resilient cooperative distributed control approach equipped with baseline cyberattack detectors is proposed. We will then discuss the vulnerability of the proposed controller and attack detection algorithms to stealthy FDI attacks. Section 11.4 analyzes the stability and attack-resilient features of the proposed cooperative controller and shows that the proposed control method guarantees stability for both voltage regulation and proportional load-sharing objectives under stealthy FDI attacks. Section 11.5 presents several case studies and results. Finally, Section 11.6 concludes the chapter.

11.2 Cyber-physical DC microgrids

A DC microgrid is a cyber-physical system wherein the coordinated control of each converter (physical networks) relies on information and communication infrastructure (cyber networks), as depicted in Figure 11.1. This section will focus on the physical and cyber framework of DC microgrids. The vulnerability of cyber layers against cyberattacks will then be discussed.

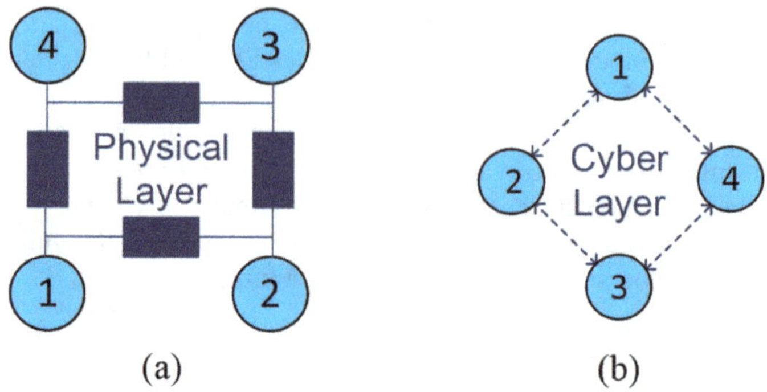

Figure 11.1 *Schematic of the (a) physical and (b) cyber layer of a DC microgrid – in this example, the physical layer is composed of* n = *4 DG units (represented as a node) connected via distribution lines (represented as an edge), and the cyber layer relies on communication links*

11.2.1 Physical networks

Consider a converter-interfaced DC microgrid composed of n DG units, which are connected via m distribution lines. Each DG unit consists of renewable energy sources with DC outputs that are simplified as an ideal DC voltage source, a DC–DC buck converter, an inductive-capacitive filter, and a local load connected at the point of common coupling (PCC). The dynamics of DG i are described by the following equations [26]:

$$\begin{aligned} L_i\dot{I}_i(t) &= -V_i(t) - r_iI_i(t) + V_{dc,i}d_i(t), \\ C_i\dot{V}_i(t) &= I_i(t) - Y_iV_i(t) - \mathbb{B}_{ij}I_{ij}(t), \end{aligned} \tag{11.1}$$

where $I_i(t)$, $V_i(t)$, $d_i(t)$, and $I_{ij}(t)$ are the current of the DC–DC converter i, the voltage at PCC i, the duty cycle of the converter i, and the current of the distribution line ij connecting DG i to DG j, respectively. In (11.1), $V_{dc,i}$ is the DC voltage of the input side of the converter i, (L_i, r_i, C_i) are the filter parameters of the DC–DC converter i, Y_i is the load conductance of DG i, and $\mathbb{B}_{ij}$ determines the direction of the line current $I_{ij}(t)$. In particular, $\mathbb{B}_{ij} = 1$ if the line current leaves DG i, whereas $\mathbb{B}_{ij} = -1$ if the line current enters DG i. If there is no distribution line between DG i and j, $\mathbb{B}_{ij} = 0$. the line dynamics can be modeled as follows:

$$L_{ij}\dot{I}_{ij}(t) = -R_{ij}I_{ij}(t) + \mathbb{B}_{ij}(V_i(t) - V_j(t)), \tag{11.2}$$

where (R_{ij}, L_{ij}) are the resistance and inductance of the line ij.

Let $\mathbf{V}(t) = [V_1(t), \ldots, V_n(t)]^T$, $\mathbf{I}(t) = [I_1(t), \ldots, I_n(t)]^T$, $\mathbf{d}(t) = [d_1(t), \ldots, d_n(t)]^T$, and $\mathbf{I}_l(t)$ is a vector of distribution line currents. The dynamics of the physical network of DC microgrids are then described by the following dynamic equations in a vector form:

$$
\begin{aligned}
[C]\,\dot{\mathbf{V}}(t) &= \mathbf{I}(t) - [Y]\,\mathbf{V}(t) - \mathbb{B}\mathbf{I}_l(t),\\
[L_l]\,\dot{\mathbf{I}}_l(t) &= -[R_l]\,\mathbf{I}_l(t) + \mathbb{B}^T\mathbf{V}(t),\\
[L]\,\dot{\mathbf{I}}(t) &= -\mathbf{V}(t) - [r]\,\mathbf{I}(t) + [V_{dc}]\,\mathbf{d}(t),
\end{aligned}
\tag{11.3}
$$

where $[C] = diag(C_1, \ldots, C_n)$, $[L] = diag(L_1, \ldots, L_n)$, $[r] = diag(r_1, \ldots, r_n)$, $[Y] = diag(Y_1, \ldots, Y_n)$, $\mathbb{B}$ is the incidence matrix of the physical network of microgrids, $[R_l]$ and $[L_l]$ are block-diagonal matrices whose diagonal elements are line resistances and inductances. Throughout this chapter, the symbol $[X]$ describes a diagonal matrix of an appropriate dimension whose diagonal elements are X_i.

According to (11.3), the physical layer of DC microgrids can be modeled by an undirected connected graph $\mathcal{G}$ with a node set $\mathcal{V} = \{1, \ldots, n\}$ and an edge set $\mathcal{E} = \{1, \ldots, m\}$. Each node in $\mathcal{V}$ represents a DG unit, whereas each edge in $\mathcal{E}$ represents a distribution line, see Figure 11.1(a).

11.2.2 Cyber networks

In a DC microgrid with n DG units, the main objectives in a lower-level control system are to regulate the average voltage of voltage signals at PCCs and to guarantee a proportional current sharing among DG units while maintaining the stability of DC microgrids. The objectives can mathematically be expressed as follows:

$$\lim_{t\to\infty} \frac{1}{n}\sum_{i=1}^{n} V_i(t) - V^* = 0, \tag{11.4}$$

$$\lim_{t\to\infty} \left(\tfrac{I_i(t)}{I_i^s} - \tfrac{I_j(t)}{I_j^s}\right) = 0, \tag{11.5}$$

for $i, j \in \mathcal{V}$, where V^* is a given reference set-point and I_i^s is the rated current of DG i.

To achieve the above-mentioned objectives in (11.4) and (11.5), cooperative distributed control algorithms are used. In a cooperative control framework, each DG is associated with a local controller, where the local controllers of DG units

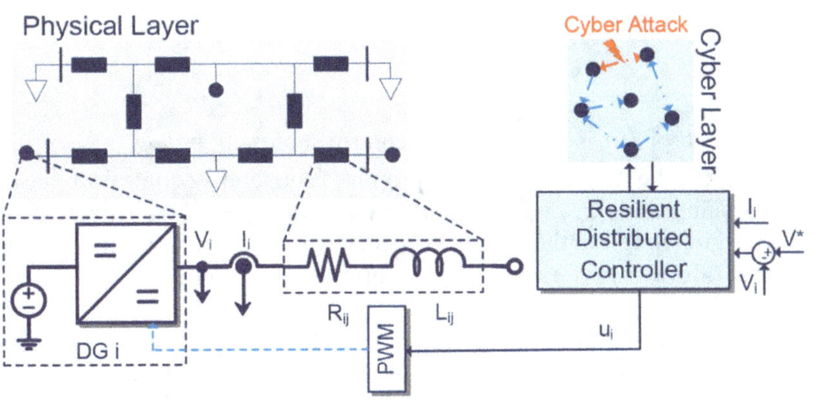

Figure 11.2 Schematic of networked DC microgrids consisting of multiple DG units equipped with distributed controllers over cyber networks

exchange data with their neighbors via a cyber network, see Figure 11.1(b), which is modeled as an undirected connected graph $\mathcal{G}_C$. Figure 11.2 shows the schematic diagram of a DC microgrid and its distributed controller.

11.2.3 Impact of cyber attacks on microgrids

A major concern in cooperative control schemes is the vulnerability to cyberattack surfaces that potentially corrupt the integrity and availability of exchanged data, sensor measurements, or actuation commands. The existence of cyberattacks in microgrids' control systems might have a detrimental impact on microgrid stability and performance and even lead to safety and reliability issues. Hence, new resilient control and monitoring technologies are required to rigorously enhance the resilience of microgrids by reducing or mitigating potential cyberattack impact on microgrid stability, voltage regulation, and power-sharing objectives.

11.3 Resilient cooperative control in DC microgrids

In this section, a cooperative distributed control strategy is developed for DC microgrids. The proposed control method is resilient to stealthy FDI attacks in cyber networks.

11.3.1 Structure of resilient cooperative controllers

The proposed resilient cooperative distributed control strategy for DC microgrids modeled in (11.3) is based on the following control law:

$$\begin{aligned}
\mathbf{u}(t) &= \left[k_1\right]\mathbf{V}(t) + \left[k_2\right]\mathbf{I}(t) + \left[k_3\right]\mathbf{v}(t) + \left[k_4\right]\mathbb{L}W\mathbf{I}(t) + \gamma\left[k_5\right]\mathbb{L}\theta(t) + \left[k_6\right]\left(\mathbf{I}(t) - \phi(t)\right)\\
\tau_v\dot{\mathbf{v}}(t) &= -\beta W\left(\mathbf{V}(t) - \mathbf{1}_n V^*\right) - K\beta W\mathbb{L}W\mathbf{I}(t) - W\alpha\beta\left(\mathbf{I}(t) - \phi(t)\right) + \gamma\beta W\mathbb{L}\theta(t) + \mathbf{d}_v(t),\\
\tau_\theta\dot{\theta}(t) &= -\mathbb{L}(\theta(t) + \gamma W\mathbf{I}(t)) + \mathbf{d}_\theta(t),\\
\tau_\phi\dot{\phi}(t) &= -\phi(t) + \mathbf{I}(t),
\end{aligned} \tag{11.6}$$

where $\mathbf{u}(t) = \left[V_{dc}\right]\mathbf{d}(t)$ is the vector of control inputs, $W = diag(\frac{1}{I_1^s}, \ldots, \frac{1}{I_n^s})$, $\mathbf{v}(t)$, $\theta(t)$, and $\phi(t)$ are the state vectors of the proposed cooperative control strategy. The control parameters are α, β, γ, K, τ_v, τ_θ, τ_ϕ, and $\left[k_j\right]$, $j = 1, \ldots, 6$. Section 11.4.5 proposes a guideline for the design of control parameters in (11.6) so that the stability and desired performance of DC microgrids under stealthy FDI attacks are guaranteed.

The cyber layer in (11.6) is based on a neighbor-to-neighbor communication graph with a graph Laplacian matrix $\mathbb{L}$. It is assumed that the communication graph in (11.6), associated with $\mathbb{L}$, is connected and undirected ($\mathbb{L} = \mathbb{L}^T$). In (11.6), $\mathbf{d}_v(t)$, $\mathbf{d}_I(t)$, and $\mathbf{d}_\theta(t)$ are uniformly bounded time-varying or constant malicious signals imposed on some DG units' local controllers and their communication links. It is assumed that $\mathbf{d_v}(t)$, $\mathbf{d_I}(t)$, and $\mathbf{d}_\theta(t)$ do not depend on the states of microgrids and cooperative controllers.

As one can observe from (11.6), the proposed cooperative controller relies on exchanging the physical states $\mathbf{I}(t)$, as well as the controller states $\theta(t)$ and $\mathbf{v}(t)$. This would increase the microgrid's vulnerability to cyberattacks $\mathbf{d_v}(t)$, $\mathbf{d_I}(t)$, and $\mathbf{d}_\theta(t)$. However, it will be shown that by a proper design of the control parameters in (11.6), the attack resilience of microgrids is enhanced.

11.3.2 Proposed baseline attack detectors

To detect the cyberattacks $\mathbf{d}_\theta(t)$ and $\mathbf{d_v}(t)$ in (11.6) and monitor the abnormality of the proposed cooperative control system in (11.6) due to the existence of cyberattacks, two baseline cyber attack detectors are developed. The anomaly detectors are co-allocated with the controller in (11.6) [27]. The detectors generate alarms to alert the unusual behavior of DC microgrids caused by cyberattacks, which is called the abnormal operation of microgrids.

11.3.2.1 Anomaly Detector I: detection of cyberattacks d_θ(t)

Anomaly Detector I for the detection of $\mathbf{d}_\theta(t)$ acts based on the following test:

$$\text{Anomaly Detector I:}\begin{cases} \mathbf{1}_n^T\theta(t) \text{ is constant, } \forall t \geq 0 : \text{normal operation,} \\ \mathbf{1}_n^T\theta(t) \text{ is not constant, } \forall t \geq 0 : \text{abnormal operation.} \end{cases} \tag{11.7}$$

In the absence of the attack $\mathbf{d}_\theta(t)$, from (11.6) one obtains that $\mathbf{1}_n^T\dot{\theta}(t) = 0$. Consequently, $\mathbf{1}_n^T\theta(t)$ is constant for $\forall t \geq 0$. However, in the presence of $\mathbf{d}_\theta(t)$, this might not be necessarily true, ultimately leading to the activation of the attack detection system in (11.7). Therefore, the detection of an abnormal operation by the Anomaly Detector I in (11.7) indicates that the microgrid's control system in (11.6) is subject to cyberattacks $\mathbf{d}_\theta(t)$.

11.3.2.2 Anomaly Detector II: detection of cyber attacks $\mathbf{d_v}$(t)

Anomaly Detector II for the detection of $\mathbf{d_v}(t)$ acts as follows:

$$\text{Anomaly Detector II:}\begin{cases} \mathbf{1}_n^T\bar{\mathbf{V}} - V^* = 0 : \text{normal operation,} \\ \mathbf{1}_n^T\bar{\mathbf{V}} - V^* \neq 0 : \text{abnormal operation.} \end{cases} \tag{11.8}$$

where $\bar{\mathbf{V}}$ is the value of $\mathbf{V}(t)$ in the steady state.

It can be shown that in the presence of $\mathbf{d_v}(t)$, the average voltage regulation objective in (11.4) might be violated. Hence, the attack $\mathbf{d_v}(t)$ can easily be detected by the attack detection technique in (11.8).

11.3.3 Attack scenarios and adversary model

Although the baseline cyberattack detectors in (11.7)–(11.8) might be able to detect simple FDI attack scenarios, they might fail to detect the presence of sophisticated cyberattacks, referred to as stealthy attacks. In case of a stealthy cyberattack, the malicious adversary might have perfect knowledge of microgrids' control systems and aims at corrupting exchanged data in the controllers by injecting false data $\mathbf{d}_\theta(t)$ and $\mathbf{d}_\mathbf{v}(t)$ so that the microgrid performance is deteriorated without being detected by the proposed attack detection algorithms in (11.7) and (11.8). Such stealthy attacks are counted as dangerous attacks, as non-stealthy attacks can be blocked by the baseline attack detectors before serious damages to microgrids are caused.

It can be shown if FDI cyberattacks $\mathbf{d}_\theta(t)$, $\mathbf{d}_\mathbf{I}(t)$, and $\mathbf{d}_\mathbf{v}(t)$ in the proposed cooperative control in (11.6) are represented as follows:

$$\mathbf{d}_\theta(t) = \mathbb{L}\delta_\theta(t), \quad \mathbf{d}_\mathbf{v}(t) = W\mathbb{L}\delta_\mathbf{v}(t), \quad \mathbf{d}_\mathbf{I}(t) = \mathbb{L}\delta_\mathbf{I}(t) \tag{11.9}$$

where $\delta_\theta(t)$, $\delta_\mathbf{v}(t)$, and $\delta_\mathbf{I}(t)$ are uniformly bounded signals, they might not be detectable by the baseline cyberattack detection algorithms in (11.7) and (11.8). The main focus of this chapter is to develop cooperative distributed controllers that are resilient to the stealthy FDI cyberattacks in (11.9).

11.3.4 Proposed model of cyber-physical DC microgrids

The interconnection of the physical network in (11.3) and the proposed cooperative control strategy in (11.6) under stealthy FDI attacks modeled in (11.9) creates a cyber-physical microgrid whose dynamics are described by following equations:

$$\begin{aligned}
[C]\dot{\mathbf{V}}(t) &= \mathbf{I}(t) - [Y]\mathbf{V}(t) - \mathbb{B}\mathbf{I}_l(t),\\
[L_l]\dot{\mathbf{I}}_l(t) &= -[R_l]\mathbf{I}_l(t) + \mathbb{B}^T\mathbf{V}(t),\\
[L]\dot{\mathbf{I}}(t) &= ([k_1] - \mathbf{I}_n)\mathbf{V}(t) + ([k_2] - [r])\mathbf{I}(t) + [k_3]\mathbf{v}(t) + [k_4]\mathbb{L}W\mathbf{I}(t) + \gamma[k_5]\mathbb{L}\theta(t)\\
&\quad + [k_6](\mathbf{I}(t) - \phi(t)) + \mathbb{L}\delta_\mathbf{I}(t),\\
\tau_v\dot{\mathbf{v}}(t) &= W\left(-\beta(\mathbf{V}(t) - \mathbf{1}_n V^*) - K\beta\mathbb{L}W\mathbf{I}(t) - \alpha\beta(\mathbf{I}(t) - \phi(t)) + \gamma\beta\mathbb{L}\theta(t) + \mathbb{L}\delta_\mathbf{v}(t)\right),\\
\tau_\theta\dot{\theta}(t) &= -\mathbb{L}(\theta(t) + \gamma W\mathbf{I}(t)) + \mathbb{L}\delta_\theta(t),\\
\tau_\phi\dot{\phi}(t) &= -\phi(t) + \mathbf{I}(t),
\end{aligned} \tag{11.10}$$

Figure 11.3 depicts the block diagram of the proposed cyber-physical DC microgrid in (11.10). The stability and attack-resilient analysis of the above system will be discussed in the next section.

11.3.5 Metrics for resilient analysis

In order to characterize the resilient properties of cyber-physical DC microgrids in (11.10) in the presence of stealthy FDI attacks in (11.9), several metrics are defined in this chapter. The first important metric is the stability of cyber-physical microgrids in (11.10) under stealthy FDI attacks. Other metrics described as follows are related to voltage regulation and proportional load sharing objectives in (11.4) and (11.5):

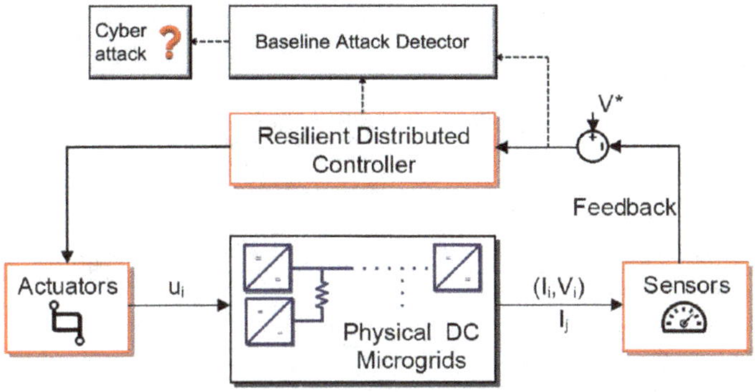

Figure 11.3 *Block diagram of cooperatively controlled DC microgrids augmented with a baseline attack detector scheme*

$$
\begin{aligned}
J_V &= \left| \lim_{t\to\infty} \mathbf{1}_n^T(\mathbf{V}(t) - \mathbf{1}_n V^*) \right|, \\
J_I &= \left\| \lim_{t\to\infty} \mathbb{L} W\mathbf{I}(t)) \right\|_\infty
\end{aligned}
\tag{11.11}
$$

As one can observe from (11.11), the smaller values of J_V and J_I indicate better resilience to cyberattacks in DC microgrids.

11.4 Stability and attack-resilience analysis

In this section, the stability of the proposed resilient cooperative control strategy in (11.6) is analyzed. To this end, first cyber-physical microgrids in (11.10) are represented in the state space framework. Then, we discuss their stability analysis using a Lyapunov-based approach.

11.4.1 State space representation of cyber-physical DC microgrids

Cyber-physical DC microgrids in (11.10) can be presented in a close-loop form as follows:

$$
\mathbf{E}\dot{\mathbf{e}}(t) = \mathbf{A}\mathbf{e}(t) + \mathbf{B}\delta(t), \tag{11.12}
$$

where

$$
\begin{aligned}
\mathbf{e}(t) &= \mathbf{x}(t) - \bar{\mathbf{x}}, \\
\mathbf{x}(t) &= \left[\mathbf{V}^T(t), \mathbf{I}^T(t), \mathbf{I}_l^T(t), \mathbf{v}^T(t), \theta^T(t), \phi^T(t)\right]^T, \\
\delta(t) &= \left[\delta_{\mathbf{I}}{}^T(t), \delta_{\mathbf{v}}{}^T(t), \delta_\theta{}^T(t)\right]^T, \\
\bar{\mathbf{x}} &= \left[\bar{\mathbf{V}}^T, \bar{\mathbf{I}}^T, \bar{\mathbf{I}}_l^T, \bar{\mathbf{v}}^T, \bar{\theta}^T, \bar{\phi}^T\right]^T,
\end{aligned}
\tag{11.13}
$$

where $\bar{\mathbf{V}}$, $\bar{\mathbf{I}}$, $\bar{\mathbf{I}}_l$, $\bar{\mathbf{v}}$, $\bar{\theta}$, and $\bar{\phi}$, respectively, are the equilibrium of $\mathbf{V}(t)$, $\mathbf{I}(t)$, $\mathbf{I}_l(t)$, $\mathbf{v}(t)$, $\theta(t)$, and $\phi(t)$ in the absence of cyberattacks, i.e., $\delta(t) = 0$. Note that $\delta(t)$ is assumed to be independent of $\mathbf{e}(t)$.

The state space matrices $(\mathbf{A}, \mathbf{B}, \mathbf{E})$ in (11.12) are defined as follows:

$$\mathbf{A} = \begin{bmatrix} -[Y] & \mathbf{I}_n & -\mathbb{B} & \mathbf{0}_n & \mathbf{0}_n & \mathbf{0}_n \\ [k_1 - 1] & ([k_2] + [k_4]\mathbb{L}W - [r] + [k_6]) & \mathbf{0}_n & [k_3] & \gamma\,[k_5]\,\mathbb{L} & -[k_6] \\ \mathbb{B}^T & \mathbf{0}_n & -[R_l] & \mathbf{0}_n & \mathbf{0}_n & \mathbf{0}_n \\ -\beta W & -\beta\,(KW\mathbb{L}W + \alpha\mathbf{I}_n) & \mathbf{0}_n & \mathbf{0}_n & \beta\gamma W\mathbb{L} & \beta W\alpha \\ \mathbf{0}_n & -\mathbb{L}\gamma W & \mathbf{0}_n & \mathbf{0}_n & -\mathbb{L} & \mathbf{0}_n \\ \mathbf{0}_n & \mathbf{I}_n & \mathbf{0}_n & \mathbf{0}_n & \mathbf{0}_n & -\mathbf{I}_n \end{bmatrix},$$

$$\mathbf{B} = \begin{bmatrix} \mathbf{0}_n & \mathbf{0}_n & \mathbf{0}_n \\ \mathbb{L} & \mathbf{0}_n & \mathbf{0}_n \\ \mathbf{0}_n & \mathbf{0}_n & \mathbf{0}_n \\ \mathbf{0}_n & W\mathbb{L} & \mathbf{0}_n \\ \mathbf{0}_n & \mathbf{0}_n & \mathbb{L} \\ \mathbf{0}_n & \mathbf{0}_n & \mathbf{0}_n \end{bmatrix}, \quad \mathbf{E} = \begin{bmatrix} [C] & \mathbf{0}_n & \mathbf{0}_n & \mathbf{0}_n & \mathbf{0}_n & \mathbf{0}_n \\ \mathbf{0}_n & [L] & \mathbf{0}_n & \mathbf{0}_n & \mathbf{0}_n & \mathbf{0}_n \\ \mathbf{0}_n & \mathbf{0}_n & [L_l] & \mathbf{0}_n & \mathbf{0}_n & \mathbf{0}_n \\ \mathbf{0}_n & \mathbf{0}_n & \mathbf{0}_n & \tau_v\mathbf{I}_n & \mathbf{0}_n & \mathbf{0}_n \\ \mathbf{0}_n & \mathbf{0}_n & \mathbf{0}_n & \mathbf{0}_n & \tau_\theta\mathbf{I}_n & \mathbf{0}_n \\ \mathbf{0}_n & \mathbf{0}_n & \mathbf{0}_n & \mathbf{0}_n & \mathbf{0}_n & \tau_\phi\mathbf{I}_n \end{bmatrix}. \tag{11.14}$$

From (11.10), it can be shown that $\bar{\mathbf{V}}$ and $\bar{\mathbf{I}}$ satisfy the following conditions:

$$\beta W((-\bar{V} + 1_n V^*) - K\mathbb{L}W\bar{I} - \gamma^2\mathbb{L}W\bar{I}) = 0 \tag{11.15}$$

By multiplying LHS of (11.15) by $\frac{1}{n}\mathbf{1}_n^T W^{-1}$ and taking into account the properties of the Laplacian matrix $\mathbb{L}$, one obtains that:

$$\frac{1}{n}1_n^T\bar{\mathrm{V}} - V^* = 0. \tag{11.16}$$

Moreover, (11.15) can be rewritten as follows:

$$\mathbb{L}W\bar{I} = \frac{1}{(K+\gamma^2)}\left(-\bar{\mathrm{V}} + 1_n V^*\right). \tag{11.17}$$

If the value of γ is chosen to be sufficiently large, from (11.17), one can obtain that $\mathbb{L}W\bar{I} \approx 0$ As a result, in the absence of cyberattacks, the proportional current sharing in (11.5) is guaranteed provided that γ is large.

11.4.2 Stability analysis in the absence of stealthy FDI attacks

To demonstrate the stability of cyber-physical DC microgrid in (11.12) in the absence of FDI cyberattacks, it is sufficient to show that $\mathbf{E}^{-1}\mathbf{A}$ in (11.12) is a Hurwitz (stable) matrix. Note that since $\mathbf{E} \succ 0$, $\mathbf{E}^{-1}$ exists.

Theorem 11.4.1. *It is assumed that the communication graph associated with the Laplacian matrix $\mathbb{L}$ in (11.10) is connected. If the parameters of the proposed resilient cooperative distributed control technique in (11.6) belong to the following set:*

$$\mathscr{Z} = \left\{ \begin{array}{c} 0 < K,\ 0 < \gamma,\ 0 < \alpha,\ 0 < \beta,\ 0 < \tau_v,\ 0 < \tau_\theta,\ 0 < \tau_\Phi \\ k_{1,i} < 1,\ k_{2,i} < r_i,\ 0 < \_3, i < \frac{\tau_v}{\beta L_i}(1 - k_{1,i})(r_i - k_{2,i}), \\ k_{4,i} = \frac{k}{I_i^s}(1 - k_{1,i}),\ k_{5,i} = \frac{1}{I_i^s}(1 - k_{1,i}),\ k_{6,i} = \alpha(k_{1,i} - 1),\ i \in \mathscr{V} \end{array} \right\} \tag{11.18}$$

$\mathbf{E}^{-1}\mathbf{A}$*in (11.12) is Hurwitz.*

Proof 11.4.2. The detailed proof of Theorem 11.4.1 can be found in [25].

11.4.3 Stability under stealthy FDI attacks

Since $\delta(t)$ in (11.12) does not depend on $\mathbf{e}(t)$, the stability of $\mathbf{E}^{-1}\mathbf{A}$ in (11.12) results in the stability of cyber-physical microgrids with the proposed cooperative controller in (11.6) in the presence of stealthy cyberattacks. Hence, if the control parameters in (11.6) are appropriately designed in the set $\mathcal{Z}$ given in (11.18), the cooperative control strategy in (11.6) guarantees the stability of DC microgrids while under cyberattacks modeled in (11.9). Moreover, the stability of $\mathbf{E}^{-1}\mathbf{A}$ in (11.12) implies the input-to-state stability of (11.12). As a result, for a potential bounded attack injection $\delta(t)$, the states of cyber-physical DC microgrids in (11.12) are bounded.

11.4.4 Stealthy FDI attack-resilience analysis

In this subsection, it will be shown that the proposed cooperative control strategy in (11.6) with parameters in (11.18) is resilient to stealthy FDI attacks in (11.9) by a proper choice of γ and β. To analyze the attack-resilience feature of the proposed distributed controller, we consider the state-space model in (11.12). The error vector $\mathbf{e}(t)$ is obtained as follows:

$$\mathbf{e}(t) = e^{(\mathbf{E}^{-1}\mathbf{A})t}\mathbf{e}(0) + \int_0^t e^{(\mathbf{E}^{-1}\mathbf{A})(t-\tau)}\mathbf{E}^{-1}\mathbf{B}\delta(\tau)\, d\tau. \tag{11.19}$$

By invoking the triangle inequality, one obtains that

$$\|\mathbf{e}(t)\| \leq \left\| e^{(\mathbf{E}^{-1}\mathbf{A})t}\mathbf{e}(0) \right\| + \left\| \int_0^t e^{(\mathbf{E}^{-1}\mathbf{A})(t-\tau)}\mathbf{E}^{-1}\mathbf{B}\delta(\tau) d\tau \right\| \tag{11.20}$$

As $\mathbf{E}^{-1}\mathbf{A}$ is a stable matrix (see Theorem 11.4.1), $\lim_{t\to\infty} \left\| e^{(\mathbf{E}^{-1}\mathbf{A})t}\mathbf{e}(0) \right\| = 0$. As a result, at the steady state, we have

$$\lim_{t\to\infty} \|\mathbf{e}(t)\| \leq \lim_{t\to\infty} \left\| \int_0^t e^{(\mathbf{E}^{-1}\mathbf{A})(t-\tau)}\mathbf{E}^{-1}\mathbf{B}\delta(\tau) d\tau \right\|. \tag{11.21}$$

Given the uniform boundedness of $\delta(t)$, there exists a constant vector Δ and a positive constant $t^\star$ such that for all $t \geq t^\star$ the following inequality holds:

$$\left\| \int_0^t e^{(\mathbf{E}^{-1}\mathbf{A})(t-\tau)}\mathbf{E}^{-1}\mathbf{B}\delta(\tau) d\tau \right\| \leq \left\| \int_0^t e^{(\mathbf{E}^{-1}\mathbf{A})(t-\tau)}\mathbf{E}^{-1}\mathbf{B}\Delta d\tau \right\|. \tag{11.22}$$

Hence, from (11.21) and (11.22), one obtains that

$$\lim_{t\to\infty} \|\mathbf{e}(t)\| \leq \lim_{t\to\infty} \left\| \int_0^t e^{(\mathbf{E}^{-1}\mathbf{A})(t-\tau)}\mathbf{E}^{-1}\mathbf{B}\Delta \, d\tau \right\| = \left\| -\mathbf{A}^{-1}\mathbf{B}\Delta \right\|. \tag{11.23}$$

It can be shown that for a sufficiently large value of γ and β, $\left\|-\mathbf{A}^{-1}\mathbf{B}\Delta\right\| \approx 0$. Thus, $\lim_{t\to\infty} \|\mathbf{e}(t)\| \approx 0$. As a result, it yields that

$$\lim_{\gamma,\beta\to\infty} \left(\lim_{t\to\infty} \|\mathbf{e}(t)\| \right) = 0. \tag{11.24}$$

The above result implies that for a sufficiently large value of γ and β, the steady-state value of $\mathbf{e}(t)$ converges to zero; therefore, the states of cyber-physical microgrids in (11.12) converge to $\bar{\mathbf{x}}$, which are the equilibria in the absence of cyberattacks. As a result

$$\begin{aligned} \lim_{\gamma,\beta\to\infty} \left(\lim_{t\to\infty} \mathbf{V}(t) \right) &= \bar{\mathbf{V}}, \\ \lim_{\gamma,\beta\to\infty} \left(\lim_{t\to\infty} \mathbf{I}(t) \right) &= \bar{\mathbf{I}}, \end{aligned} \tag{11.25}$$

where $(\bar{\mathbf{V}}, \bar{\mathbf{I}})$ satisfy (11.16) and (11.17). From (11.16) and (11.17), a conclusion is drawn that the average voltage regulation in (11.4) is guaranteed; furthermore, the proportional current sharing in (11.5) is achieved if γ is chosen to be sufficiently large.

11.4.5 Design of resilient cooperative control parameters

This section highlights the design procedure of the proposed resilient cooperative distributed control scheme for DC microgrids that are resilient against stealthy FDI cyber attacks in communication links. DG unit i is equipped with the following distributed controller:

$$\begin{aligned} u_i(t) &= k_{1,i}V_i(t) + k_{2,i}I_i(t) + k_{3,i}v_i(t) + k_{4,i}\textstyle\sum_{j\in N_i} \eta_{i,j}\left(\frac{I_i(t)}{I_i^s} - \frac{I_j(t)}{I_j^s}\right) \\ &\quad + k_{5,i}\gamma \textstyle\sum_{j\in N_i} \eta_{i,j}\left(\theta_i(t) - \theta_j(t)\right) + k_{6,i}\left(I_i(t) - \phi_i(t)\right), \\ \tau_v \dot{v}_i(t) &= \frac{\beta}{I_i^s}\left(-V_i(t) + V^*\right) + \frac{\gamma\beta}{I_i^s}\sum_{j\in N_i} \eta_{i,j}\left(\theta_i(t) - \theta_j(t)\right) - \frac{\alpha\beta}{I_i^s}\left(I_i(t) - \phi_i(t)\right) \\ &\quad - \tfrac{K\beta}{I_i^s}\textstyle\sum_{j\in N_i} \eta_{i,j}\left(\frac{I_i(t)}{I_i^s} - \frac{I_j(t)}{I_j^s}\right), \\ \tau_\theta \dot{\theta}_i(t) &= -\gamma \textstyle\sum_{j\in N_i} \eta_{i,j}\left(\frac{I_i(t)}{I_i^s} - \frac{I_j(t)}{I_j^s}\right) - \textstyle\sum_{j\in N_i} \eta_{i,j}\left(\theta_i(t) - \theta_j(t)\right), \\ \tau_\phi \dot{\phi}_i(t) &= -\phi_i(t) + I_i(t), \end{aligned} \tag{11.26}$$

where N_i is the set of neighboring DG units of DG i. Table 11.1 summarizes the design criteria for the control parameters in (11.26) such that the stability and attack resilience of DC microgrids are guaranteed.

As one can observe from (11.26), the design of local controllers of DG units does not rely on the global-level information in microgrids, and only local knowledge of DG units' parameters or the estimate of their parameters is required. Moreover, the control design does not need the knowledge of distribution lines, which is widely

Table 11.1 Design criteria for resilient controller parameters in (11.26)

Control parameter	Criterion	Control parameter	Criterion
α	$\alpha > 0$	$k_{1,i}$	$k_{1,i} < 1$
β	$\beta > 0$*	$k_{2,i}$	$k_{2,i} < r_i$
K	$K > 0$	$k_{3,i}$	$0 < k_{3,i} < \frac{\tau_v I_i^s}{\beta L_i}(1 - k_{1,i})(r_i - k_{2,i})$
γ	$\gamma > 0$†	$k_{4,i}$	$k_{4,i} = -K(1 - k_{1,i})$
τ_v	$\tau_v > 0$	$k_{5,i}$	$k_{5,i} = 1 - k_{1,i}$
τ_θ	$\tau_\theta > 0$	$k_{6,i}$	$k_{6,i} = \alpha(k_{1,i} - 1)$
τ_ϕ	$\tau_\phi > 0$		

*Sufficiently large.
†Sufficiently large.

uncertain or unknown. These important features of the proposed resilient cooperative distributed control strategy facilitate the plug-and-play capability in DC microgrids.

11.5 Simulation results

In this section, a series of simulation case studies is conducted to illustrate the previous analysis and results. We consider a 48 DC microgrid composed of $n = 4$ converter-interfaced DG units that are connected via resistive-inductive power lines in loop topology, as shown in Figure 11.1(a). Each DG feeds a resistive load of 10 and is controlled via a distributed controller whose structure is given in (11.26). The parameters of buck converters and lines, as well as the designed control parameters are given in Table 11.2.

In the first case study, it is assumed that constant stealthy false data $\delta_\theta(t)$ are injected to the controller at $t = 1$. The voltage signals at PCCs, the current of DC–DC converters, and the average value of PCC voltages are plotted in Figure 11.4. As one can observe from this figure, the proposed resilient controller is able to provide an accepted level of operational normality in DC microgrids in response to constant stealthy FDI cyberattacks.

To show the resilience of the proposed distributed control approach to time-varying cyberattacks, it is assumed that the cyber-physical DC microgrid is subject to time-varying stealthy FDI attacks in (11.9) launched at $t = 3$, where $\delta_\theta(t)$ and $\delta_v(t)$ are a mixture of constant, sinusoidal, and sawtooth signals of different magnitudes and frequencies. Figure 11.5 highlights the impact of the resilience parameters (β, γ) on the resilient metric J_I in (11.11). Note that since the stealthy FDI attacks in (11.9) do not affect the voltage regulation objective in (11.4), the resilient metric J_V in (11.11) has not been considered. As one can see from Figure 11.5, increasing the values of (β, γ) leads to better attack resilience in DC microgrids. However, the

Table 11.2 Parameters of the DC microgrid and its resilient distributed control in section 11.5

Microgrid parameters	Value	Control parameters	Value
L_i	2.64μH	K	10
C_i	2.2mF	τ_θ	0.01
I_i^s	$I_1^s = I_2^s = 1,\ I_3^s = I_4^s = 2$	τ_ϕ	0.01
R_{ij}	0.5Ω	τ_v	0.01
L_{ij}	2μH	α	10
Y_i	$0.1\Omega^{-1}$	$k_{1,i}$	−10
V^*	48V	$k_{2,i}$	−10
		$k_{3,i}$	5

high values of these parameters in (11.6) might adversely affect transient behavior in voltage regulation and current-sharing responses.

Figure 11.6 depicts the voltage and current trajectories for $(\beta, \gamma) = (100, 100)$. Upon launching cyberattacks at $t = 3$, the resilient controller mitigates the effects of the attacks on voltage regulation and current-sharing objectives.

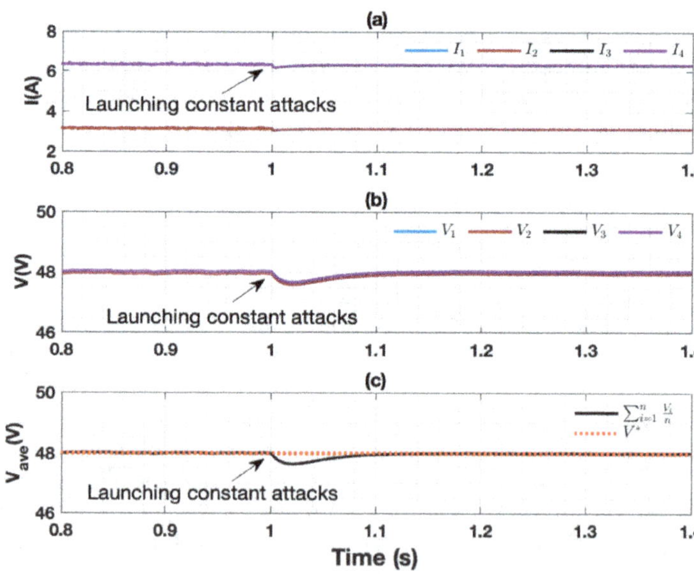

Figure 11.4 Performance of the proposed resilient cooperative controller in (11.6) to stealthy constant FDI cyberattacks launched at t = *1s: (a) current signals of DG units, (b) voltage signals at PCCs, and (c) average PCC voltages*

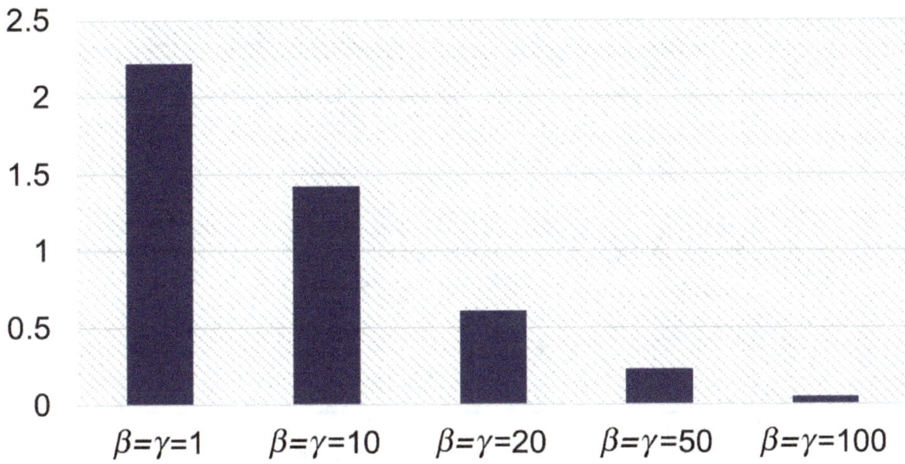

Figure 11.5 *Resilient metric* J_I *in (11.11) for different values of* (β, γ)

The next case study in Figure 11.7 shows the performance of the proposed resilient distributed control strategy in (11.6) to a load change, where the load of DG 4 is suddenly doubled at $t = 2$. Moreover, cyberattacks are launched on the duty cycles of all four buck converters at $t = 2.5$. Figure 11.7 shows the voltage signals at PCCs,

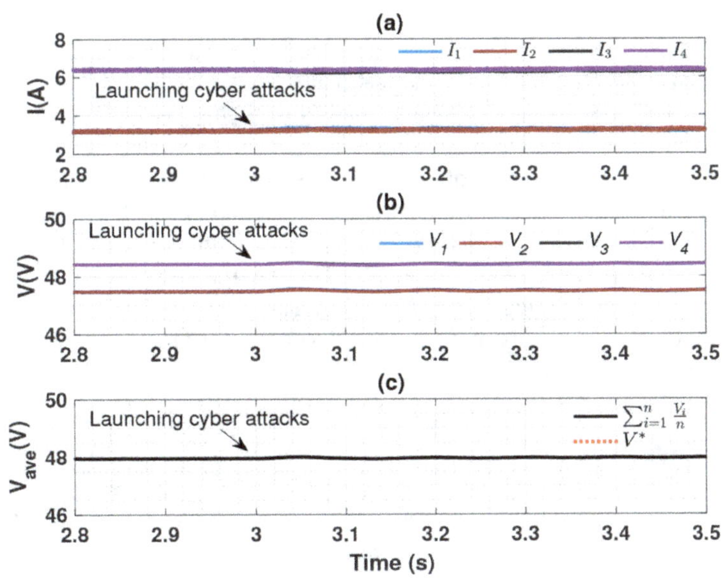

Figure 11.6 *Performance of the proposed resilient cooperative controller in (11.6) to time-varying stealthy FDI cyberattacks launched at* t = *3s: (a) current signals of DG units, (b) voltage signals at PCCs, and (c) average PCC voltages*

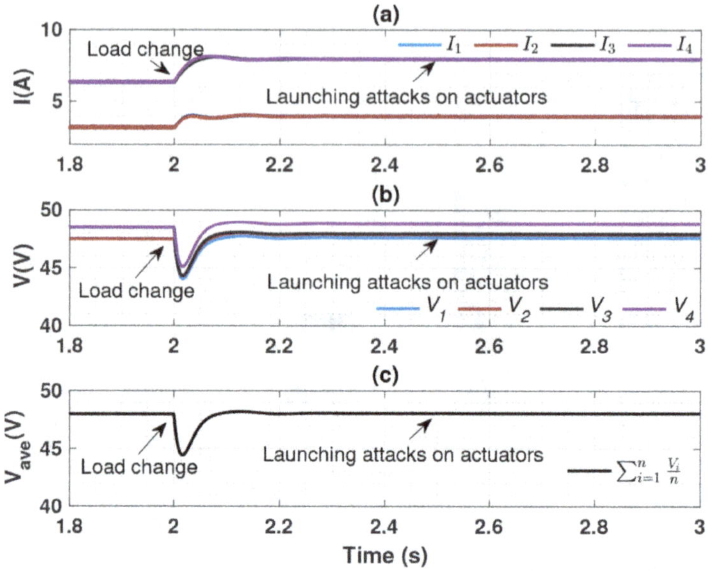

Figure 11.7 Performance of the proposed resilient cooperative controller in (11.6) to a load change at t = *2s and actuator attacks launched at* t = *2.5s: (a) current signals of DG units, (b) voltage signals at PCCs, and (c) average PCC voltages*

the current of DC–DC converters, and the average value of PCC voltages under the load uncertainty and actuator attacks. This figure highlights the robustness of the proposed controller to load changes and its resilience to actuator attacks.

In the final case study, the robustness of the proposed resilient distributed controller in (11.6) is assessed under uncertainties in the topology of DC microgrids. To this end, it is assumed that the distribution line connecting DG 1 to DG 4 is disconnected at $t = 1$ and reconnected at $t = 1.25$. The voltage and current trajectories of the DC microgrid are depicted in Figure 11.8. The results of this case study reveal that the proposed resilient distributed controllers are robust to uncertainties affecting the topology of DC microgrids.

11.6 Conclusion

This chapter has investigated the resilient distributed control of DC microgrids under cyber attacks. Baseline attack detection algorithms were proposed in this chapter. However, it was shown that the attack detection algorithms are not able to detect stealthy FDI attacks on DC microgrids' control systems. A cooperative resilient distributed control approach was then presented that enables DC microgrids to achieve average voltage regulation and proportional current sharing in the presence of stealthy FDI attacks.

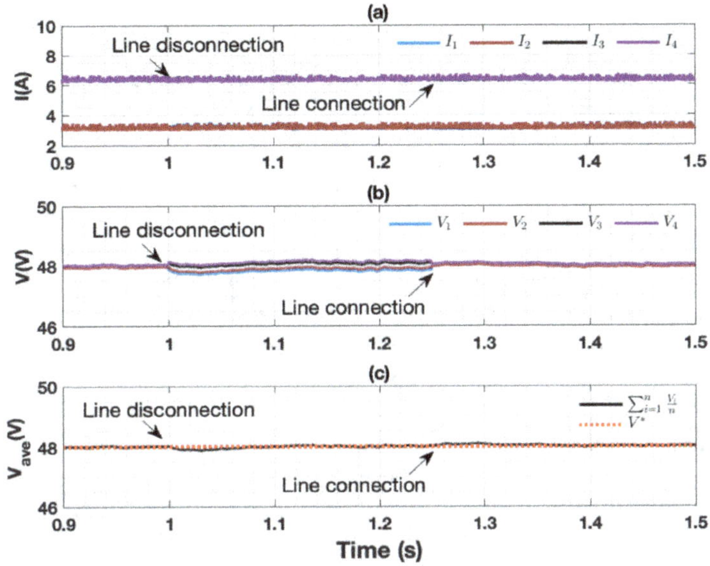

Figure 11.8 *Performance of the proposed resilient cooperative controller in (11.6) to a line disconnection at* t = *1 and its reconnection at* t = *1.25: (a) current signals of DG units, (b) voltage signals at PCCs, and (c) average PCC voltages*

Although the main focus of this chapter is on DC microgrids, we refer interested readers to [15] for general information on cybersecurity issues in AC microgrids and the design of resilient secondary controllers.

The existing resilient distributed control approaches in the literature dealing with potential cyberattacks DC microgrids, such as the proposed method in this chapter, are generally point solutions that focus only on a small part of cybersecurity issues in DC microgrids. They rely upon several assumptions: (a) the specific type of cyberattacks in DC microgrids' control systems (e.g., FDI, DoS, replay attacks), (b) only one or several vulnerable locations in controllers (e.g., sensors, actuators, and/or communication links), and (c) the number of attacked controller units. Furthermore, cyberattack detection and mitigation using the existing detection/mitigation techniques might require significant computational burden. Hence, it remains a painful and/or infeasible exercise to mitigate the adverse impact of sophisticated combinatory cyberattacks in DC microgrids. As a result, it is essential to deploy a cybersecurity layer in DC microgrids' control systems, combined with different detection and mitigation schemes and operating at different levels, in order to decrease the vulnerabilities of DC microgrids to complex and unknown cyberattacks.

References

[1] Nejabatkhah F., YW L., Liang H., Ahrabi R.R. 'Cyber-security of smart microgrids: a survey'. *Energies*. 2021;14(1):1–27.

[2] Mohan A., Brainard G., Khurana H., Fischer S. 'A cyber security architecture for microgrid deployments'. *International Conference on Critical Infrastructure Protection (ICCIP 2015)*; Arlington, VA Nov, 2015. pp. 245–59.

[3] Pal R., Hui P., Prasanna V. 'Privacy engineering for the smart Micro-Grid'. *IEEE Transactions on Knowledge and Data Engineering*. 2019;31(5):965–80.

[4] Beg O.A., Johnson T.T., Davoudi A. 'Detection of false-data injection attacks in cyber-physical DC microgrids'. *IEEE Transactions on Industrial Informatics*. 2017;13(5):2693–703.

[5] Sahoo S., Peng J.C.-H., Devakumar A., Mishra S., Dragičević T. 'On detection of false data in cooperative DC microgrids—a discordant element approach'. *IEEE Transactions on Industrial Electronics*. 2020;67(8):6562–71.

[6] Gallo A.J., Turan M.S., Boem F., Parisini T., Ferrari-Trecate G. 'A distributed cyberattack detection scheme with application to DC microgrids'. *IEEE Transactions on Automatic Control*. 2020;65(9):3800–15.

[7] Beg O.A., Nguyen L.V., Johnson T.T., Davoudi A. 'Signal temporal logic-based attack detection in DC microgrids'. *IEEE Transactions on Smart Grid*. 2019;10(4):3585–95.

[8] Zuo S., Altun T., Lewis F.L., Davoudi A. 'Distributed resilient secondary control of DC microgrids against unbounded attacks'. *IEEE Transactions on Smart Grid*. 2020;11(5):3850–9.

[9] Abhinav S., Modares H., Lewis F.L., Davoudi A. 'Resilient cooperative control of DC microgrids'. *IEEE Transactions on Smart Grid*. 2019;10(1):1083–85.

[10] Habibi M.R., Baghaee H.R., Dragičević T., Blaabjerg F. 'Detection of false data injection cyberattacks in DC microgrids based on recurrent neural networks'. *IEEE journal of emerging and selected topics in power electronics*. 2020:1–1.

[11] Abhinav S., Modares H., Lewis F.L., Ferrese F., Davoudi A. 'Synchrony in networked microgrids under attacks'. *IEEE Transactions on Smart Grid*. 2018;9(6):6731–41.

[12] Chlela M., Mascarella D., Joós G., Kassouf M. 'Fallback control for Isochronous energy storage systems in autonomous microgrids under denial-of-service cyberattacks'. *IEEE Transactions on Smart Grid*. 2018;9(5):4702–11.

[13] Zuo S., Beg O.A., Lewis F.L., Davoudi A. 'Resilient networked AC microgrids under unbounded cyberattacks'. *IEEE Transactions on Smart Grid*. 2020;11(5):3785–94.

[14] Chen Y., Qi D., Dong H., Li C., Li Z., Zhang J. 'A FDI attack-resilient distributed secondary control strategy for islanded microgrids'. *IEEE Transactions on Smart Grid*. 2020;12(3):1929–38.

[15] Sadabadi M.S., Sahoo S., Blaabjerg F. 'A fully resilient cyber-secure synchronization strategy for AC microgrids'. *IEEE Transactions on Power Electronics*. 2021;36(12):13372–8.

[16] Danzi P., Angjelichinoski M., Stefanovic C., Dragičević T., Popovski P. 'Software-defined microgrid control for resilience against denial-of-service attacks'. *IEEE Transactions on Smart Grid*. 2019;10(5):5258–68.

[17] Hu S., Yuan P., Yue D., Dou C., Cheng Z., Zhang Y. 'Attack-resilient event-triggered controller design of DC microgrids under DOS attacks'. *IEEE Transactions on Circuits and Systems I: Regular Papers*. 2020;67(2):699–710.

[18] Sahoo S., Mishra S., Peng J.C.-H., Dragicevic T. 'A stealth cyberattack detection strategy for DC microgrids'. *IEEE Transactions on Power Electronics*. 2019;34(8):8162–74.

[19] Sahoo S., Dragičević T., Blaabjerg F. 'An event-driven resilient control strategy for DC microgrids'. *IEEE Transactions on Power Electronics*. 2020;35(12):13714–24.

[20] Sahoo S., Yang Y., Blaabjerg F. 'Resilient synchronization strategy for AC microgrids under cyberattacks'. *IEEE Transactions on Power Electronics*. 2021;36(1):73–7.

[21] Sahoo S., Peng J.C.-H., Mishra S., Dragicevic T. 'Distributed screening of hijacking attacks in DC microgrids'. *IEEE Transactions on Power Electronics*. 2020;35(7):7574–82.

[22] Sahoo S., Dragičević T., Blaabjerg F. 'Multilayer resilience paradigm against cyberattacks in DC microgrids'. *IEEE Transactions on Power Electronics*. 2021;36(3):2522–32.

[23] Zhou Q., Shahidehpour M., Paaso A., Bahramirad S., Alabdulwahab A., Abusorrah A. 'Distributed control and communication strategies in networked microgrids'. *IEEE Communications Surveys & Tutorials*. 2020;22(4):2586–633.

[24] Rieger C.G., Gertman D.I., McQueen M.A. 'Resilient control systems: next generation design research'. *2009 2nd Conference on Human System Interactions*; Catania, Italy, 21-23 May; 2009. pp. 632–6.

[25] Sadabadi M.S., Sahoo S., Blaabjerg F. 'Stability oriented design of cyberattack resilient controllers for cooperative DC microgrids'. *IEEE Transactions on Power Electronics*. 2021;37(2):1–21.

[26] Sadabadi M.S. 'Line-independent plug-and-play voltage stabilization and L_2 gain performance of DC microgrids'. *IEEE Control Systems Letters*. 2021;5(5):1609–14.

[27] Sandberg H. 'Stealthy false data injection attacks in feedback systems revisited. M FR, H TAM, editors'. *Safety, Security and Privacy for Cyber-Physical Systems*. Springer, Cham; 2021. pp. 61–78.

Chapter 12

Cyber-physical testbeds for smart grid and electric vehicle-charging infrastructure

Kirti Gupta[1], Bijaya Ketan Panigrahi[1], Subham Sahoo[2], and Frede Blaabjerg[2]

To enhance the reliability, resiliency and sustainability of a microgrid (MG) in the presence of several distributed energy resources (DERs) and loads, communication systems play a vital role, forming it into a cyber-physical system (CPS). However, integration of information and communication technology (ICT) in the MG architecture exposes MG to potential malicious cyber attacks as well. Similarly, electric vehicles (EVs) also come with a host of many cybersecurity risks. The integration of power, communication and transportation sectors empowers the intruders to exploit the vulnerability and disrupt the operation of multiple sectors. Testbed plays a vital role in the study of the threat landscape and the impact of cyber attacks (such as denial of service (DoS) attacks and false data injection attacks (FDIA)) in such cyber-physical infrastructure. This study would further aid to develop and verify the detection and mitigation schemes against cyber attacks.

This work presents a detailed study on the development of real-time (RT) cosimulation testbed for inverter-based MG and RT testbed for EV-charging infrastructure. The MG testbed consists of OP5700 RT simulator to emulate cyber-physical layer through HYPERSIM software; and SEL-3530 Real-Time Automation Controller (RTAC) hardware configured with ACSELERATOR RTAC SEL-5033 software. A human–machine interface (HMI) is further developed in ACSELERATOR Diagram Builder SEL-5035 software for local/remote monitoring and control. Furthermore, communication protocols such as Modbus, sampled measured values (SMVs), generic object-oriented substation event (GOOSE) and distributed network protocol 3 (DNP3) on an Ethernet-based interface were established to link the cyber and physical layers. Furthermore, the testbed for EV-charging infrastructure consists of AC and DC emulators to emulate the respective charging station (AC/DC) and EV. It also constitutes various charging cables supported for different charging protocols such as combined charging system (CCS), GB/T and CHAdeMO. To

[1]Department of Electrical Engineering, Indian Institute of Technology, Delhi, India
[2]Department of Energy, Aalborg University, Aalborg, Denmark

monitor the message exchanges, charging discovery system (CDS) hardware along with its software is used. Few of the test cases have been presented to demonstrate the ability of these testbeds.

12.1 Introduction

Towards the path of smart grid development, India aims to achieve a target of 450 GW renewable energy installed capacity by 2030 [1]. MG is the elementary unit for the smart grid, which aids in accommodating various DERs through power electronic converters. The hierarchical control architecture is the standard for the desired control and operation of the system. This is a three-layer architecture comprising primary, secondary and tertiary control layers. For an islanded mode of operation, MG is integrated with primary and secondary control layers. MGs integrated with distributed secondary control are reliable alternatives over the centralised control that was affected by single point-of-failure and required high communication bandwidth. With the inclusion of communication network in the distributed control architecture, MGs have evolved into a CPS. The reliance on the information exchanged among the power electronic devices (such as inverters) makes them vulnerable to cyber intrusions. Any infiltration in the cyber layer can also affect the physical security of the CPS due to the tight integration of cyber and physical layers.

EVs can play a major role in alleviating the reliance on fossil fuels for transport and electricity systems, thus evolving the power grid to a smart grid. This has made EVs increasingly popular, and their rapid growth is expected in the coming decades from 3 to 125 million by 2030 [2, 3]. An EV-charging ecosystem consists of EV, EV user, electric vehicle supply equipment (EVSE, commonly termed as charger), charging station, EVSE service provider and energy aggregator. These terms would be briefly explained in the later section. The bidirectional power and information exchange between the EV and EVSE is the key feature for smart charging of EVs through public/private charging stations. The involvement of communication infrastructure comes with the advantages as well as its security concerns and threats. The intruders, after diagnosing the vulnerabilities of these communication protocols and the different entities involved, can launch the cyber attacks. The impact of these cyber attacks can range from moderate to severe. The moderate level impact may consist of DoS, in which an EVSE would refuse to charge an EV. The extensive level of impact can be the damage and destruction of the hardware entities involved. At last, the severe level of impact would be human safety concerns and widespread impact to the electrical grid. Many countries around the globe set their target in adoption of EV, but a range of challenges restrains most of them from achieving the specified target. As described above, with the advancements in control architectures and inclusion of ICT, modern CPSs have become more reliant on the information exchanges. This raises security concerns and mandates developing an extensive vulnerability assessment platform. A testbed plays a vital role and suffices this purpose. In this regard, this work presents two different testbeds each for MG- and

EV-charging infrastructure. These testbeds consist of both physical and cyber layers that would be briefly explained in the later sections.

12.2 Brief description of MG

An MG is a CPS consisting of power-electronic devices (such as inverters), lines, loads, circuit breakers (CBs), sensors, actuators and controllers in the physical layer. The devices responsible for information exchanges such as switches, routers and communication links for wired and wireless communication in the cyber layer. The control layer residing at the junction of these two layers, consists of primary controllers (PC) and secondary controllers (SC) for an islanded MG. The PC consists of fast inner current and voltage control loops with droop control. The SC is implemented over distributed control architecture with the objective of frequency restoration, proportional active and reactive power sharing with m_p and n_q being the droop coefficients for sharing of active power (P) and reactive power (Q), respectively, and are represented as:

1. Frequency restoration:

$$\lim_{t \to \infty} \omega^i(t) = \omega_{ref} \tag{12.1}$$

2. Proportional active power sharing:

$$\lim_{t \to \infty} [m_p^j P^j(t) - m_p^i P^i(t)] = 0 \tag{12.2}$$

3. Proportional reactive power sharing:

$$\lim_{t \to \infty} [n_q^j Q^j(t) - n_q^i Q^i(t)] = 0 \tag{12.3}$$

where $j \in N_i$. That is, all the immediate neighbours of ith DER.

Further details regarding the equations involved in the primary and secondary control layers can be obtained from [4]. The PC after receiving the measurements from sensors, transmits the control signal to the respective DER. Furthermore, to improve the steady-state performance, SC receives the information regarding frequency and power from its and neighbouring DERs to generate the correction signal to the respective PC. These message exchanges can be wired/wireless, which can be the entry point to the cyberattacks, as shown in Figure 12.1.

12.3 Vulnerabilities associated with MG

Attacks on power sector have diverse impacts in recent decades [5, 6], with some of them being unnoticed while others having severe impacts on both economy and human life. The Stuxnet attack in Iran was first of its kind, endangering the power sector [7]. The cyber attacks in Kyiv, Ukraine, in two consecutive years in 2015 [8]

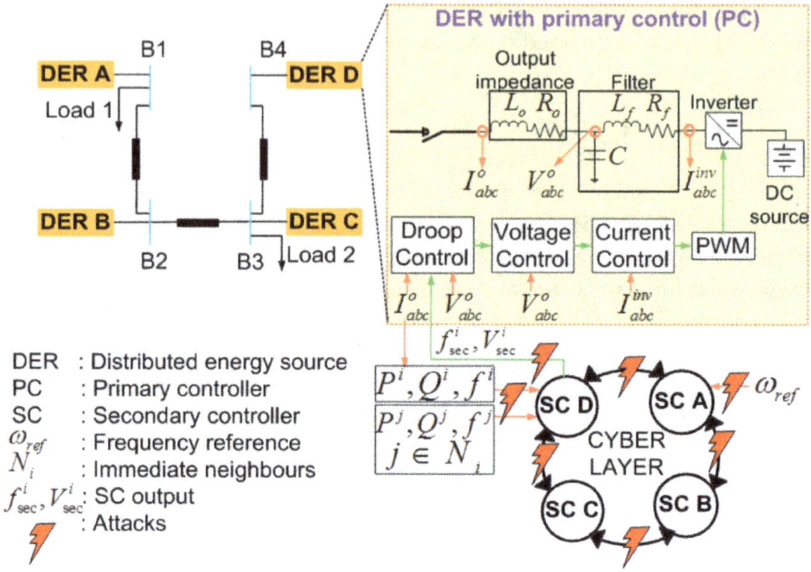

Figure 12.1 Test MG with four DERs

and in 2016 [9] are other examples featuring the vulnerabilities in the power sector. The recent case study by Recorded Future's Insikt Group revealed the suspected network intrusions by RedEcho in India's power sector, targeting four out of five Regional Load Dispatch Centres (RLDCs) [10]. Recently, integrated wind company Vestas also experienced cyber intrusion, disrupting the communication with the consumer systems [11]. In addition, German grid operators also confronted a communication breakdown due to fault in the satellite connection, which was suspected to be the consequence of cyber attack. This resulted in the switching over of solar and wind power plants automatically to autopilot mode [12]. In the report of National Institute of Standards and Technology Internal Reports (NISTIR) 7628, MG cybersecurity was considered as a major threat scenario [13]. The growing events of such cyber intrusions signify that the threat is real and mandates the diagnosis of such events [14].

A pictorial representation of plausible attack vectors in a MG environment is represented in Figure 12.2. The control layer sends the control signal to the actuators after receiving the information through sensors. These information exchanges are carried out over the cyber layer, which may expose MG to cyber intrusions. The vulnerabilities associated with each entity in an MG are as follows:

- **Secondary controller:** The distributed control can be implemented on PLCs (programmable logic controllers) with communication interfaces making it vulnerable to cyber intrusions [15, 16].
- **Human machine interface:** The attackers can exploit the vulnerabilities and penetrate HMI remotely to disrupt the operation, inject malware, etc.

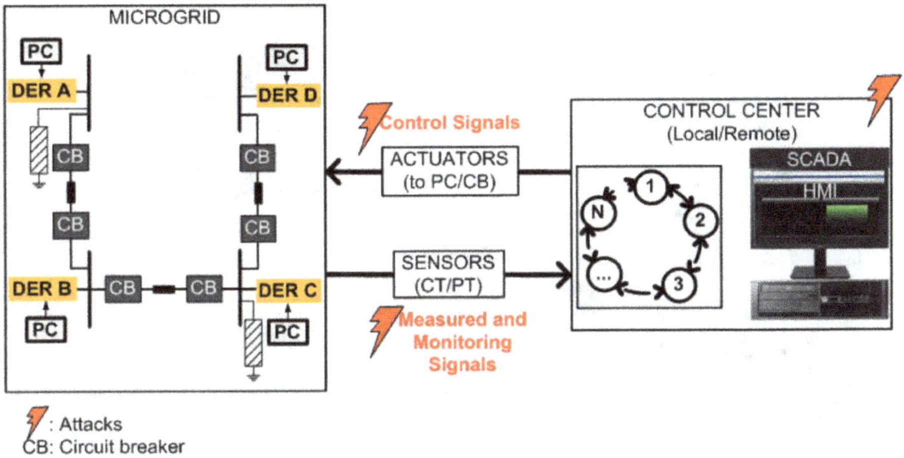

Figure 12.2 Attack pathways in an MG CPS

- **Communication links:** The wired or wireless links can be either destroyed physically or infiltrated by attackers to manipulate the signals.
- **Routers and gateways:** The attackers after gaining access can launch several attacks such as DoS attacks, routing table poisoning and packet mistreating attacks.
- **Communication protocols:** The vulnerability of the protocols can be exploited to inject the attack. IEC 61850 is an interoperability standard making it suitable for many applications but since integrated over Ethernet interface also makes it vulnerable to cyber attacks.
- **Merging units:** The voltage and current information carried from merging units to protection devices can be manipulated by the attackers after getting access to the network.
- **Relays:** The attacker after getting unauthorised access to the network can disrupt the messages exchanged among the relays and between relay and CB.

As a result, power electronic systems security becomes a key driver for protection against such threats. For this, a cosimulation testbed platform is presented for vulnerability assessment of various entities involved in the MG environment, which is explained the next section.

12.3.1 Overview of the testbed for MG

Each testbed has its particular objective to study various aspects such as cybersecurity, control and stability. The testbed described here is a cybersecurity-oriented testbed, with physical and cyber layers developed to study the impact of various cyber intrusions. The work presents a cosimulation platform, integrating two devices such as OP-5700 to emulate physical layer, control layer and partial cyber layer; and RTAC to emulate the cyber layer responsible for supervisory control and monitoring. The information exchanges are

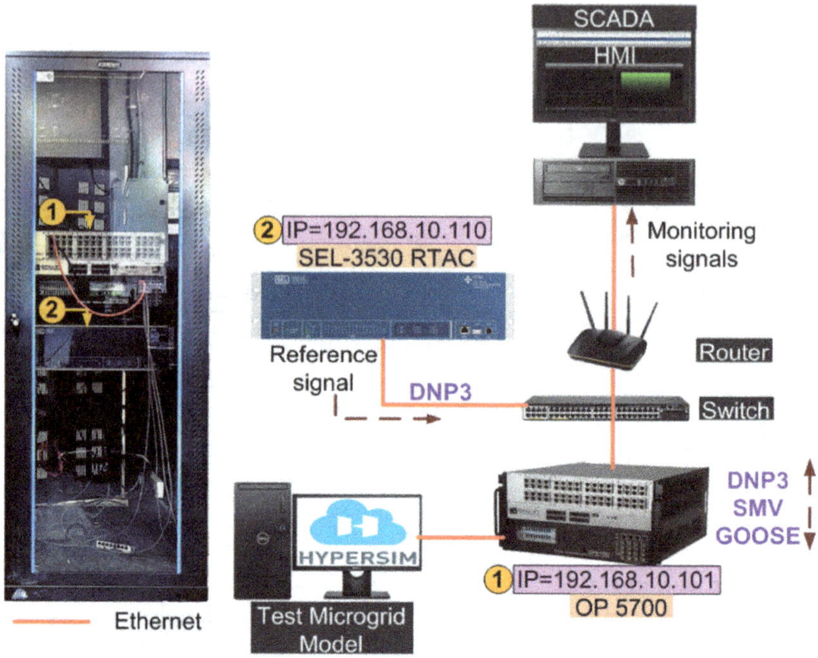

Figure 12.3 Cosimulation testbed for MG cybersecurity

carried out over the Ethernet interface. Although several protocols have been integrated by researchers in secondary control layer such as DDS, GSE [17] and ZeroMQ [18], there was a need to integrate the heterogeneous sources through an interoperability standard. Hence, IEC 61850 was the solution to this, where SMV protocol is used for information exchanges between the secondary controllers at a sampling frequency of 4 kHz. Furthermore, DNP3 was used for monitoring and supervisory control in HMI, as shown in Figure 12.3. In protection architecture, SMV messages transmit voltage and current information through merging units (MUs) to protection devices. Additionally, GOOSE is utilised for information exchanges among the relays and between relay and CB. The message exchanges captured over Wireshark is shown in Figure 12.4.

A simple case on a four-bus network with inverter-based sources is tested on the developed cosimulation platform. The different regions (R) mark different events such as normal operation (R-I), load change (R-II), stealthy FDIA (R-III) and FDIA with unfair attack value (R-IV). In Figure 12.5, a load change at 2 seconds is followed by a stealthy FDIA at 5 seconds, and on increasing the attack vector to an unfair value at 8 seconds, the system becomes unstable. More details regarding formulation of FDIA can be obtained from [19, 20].

12.3.2 Features of testbed

The features of the developed testbed are elaborated in this section. Apart from providing scalability for different test cases, it provides a platform to integrate different

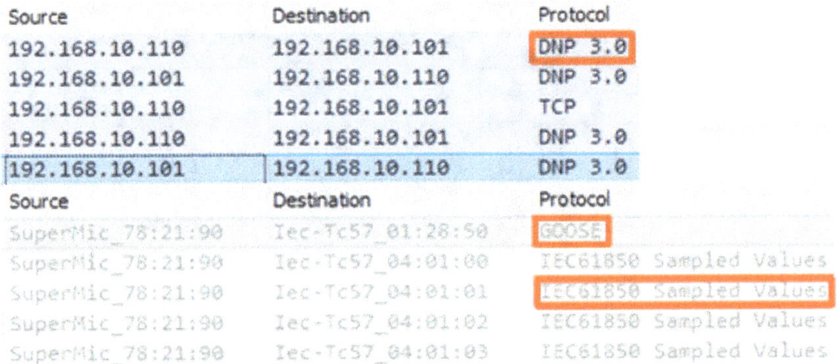

Figure 12.4 Message exchanges on wireshark

standard and non-standard protocols to test their efficiency. Therefore, it provides a realistic extension to various experiments conducted involving different networks, scenarios. Additionally, it provides a vulnerability assessment platform necessary to observe the vulnerability associated with each entity closely, necessary to develop detection and mitigation schemes against the cyber attacks. The prominent features of the developed testbed are:

> Notable features of MG cybersecurity testbed: The testbed developed for MG cybersecurity analysis provides a platform scalable to different networks or control models, several attack-scenario generation. It also facilitates to integrate various communication protocols to test their feasibility and efficiency depending on the use case. The developed testbed can be further extended to more realistic scenarios by integrating actual relay, controllers, etc. Additionally, it provides a platform for vulnerability assessment associated with entities involved to develop their countermeasures.

12.3.2.1 Scalability

The testbed provides the facility to design different network architectures incorporated with several different control algorithms to find the best suited control algorithm based on its performance. The result of a four-bus test MG consisting of four DERs is presented here for instance, but this can further be extended to integrate several DERs for different network architectures.

12.3.2.2 Communication protocols

Several inbuilt communication protocols such as C37.118 for wide area applications, DNP3 for monitoring applications, IEC 61850 for time-critical applications and many more, as shown in Figure 12.6, are available to incorporate in a network, depending on the applications. In addition, the developed testbed is not limited to

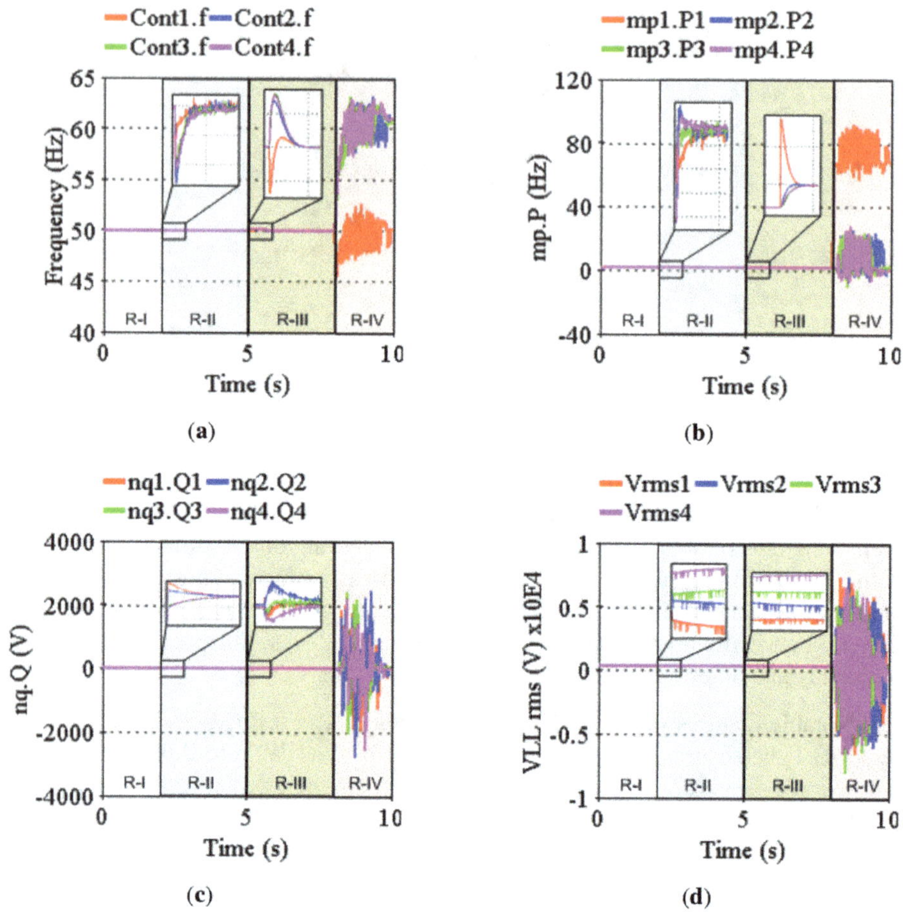

Figure 12.5 *Real-time simulation results on HYPERSIM with load change at 2 seconds, FDIA initiated at 5 seconds with small value and increased to unfair value at 8 seconds (a) frequency restoration; (b) proportional active power sharing; (c) proportional reactive power sharing; and (d) voltage across buses*

these inbuilt communication protocols but can further be extended to integrate several other communication protocols externally.

12.3.2.3 Generation of attack scenarios

Various cyber attacks such as DoS attacks, FDIAs, hijacking attacks and replay attacks can be modelled to study the impacts of these cyber intrusions. This analysis would help to develop the respective countermeasures and test its effectiveness.

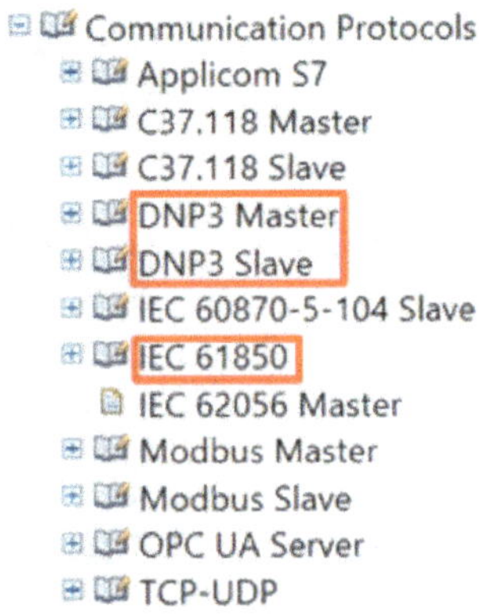

Figure 12.6 Communication protocols available in MG testbed

12.3.2.4 Extension to more realistic scenarios

The developed testbed can be extended to more realistic environment by integrating hardware. This may include inclusion of actual relays, PLCs or controllers with control algorithm in the network. This would enhance the effectiveness of the proposed schemes.

12.4 Brief description of EV-charging infrastructure

The power exchange between grid and EV can be bidirectional through EVSE. The grid-to-vehicle (G2V) power exchange is more common, whereas vehicle-to-grid (V2G) is limited to small-scale pilot projects [21, 22]. However, aiming at decarbonisation goals, V2G energy transfer would expand rapidly in near future. An EV-charging CPS consists of EV, EVSE and charging station with several EVSE in the physical layer. In order to facilitate the integration of EVs, the chargers connect EVs to the grid using a charging device with a requirement of information exchanges for data regarding state of charge (SoC), charging status, transactions, control commands between the various entities in an EV ecosystem through either low-level communication or high-level communication in the cyber layer. The low-level communication is carried over pulse width modulation (PWM) between EV and EVSE. The IEC 61851–1 specifies the meaning of duty cycle values. The duty cycle is controlled by EVSE, whereas the voltage of the signal is controlled by the EV. The high-level communication, on the other hand, is during DC charging between EV and EVSE from the start, after the plug-in of the charging cable. It is used to exchange parameters related to charging such as voltage, current, SoC and charging time remaining.

According to the voltage and power transfer ratings, EVSE is divided into three levels, namely Level 1 (L1), Level 2 (L2) and Level 3 (L3). L1 EVSE is of 120 V, 12–16 A single-phase AC supply with power transfer rate of 1.44–1.92 kW dominant in the United States. L2 EVSE is single-phase 240 V AC or three-phase 400 V AC up to 80 A and rated

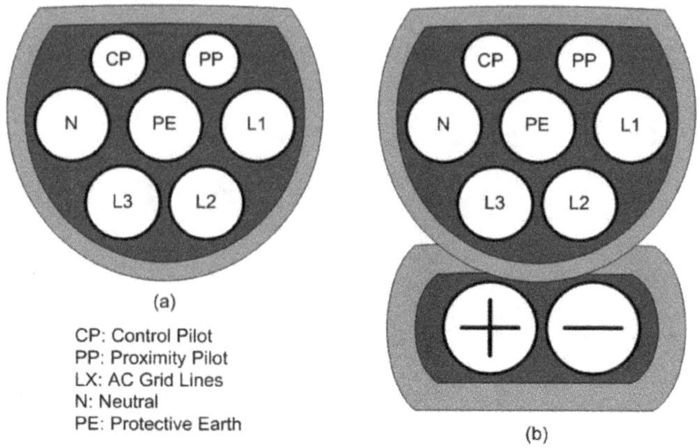

Figure 12.7 Pin configuration in (a) AC and (b) DC chargers

power up to 55 kW dominant outside the United States. However, L3 EVSE is DC fast chargers bulkier than their AC counterpart due to the presence of AC to DC converters. The pin configuration of AC and DC chargers are shown in Figure 12.7. The communication between EVSE and EV is provided through a pilot signal found on one of the pins (i.e., control pilot (CP)) in the cord and connector. The charging station sends a square wave signal of 1 kHz frequency on CP to describe the maximum current available with the help of PWM signal. There are a total of six states out of which states A, B and C are general states, whereas states D, E and F are rare states. The description and voltage level corresponding to each state is mentioned in Table 12.1. Proximity pilot is used by EV to detect the plug. It prevents driving away of vehicle while connected and also can detect the plug removal. The AC grid lines, neutral pins are for the power supply connections, whereas protective earth is present for the protection purpose.

IEC 61851–1 specification is the world standard for AC charging. It uses low-level communications. However, there are regional standards for DC charging. It requires high-level communication. Among different standards, SAE J1722, VDE-AR-E 2623-2-2 and Tesla Destination are for L1 and L2 EVSE. For L3 EVSE, there are CCS, GB/T, CHAdeMo and Tesla Supercharger. DIN SPEC 70121:2014–12 is

Table 12.1 Signal states of control pilot

State	Pilot high voltage	Pilot low voltage	Description
State A	12 V	NA	Not connected
State B	9 V	−12 V	EV connected, ready to charge
State C	6 V	−12 V	EV charging
State D	3 V	−12 V	EV charging, ventilation required
State E	0 V	0 V	Error
State F	NA	−12 V	Unknown error

a preliminary standard defining the communication between the EVSE and the EV for DC charging. This standard is based on requirements specified in ISO 15118–2. Moreover, the CCS is a global standard system that combines all established AC charging solutions with ultrafast DC charging in a single system.

12.5 Vulnerabilities associated with EV-charging infrastructure

With the wide deployment of public charging infrastructures and EVs, attack vectors are further escalated. This raises concerns regarding the breaching of information and further affecting the performance and stability of the overall system. According to a report by Kaspersky Lab, the data of a user stored on RFID card (used to complete the transaction) can be replicated to accomplish unauthorised transactions. In addition, man-in-the-Middle (MitM) attack can be initiated on EVSE by exploiting USB ports, accessing logs and critical data to inject malicious firmware [23]. In a similar manner, attacker can penetrate into the private home chargers and gain access to manipulate the EV-charging process [24]. Apart from EVSE, the vulnerabilities of EVs can also be exploited through cyber intrusions. According to a report, an intruder remotely gained access to control a Nissan Leaf using vehicle identification number [25]. The presence of these vulnerabilities necessitates a comprehensive vulnerability assessment to reveal the threats involved and the gaps, jeopardising the deployment of EV technology.

Figure 12.8 represents the possible attack pathways in the EV-charging environment. The message exchanges between the various entities can act as entry point for cyber attacks. This can lead to cyber intrusions, hence instigate data privacy concerns. These attacks may include unauthorised access to information (such as consumer payment and location details), tampering of data (such as energy consumption), DoS (refusing to charge). The diversity of the players and communication technologies involved further aggravates the situation. The vulnerabilities associated with each entity in an EV ecosystem are as follows:

- **EV:** An attacker can invade into the functioning and management system of an EV by penetrating the communication involved and can have severe impacts.
- **EV user:** The private data of an EV user, such as location data, transaction information and banking details, can be maliciously used by the attacker for his/her personal gains.
- **EVSE:** The EVSE can be in the hands of someone with malicious intent, which can lead to system compromise with cyber/physical access to the hardware.
- **Utility:** The attacker can manipulate the charging/discharging commands, which can have devastating impacts and can also cause instability in the grid.
- **Communication protocols:** The vulnerability of the communication standards incorporated for exchanging data among various entities (such as EV, EVSE, energy aggregator and EVSE service provider) can be exploited, causing catastrophic impacts.

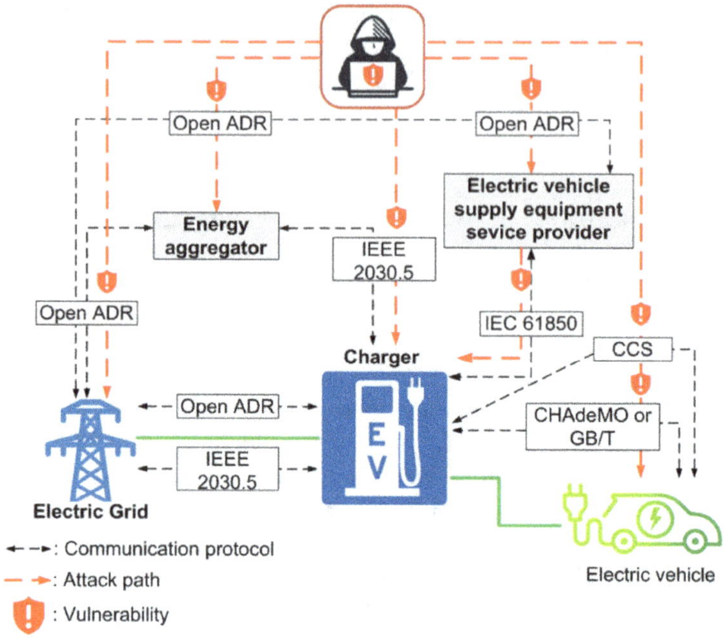

Figure 12.8 Attack pathways in an EV-charging infrastructure

According to the recent literature, wired communication layer of a CAN bus or power line communication (PLC) using communication protocol ISO 15118 are vulnerable to cyber attacks [26–28]. In [27], possibility of remote eavesdropping of messages in 54 EVSE with CCS connectors was demonstrated. In [29], a data-driven attack on power grid by manipulating EVSE demands was presented that caused frequency instability. Such vulnerabilities mandate a platform to study the effects of such cyber intrusions, to develop an efficient detection and mitigation schemes against these cyber attacks and to develop a trustworthy and secure ecosystem crucial for both the consumers and the grid operators. In this regard, this research work presents a testbed for EV-charging infrastructure with different charging protocols.

12.5.1 Overview of the testbed for EV-charging infrastructure

The security of EV-charging infrastructure has gained great research interest in recent years. Earlier in 2013, a generic study on EV-charging security was presented in [30]. The impact of EV on grid was investigated in [31]. Most of the research studies pertaining to the grid integration are at the offline simulation stage, which may not include the non-linear models of power electronic devices and hence may not include all the dynamics involved with the actual system. The testbed presented consists of actual EVSE (AC and DC) and EVs along with a monitoring device CDS, with its software to monitor the message exchanges between EV and EVSE. Additionally, EVSE and EVs of different types can be emulated by adjusting the parameters accessible to the user. The EV and EVSE are connected in the testbed through the actual charging cables, and the information is exchanged between them through the real charging protocols (such as CCS, GB/T and CHAdeMO).

12.5.2 Features of testbed

The salient features of the developed testbed are explained in this section. The testbed in Figure 12.9 is scalable to design different EVSE and EVs by defining the parameters accessible to the users. Furthermore, it provides a platform to integrate EVSE and EVs through the standard communication platform. The testbed presented for EV-charging infrastructure can operate in three testing modes, namely EV test mode, EVSE test mode and MitM test mode. The usage of each test mode is explained in this section. It also facilitates insulation resistance emulation. On the whole, this provides a realistic experience for charging EV integrated through standard charging protocols. Additionally, vulnerability associated with each entity and charging protocol can be tested in this testbed against different cyber intrusions. This would facilitate to develop detection and mitigation schemes for a secure operation of EV-charging infrastructure. The notable features of the EV-charging testbed are as follows:

> Notable features of EV-charging infrastructure testbed: The testbed developed for EV-charging infrastructure provides a platform to integrate various EVSE (AC/DC) with EVs through standard charging protocols. It also facilitates to emulate different EVSE and EVs by defining their associated parameters. The testbed incorporates three different testing modes, namely EV test mode, EVSE test mode and MitM test mode according to the type of EVSE and EV connected (actual or emulated) for the applications of designing EVs, EVSE compatible with the existing ones and monitoring message exchanges. Furthermore, it provides a vulnerability assessment platform to investigate the vulnerability associated with different entities. This would accordingly help in developing their respective mitigation schemes for different cyber intrusions.

12.5.2.1 Scalability

The developed testbed for EV-charging system is scalable to user-defined specifications to EVSE and EV. With the aid of emulators, we can model any EVSE or EV present in the market or even specify a new specification for them. It would prove to be very interesting to observe the behaviour of each entity with the variation of parameters.

12.5.2.2 Charging protocols

The testbed is supported with different charging protocols for both AC and DC charging. The different protocols available are shown in Figure 12.10. In this testbed, we have presented results with the integration of IEC 61851-1:2010 and DIN SPEC 70121:2014–12 for AC and DC charging, respectively.

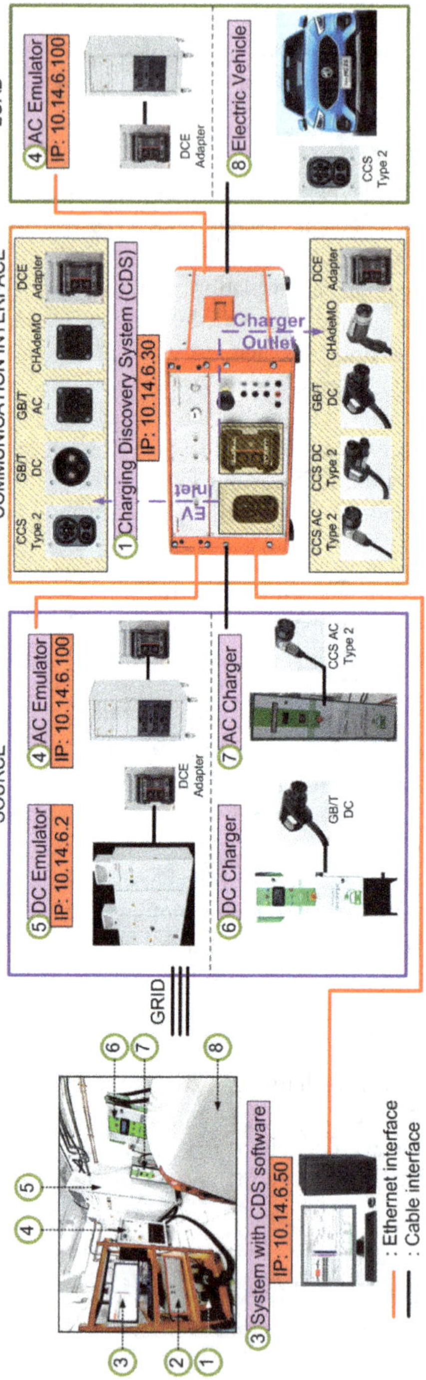

Figure 12.9 Testbed for EV-charging infrastructure

IEC 61851-1:2010
SAE J1772:2010
DIN SPEC 70121:2014-12
ISO 15118:2013 Ed.1
ISO 15118-20 DIS
GB/T 27930-2011
GB/T 27930-2015
GB/T 18487-2015
CHAdeMO 0.9, 0.9.1
CHAdeMO 1.0.0, 1.0.1, 1.1, 1.2
CHAdeMO 2.0

Figure 12.10 Protocols available in EV-charging testbed

12.5.2.3 Insulation resistance emulation

The testbed facilitates to test insulation monitoring function of EV or EVSE. It is used to emulate insulation fault systematically. A variable resistance between DC+ and PE and DC− and PE is connected to CDS for conducting this test. The arrangement for performing this test is shown in Figure 12.11.

12.5.2.4 Modes of testing

A total of three modes of testing are present in the developed testbed. Each of them can be used depending on the application. The different modes of testing available in the EV-charging testbed are as follows:

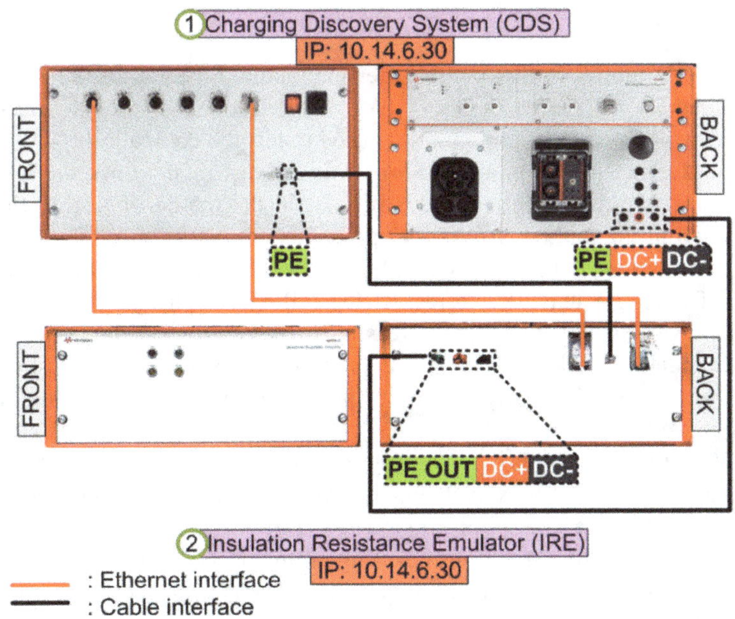

Figure 12.11 Connection diagram for insulation resistance emulation

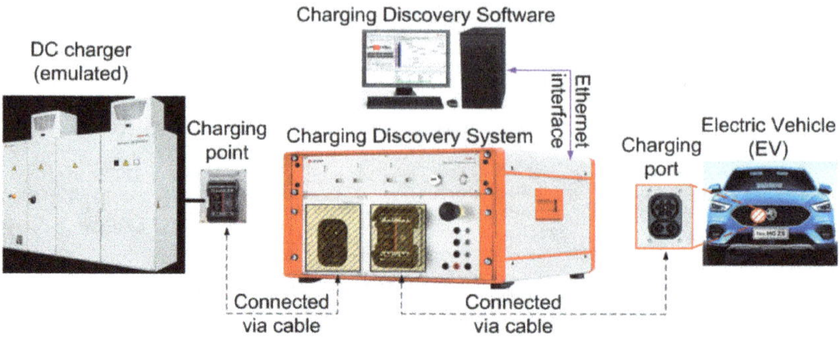

Figure 12.12 EV test mode during DC charging

1. **EV test mode:** During this mode of testing, there are actual EVs connected to emulated EVSE. Various EVSE such as AC or DC can be emulated through the emulators. This test is conducted to test the compatibility of a EV with different chargers (emulated). It is used to validate different charging protocols of various chargers with actual EV. Once an EV has passed different test cases in this mode of testing, one can launch EV in the market. Hence, it provides an easy solution without the necessity to have all actual EVSE at a place to test its feasibility to an EV. One can emulate different chargers and can easily check its compatibility to the EV. Figure 12.12 shows setup for EV mode of testing in DC charging system. Similarly, it can be configured for AC charging and different EVs as well.
2. **EVSE test mode:** In this mode of testing, actual EVSE is used to charge the emulated EVs. Different types of EVs can be modelled through user-defined blocks available in the software. It is used to test the compatibility of an EVSE with different EVs (emulated). Similar to EV test mode, once the EVSE has passed all the available test cases present in this mode of testing, it can be launched in the market. It provides a convenient platform to test the compatibility of various emulated EVs with developed EVSE. It would hence cut the cost of having actual EVs to test its feasibility to a charger. One can simply emulate various EVs and test the feasibility of the developed charger to different emulated EVs. Figure 12.13 shows setup for EVSE mode of testing in DC charging system. Similarly, it can be configured for AC charging and different EVs as well.
3. **MitM test mode:** In this mode of testing, both EVSE and EV used are real. One can capture the messages exchanged during the process of charging/discharging. It can act as a great tool to observe the information being exchanged and test the vulnerability of various charging protocols. Once this test is conducted, one can find the entry point of attackers and hence can work accordingly for the defence strategy. Figure 12.14 shows setup for MitM mode of testing in DC charging system. Similarly, it can be configured for AC charging and different EVs as well.

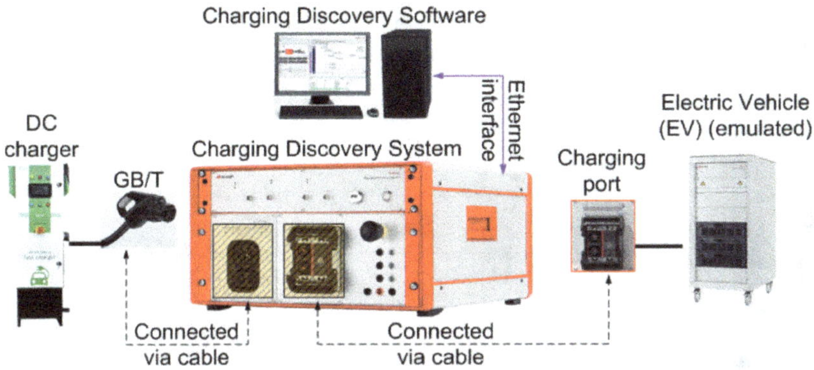

Figure 12.13 EVSE test mode during DC charging

In the developed testbed for EV-charging system, two test cases for EV mode of testing are presented here. The first case is EV test mode during AC charging with IEC 61851–1: 2010 standard. The variation of parameters such as voltage across control pilot, SoC of the battery, current and voltage during the process of charging obtained through CDS is shown in Figure 12.15. Furthermore, the message exchanges captured over CDS shown in Figure 12.16.

The second case is EV test mode during DC charging with DIN SPEC 70121:2014–12 standard. The variation of parameters such as voltage across control pilot, SoC of the battery, current and voltage during the process of charging obtained through CDS is shown in Figure 12.17. The message exchanges captured over CDS are shown in Figure 12.18, respectively.

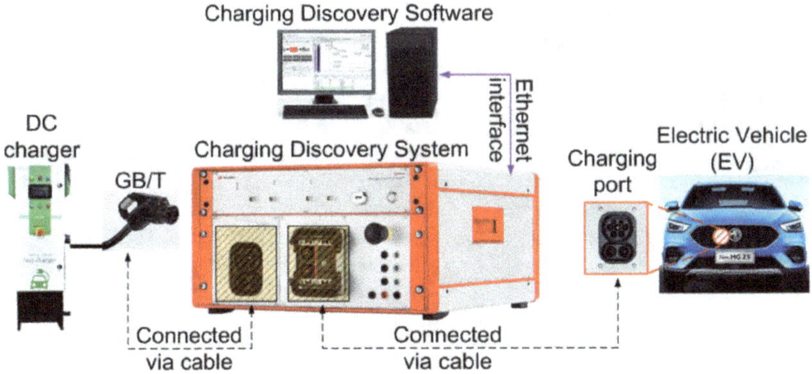

Figure 12.14 MitM test mode during DC charging

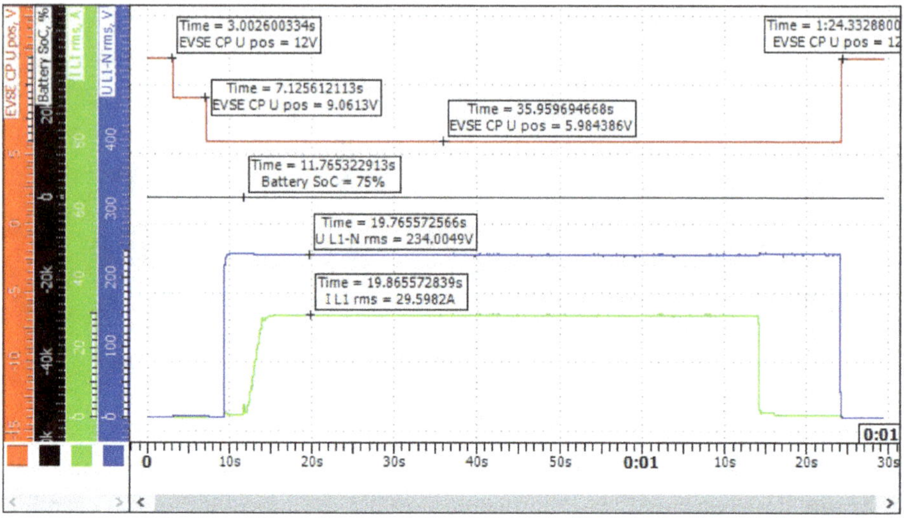

Figure 12.15 *Variation of different parameters for EV test mode during AC charging over IEC 61851-1:2010 standard obtained through CDS*

Time	Source	Message type	Message Name	Message Details
00:00:00.000200688	System	ID_UseCaseInfo		
00:00:00.002187412	System	Version	System information	System information: IP addr...
00:00:00.002229568	System	Version	Firmware version	Version information: Applica...
00:00:00.002257072	System	Version	Source sink information	Power Source information: V...
00:00:00.002273248	System	Version	Client version	Client information: Version: ...
00:00:00.130352721	System	LOGMESSAGE	INFORMATION	SECC: Initializing component...
00:00:00.132673621	System	LOGMESSAGE	INFORMATION	SECC: Initialization completed!
00:00:00.133032506	System	LOGMESSAGE	INFORMATION	Starting "EVSE Controller
00:00:00.134771561	System	Event	Statemachine transition	CPState=State A1
00:00:00.260522442	System	Event	Statemachine transition	CPState=State B1
00:00:01.275230906	System	Event	Statemachine transition	CPState=State B2
00:00:06.485850703	System	Event	Statemachine transition	CPState=State C2
00:02:01.737512203	System	LOGMESSAGE	INFORMATION	Stopping system (Source: "...
00:02:01.754613674	System	Event	Statemachine transition	CPState=State C1
00:02:03.852119867	System	Event	Statemachine transition	CPState=State B1
00:02:03.881675801	System	Event	Statemachine transition	CPState=State A1
00:02:03.966490011	System	Event	Statemachine transition	CPState=Finished
00:02:04.174481781	System	LOGMESSAGE	INFORMATION	SECC: Shutdown components!
00:02:09.714953741	System	ID_EndOfFile		

Figure 12.16 *Message exchanges during EV test mode during AC charging obtained through CDS*

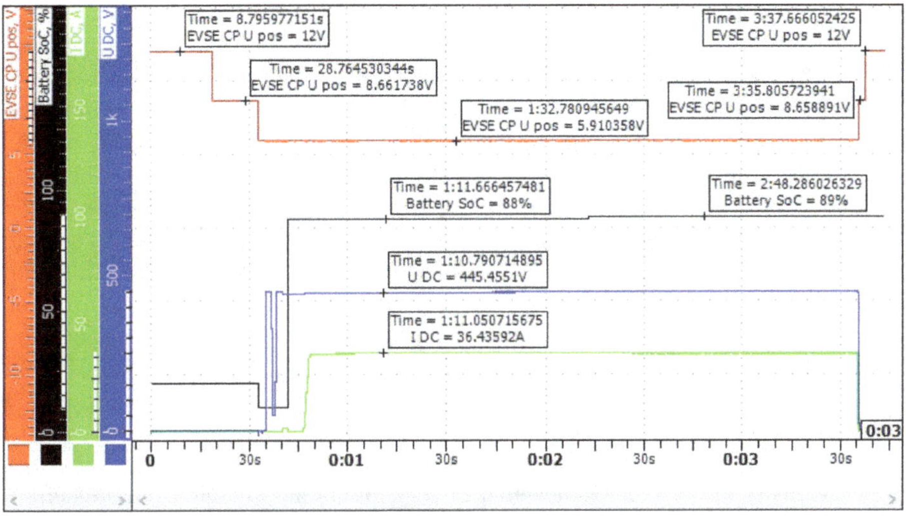

Figure 12.17 Variation of different parameters for EV test mode during DC charging over DIN SPEC 70121:2014–12 standard obtained through CDS

Time	Source	Message type	Message Name	Message Details
00:00:10.279412267	System	Event	Statemachine transition	CPState=State A1
00:00:15.652013785	System	Event	Statemachine transition	CommState=SLAC running
00:00:15.652194973	System	Event	Statemachine transition	CPState=State B1
00:00:15.671462422	System	Event	Statemachine transition	CPState=State B2
Time	Source	Message type	Message Name	Message Details
00:00:34.666722641	System	Event	Statemachine transition	CommState=Wait for PreC...
Time	Source	Message type	Message Name	Message Details
00:00:38.393590371	System	Event	Statemachine transition	CommState=Wait for Curre...
Time	Source	Message type	Message Name	Message Details
00:03:03.227563605	System	Event	Statemachine transition	CommState=Wait for Weldi...
00:03:03.281628441	EV	V2G	WeldingDetectionReq	EVReady=true, EVErrorCo...
00:03:03.284157094	System	Event	Statemachine transition	CPState=State B2
Time	Source	Message type	Message Name	Message Details
00:03:06.113341959	System	Event	Statemachine transition	CPState=State A1
00:03:06.202792987	System	Event	Statemachine transition	CommState=Finished
00:03:06.204090332	System	Event	Statemachine transition	CPState=Finished
00:03:06.414443778	System	LOGMESSAGE	INFORMATION	SECC: Shutdown components!

Figure 12.18 Few message exchanges during EV test mode during DC charging obtained through CDS

12.6 Conclusion

The work demonstrated two testbeds for two CPS, namely MG- and EV- charging system. The testbed for MG is developed on a cosimulation platform with standard communication protocols in the cyber layer. Similarly, the EV-charging testbed is developed having the facility to integrate different charging protocols between EV and EVSE (both real and emulated). It presents possible entry points of cyber attacks in both the testbeds. To summarise, development of such testbeds can help the researchers to exploit the vulnerabilities and work to develop the mitigation strategies against these cyber intrusions. These developed schemes can further be tested and validated on the same testbed for effectiveness.

References

[1] Ministry of New and Renewable Energy [online]. Available from https://pib.gov.in/PressReleaseIframePage.aspx?PRID=1762960&RegID=3&LID=1 [Accessed 04 Mar 2022].

[2] 'EEI electric vehicle sales: facts & figures'. *EEI Electric Transportation IEEE*. 2019.

[3] Bunsen T. 'Global ev outlook 2018: towards cross-modal electrification'. *IEA*. 2018.

[4] Gupta K., Sahoo S., Panigrahi B.K., Blaabjerg F., Popovski P. 'On the assessment of cyber risks and attack surfaces in a real-time co-simulation cybersecurity testbed for inverter-based microgrids'. *Energies*. 2021;14(16):4941.

[5] Loukas G. *Cyber-physical attacks: a growing invisible threat [online]*. 2022. Available from https://www.elsevier.com/books/cyberphysical-attacks/loukas/978-0-12-801290-1 [Accessed 4 Mar 2022].

[6] Liu C.-C., Stefanov A., Hong J., Panciatici P. 'Intruders in the grid'. *IEEE Power and Energy Magazine*. 2012;10(1):58–66.

[7] Stuxnet. 2015. Available from https://en.wikipedia.org/wiki/Stuxnet [Accessed 4 Mar 2022].

[8] Case D.U. '*Analysis of the cyberattack on the ukrainian power grid*'. *Electricity Information Sharing and Analysis Center (E-ISAC)*. 388. Washington, DC; 2016.

[9] Condliffe J. *Ukraine's power grid gets hacked again, a worrying sign for infrastructure attacks* [MIT Technology Review [online]. 2016. Available from https://www.technologyreview.com/2016/12/22/5969/ukraines-power-grid-gets-hacked-again-a-worrying-sign-for-infrastructure-attacks/

[10] *Redecho targets the Indian power sector amid heightened border tensions [online]*. Available from https://www.recordedfuture.com/redecho-targeting-indian-power-sector/ [Accessed 4 Mar 2022].

[11] *Vestas recovers from cyber attack and data breach [online]*. Available from https://www.power-technology.com/news/company-news/vestas-cyber-attack-data-breach-wind/ [Accessed 4 Mar 2022].

[12] *Satellite cyber attack paralyzes 11gw of German wind turbines [online].* Available from https://www.pv-magazine.com/2022/03/01/satellite-cyber-attack-paralyzes-11gw-of-german-wind-turbines/ [Accessed 4 March 2022].

[13] *Introduction to NISTIR 7628 guidelines for smart GRID cyber security [online].* Available from https://www.nist.gov/system/files/documents/smartgrid/nistir-7628_total-2.pdf [Accessed 4 Mar 2022].

[14] Gupta K., Sahoo S., Mohanty R., Panigrahi B.K., Blaabjerg F. 'Decentralized anomaly characterization certificates in cyber-physical power electronics based power systems'. *IEEE 22nd Workshop on Control and Modelling of Power Electronics (COMPEL)*; Cartagena, Colombia, 2-5 Nov; 2021. pp. 1–6.

[15] Wang Y., Mondal S., Deng C., Satpathi K., Xu Y., Dasgupta S. 'Cyber-resilient cooperative control of bidirectional interlinking converters in networked AC/DC microgrids'. *IEEE Transactions on Industrial Electronics*. 2021;68(10):9707–18.

[16] Sahoo S., Dragičević T., Blaabjerg F. 'Cyber security in control of grid-tied power electronic converters—challenges and vulnerabilities'. *IEEE Journal of Emerging and Selected Topics in Power Electronics*. 2021;9(5):5326–40.

[17] Starke M., Herron A., King D., Xue Y. 'Implementation of a publish-subscribe protocol in microgrid Islanding and resynchronization with self-discovery'. *IEEE Transactions on Smart Grid*. 2019;10(1):361–70.

[18] Du Y., Tu H., Lukic S. 'Distributed control strategy to achieve synchronized operation of an Islanded MG'. *IEEE Transactions on Smart Grid*. 2019;10(4):4487–96.

[19] Sahoo S., Yang Y., Blaabjerg F. 'Resilient synchronization strategy for AC microgrids under cyberattacks'. *IEEE Transactions on Power Electronics*. 2021;36(1):73–7.

[20] Sadabadi M.S., Sahoo S., Blaabjerg F. 'A fully resilient cyber-secure synchronization strategy for AC microgrids'. *IEEE Transactions on Power Electronics*. 2021;36(12):13372–8.

[21] *E.ON and nissan are launching new V2G Trial project in the UK [online].* Available from https://insideevs.com/news/437785/eon-nissan-v2g-trial-project-uk/ [Accessed 4 Mar 2022].

[22] *Vehicle2 grid [online].* Available from https://amsterdamsmartcity.com/projects/vehicle2grid [Accessed 4 Mar 2022].

[23] *Don't be sure charging your electric car is secure enough [online].* Available from https://www.kaspersky.com/blog/electric-cars-chargingproblems/20652/ [Accessed 4 Mar 2022].

[24] *Remotely controlled EV home chargers – the threats and vulnerabilities [online].* Available from https://securelist.com/remotely-controlled-ev-home-chargers-the-threats-and-vulnerabilities/89251/ [Accessed 4 Mar 2022].

[25] *Controlling vehicle features of Nissan LEAFs across the globe via vulnerable APIs [online].* Available from https://www.troyhunt.com/controlling-vehicle-features-of-nissan/ [Accessed 4 Mar 2022].

[26] Harnett K. 'DOE/DHS/DOT volpe technical meeting on electric vehicle and charging station cybersecurity report'. John Volpe Nat. Transp. Syst. Center (US), Cambridge, MA: Tech. Rep. DOT-VNTSC-DOE; 2018.

[27] Baker R., Martinovic I. 'Losing the car keys: wireless PHY-layer insecurity in EV charging'. Proceedings of the 28th USENIX Security Symposium; Santa Clara, CA, 14–16 Aug; 2019. pp. 407–24.

[28] Bao K., Valev H., Wagner M., Schmeck H. 'A threat analysis of the vehicle-to-grid charging protocol ISO 15118'. *Computer Science – Research and Development*. 2018;33(1-2):3–12.

[29] Acharya S., Dvorkin Y., Karri R. 'Public plug-in electric vehicles + grid data: is a new cyberattack vector viable?' *IEEE Transactions on Smart Grid*. 2020;11(6):5099–113.

[30] Mustafa M.A., Zhang N., Kalogridis G., Fan Z. 'Smart electric vehicle charging: security analysis'. *IEEE PES Innovative Smart Grid Technologies Conference (ISGT)*; Washington, DC, 24–27 Feb; 2013. pp. 1–6.

[31] Carryl C., Ilyas M., Mahgoub I., Rathod M. 'The PEV security challenges to the smart grid: analysis of threats and mitigation strategies'. *International Conference on Connected Vehicles and Expo (ICCVE)*. IEEE; Las Vegas, NV, 2–6 Dec; 2013. pp. 300–5.

Chapter 13

Event-driven resiliency of microgrids against cyber attacks

Subham Sahoo[1] and Frede Blaabjerg[1]

Though recent advancements in microgrids are largely based on distributed control strategies to enhance reliability, their susceptibility to cyber attacks still remains a challenging issue. Additionally in converter-dominated microgrids, mitigation of cyber attacks upon detection in a timely manner is the need of the hour to prevent the system from immediate shutdown. Since most of the existing research is primarily focused on detection of cyber attacks in microgrids without giving prior attention to comprehensive steps of mitigation, this chapter classifies cyber attacks as *events* and introduces an event-driven cyber attack resilient strategy for microgrids, which immediately replaces the attacked signal with a *trusted* event-driven signal constructed using true transmitted measurements. This mechanism not only disengages the attack element from the control system but also replaces it with an event-triggered estimated value to encompass normal consensus operation during both steady-state as well as transient conditions even in the presence of attacks. Finally, the event detection criteria and its sensitivity are theoretically verified and validated using different conditions in the presence of stealth attacks in AC and DC systems.

13.1 Introduction

The rapid development of microgrids can be ascribed to their high flexibility in integrating renewable energy sources, storage devices, and modern electronic loads, in both grid-connected and autonomous modes of operation [1]. In this regard, distributed control structures have been trending for microgrids as they facilitate scalability, reliability, and automation in contrast to the centralized controllers, which are highly vulnerable to single point of failure [2]. Moreover, they ensure robust performance under cyber imperfections such as communication delays, link failures, and data packet losses [3]. However, this coordination philosophy cannot be

[1]Department of Energy, Aalborg University, 9220 Aalborg, Denmark

designated to be fully *reliable* owing to the availability of information of only neighboring units. This increases the vulnerability of microgrids to illegitimate sensors and actuators tampering in the form of cyber attacks [4]. As microgrids are a key component of mission critical applications such as military bases, hospitals, and industrial plants [5], it is crucial to ensure their security against adversarial attacks.

To detect the presence of cyber attacks, bad-data detection tests are usually performed to identify residual element(s) between estimated and measured states [6]. These attacks can be classified differently into a primary superset with false data injection attacks [7], denial of service [8], and replay attacks as its elements [9]. These are some of the prominent attacks that have precipitated in real-time applications. More details on other critical intrusion approaches, which are the subsets of the conventional attacks, can be found in [11]. Furthermore, these attacks can also be *coordinated*, where the attacker attains sufficient knowledge about the cyber-physical architecture to create attack vectors, which can easily bypass the abovementioned bad-data detection tests [12]. Such attacks are commonly termed as *stealth* attacks [13]. In other words, these attacks introduce zero dynamics in distributed control systems, which goes unnoticeable. The attacker can use this *discreet* behavior to collect more system information by penetrating into the control system at first and then attack microgrids to cause instability later in unforeseeable ways. Specifically in power electronics based systems, the attack mitigating action needs to be fast, otherwise the network can become unstable or even lead to shutdown.

Considerable efforts have been put recently in modeling and detection of stealth cyber attacks on various elements (sensors, communication links) in microgrids [14, 15]. Further studies on differentiating between sensor faults and cyber attacks are carried out in [16]. However, these papers are limited to detection without providing any comprehensive steps of countermeasures for normal system operation to remove the attack element(s). Additionally, the information received from the attacked unit(s) is discarded as an elementary approach to prevent the propagation of attack into the system [17]. As a result, the network connectivity is affected, which leads to disruption in the consensus theory. Another well-defined mitigation approach is to employ an observer for each unit to operate with the estimated states using the preattack points upon detection of attack. Even though these approaches are quite efficient, they have model-intensive requirements, where their performance is highly prone to model uncertainties. Moreover, the design of observer can be complex, while its real-time execution may require heavy computational resources. Additionally, an upper bound–based mitigation condition is also proposed in [19] where the mitigation strategy is selectively determined based on the total number of compromised units, termed as F-total, or the local compromised agents in the neighborhood of each unit, termed as F-local. Although it counteracts against attacks on sensors, actuators, and communication links, it might affect the cyber graph connectivity by unnecessarily abandoning neighbor's information during a load change even when there is no attack. As a result, its operation becomes a point of serious concern for stealth attacks, which entail zero dynamics in distributed networks. Hence, since the abovementioned approaches are based on restrictive assumption on the information exchange in the cyber network, a self-healing mitigation strategy needs to be developed, which provides maximum resilience for the system to recover without losing the cyber network connectivity. In [28], a partial primal-dual algorithm is determined to detect the presence of *stealth* node and link attacks. However, it is limited to detection only, without providing any comprehensive steps of countermeasures to mitigate the attack element(s). In [29], a resilient framework for unbounded cyber attacks in AC microgrid is presented. However, the resilient control update is introduced using a hidden control layer, which can still be infiltrated by an attacker conducting a *stealth* attack with sufficient system information.

To address these issues, this chapter proposes an event-driven cyber attack resilient strategy for microgrids inspired from [32–37], where only the presence of a cyber attack element in a given agent is classified as an *event*. As compared with the conventional event-triggered schemes [20, 21], attack detection criteria are used as triggering mechanisms for the proposed countermeasures to operate immediately [22, 23]. As already discussed in [14, 15], since stealth attacks involve compromised sensors, actuators, and communication links, activation of these events is defined using the detection criteria for each unit. Furthermore, the activation status of these events in the attacked agent is also transmitted to noncompromised neighbors to authenticate the transmitted measurements as false. Since the cyber and control layers are closely coupled, this action first ensures that the attacked measurement is discarded locally and in the neighboring agents. Second, as long as the events are activated in the attacked unit, an event-triggered signal is constructed using the measurement(s) of *trusted* neighbors (with authentication signal labeled as true). The signal reconstruction is done by using detection criteria as triggering mechanism to operate within prespecified thresholds. Unlike observers, the proposed mechanism does not involve any integration operation to remove the attack elements, thereby making it computationally viable. By doing so, it is ensured that the system continues to operate normally during both steady-state and transient conditions. Finally, different avenues of system operation are discussed in detail corresponding to the severity of stealth attacks on both voltage and current measurements in microgrids. As opposed to the existing resilient schemes, the proposed strategy offers $N-1$ and full scale of resilience for stealth attacks on currents, voltages, and frequency in a DC and AC microgrid comprising of N converters, respectively. Furthermore, no model-intensive design requirements and flexibility to operate without disabling the compromised cyber links are additional advantages of the proposed approach, which showcases a new norm of resiliency of cooperative microgrids against cyber attacks.

13.2 Preliminaries of cyber-physical microgrids

13.2.1 DC microgrids

An exemplary autonomous DC microgrid considered in this work is shown in Figure 13.1. N DC sources connected via DC/DC buck converters of equal power rating are interconnected to each other via tie-lines forming the physical layer of

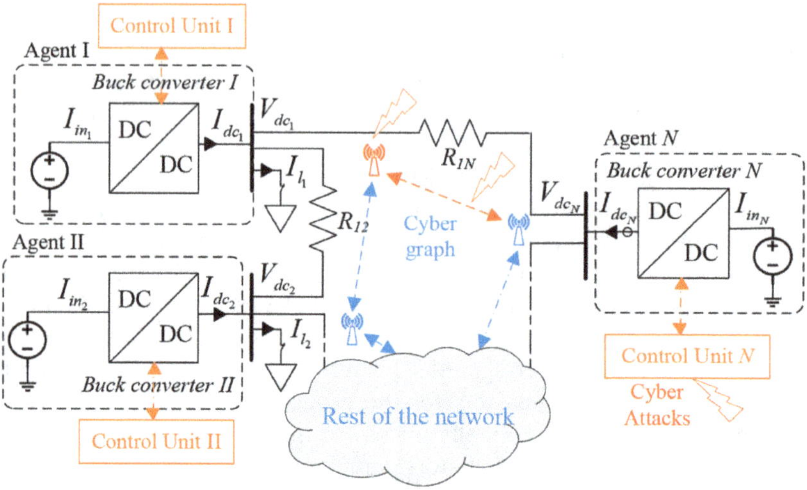

Figure 13.1 Generic cyber-physical model of DC microgrid with N agents: blue arrows represent the cyber layer, and black lines represent the physical circuit. Furthermore, red lines represent the attacked component(s) in the cyber-physical DC microgrid

the microgrid. Each converter is operated in voltage-controlled mode. Cooperative secondary controllers are employed to improve current sharing performance and voltage regulation of these sources [24]. These controllers are made possible using a distributed communication layer, which shares information only between the neighboring units. Each unit, represented as an *agent* in the cyber layer, sends and receives $x_j = \{\bar{V}_{dc_j}, I_{dc_j}^{pu}\}$ from the neighboring agent(s) to achieve secondary control objectives namely, average voltage regulation and proportionate current sharing, where $\bar{V}_{dc_j}$ and $I_{dc_j}^{pu}$ denote the average voltage estimate and per unit output current of the neighboring agents, respectively.

Each agent is represented via a node, and a communication digraph is represented via edges. They constitute an adjacency matrix $\mathbf{A} = [a_{ij}] \in R^{N \times N}$, where the communication weights are given by $a_{ij} > 0$, if $(\psi_i, \psi_j) \in \mathbf{E}$, where $\mathbf{E}$ is an edge connecting two nodes, with ψ_i and ψ_j being the local and neighboring nodes, respectively. Otherwise, $a_{ij} = 0$. Furthermore, the incoming cyber information matrix can be denoted by $\mathbf{Z}_{in} = \sum_{i \in N} a_{ij}$. Hence, if $\mathbf{A}$ and $\mathbf{Z}_{in}$ match each other, the Laplacian matrix $\mathbf{L}$ is *balanced*, where $\mathbf{L} = \mathbf{Z}_{in} - \mathbf{A}$.

Using the preliminaries of the communication graph, the local control input of the cooperative secondary controller can be written as

$$u_i(t) = \sum_{j \in M_i} \underbrace{a_{ij}(x_j(t) - x_i(t))}_{e_i(t)} \tag{13.1}$$

where $u_i = \{u_i^V, u_i^I\}$, $e_i = \{e_i^V, e_i^I\}$, respectively, as per the elements in x, and M_i is the set of neighbors of ith agent.

***Remark I**: As per the synchronization law [26], all the agents participating in distributed control will achieve consensus using $\dot{\mathbf{x}} = -\mathbf{L}\mathbf{x}$ for a well-spanned Laplacian matrix* $\mathbf{L}$ *such that* $\lim_{t\to\infty} x_i(t) = c,\ \forall\, i \in N$, *where* c *is the steady-state reference, and* N *is the number of agents.*

Using (13.1), the control inputs to achieve average voltage regulation and proportionate current sharing can be obtained from secondary sublayers I and II, respectively, by using the following voltage correction terms for i^{th} agent:

$$\textbf{Sublayer I: } \Delta V_{1_i} = H_1(s)(V_{dc_{ref}} - \bar{V}_{dc_i}) \tag{13.2}$$

$$\textbf{Sublayer II: } \Delta V_{2_i} = H_2(s)(I_{dc_{ref}} - u_i^I) \tag{13.3}$$

where $\bar{V}_{dc_i} = V_{dc_i} + \int_0^\tau \sum_{i\in M_i}(u_i^V d\tau)$, while $H_1(s)$ and $H_2(s)$ are PI controllers. Furthermore, $V_{dc_{ref}}$ and $I_{dc_{ref}}$ are the global reference voltage and current quantities for all the agents, respectively. It should be noted that $I_{dc_{ref}} = 0$ for proportionate current sharing between the agents. The correction terms obtained in (13.2) and (13.3) are finally added to the global reference voltage to achieve local voltage references for i^{th} agent using:

$$V^i_{dc_{ref}} = V_{dc_{ref}} + \Delta V_{1_i} + \Delta V_{2_i} \tag{13.4}$$

Using (13.4) as the local voltage reference for i^{th} agent, the abovementioned secondary objectives are achieved.

Using the distributed consensus algorithm for a well-connected cyber graph in a DC microgrid, the system objectives for DC microgrids using (13.1) to (13.4) shall converge to:

$$\lim_{t\to\infty} v_i(t) = V_{dc_{ref}},\ \lim_{t\to\infty} u_i^I(t) = 0 \ \ \forall i \in N \tag{13.5}$$

where, $v_i(t) = V_{dc_i}(t) + \int_{j\,\in\, M_i} u_i^V(t)$ with V_{dc_i} denoting the measured output voltage of ith agent.

13.2.1.1 Modeling of stealth attacks in DC microgrids

As shown in Figure 13.1, cyber attackers may inject unknown exogeneous signals in many ways such as, injection of false data into the controllers, sensors, communication links, etc. to disrupt the system objectives in (13.5). These attacks can be conducted in a coordinated manner to deceive the system operator in the presence of any such attack elements using the following modified inputs in (13.1), given by

$$\mathbf{u}^a(t) = \mathbf{L}\mathbf{x}(t) + \mathbf{W}\mathbf{x}_{attack} \tag{13.6}$$

where $\mathbf{u}^a$, $\mathbf{x}$, and $\mathbf{x}_{attack}$ denote the vector representation of the attacked control input $u_i^a = \{u_i^{Va}, u_i^{Ia}\}$, the states, and the attack elements $x_{attack_i} = [x^V_{attack_i}, x^I_{attack_i}]^T$, respectively. Furthermore, $\mathbf{W} = [w_{ij}]$ denotes a row-stochastic matrix with its elements given in

$$w_{ij} = \begin{cases} \dfrac{1}{M_i+1}, j \in M_i \\ 1 - \sum_{j \in M_i} w_{ij}, j = i \\ 0, j \notin M_i, j \neq i \end{cases} \tag{13.7}$$

It should be further noted that $\mathbf{x}_{attack}$ is bounded, such that the protection measures for the DC/DC converter do not operate as soon as the following holds true:

$$\mathbf{V}_{dc_{min}} < \mathbf{V}_{dc} < \mathbf{V}_{dcmax} \tag{13.8}$$

$$\mathbf{I}_{dc_{min}} < \mathbf{I}_{dc} < \mathbf{I}_{dcmax} \tag{13.9}$$

where $\mathbf{I}_{dc_{min}}$, $\mathbf{I}_{dcmax}$, $\mathbf{V}_{dc_{min}}$, and $\mathbf{V}_{dcmax}$ denote the vector representation of minimum and maximum threshold for output current, minimum, and maximum threshold for output voltages.

Remark II: Using the lemma of consensus in [26] despite of any initial conditions of $x(0)$, it can be concluded that the solution achieved for:

$$\lim_{t\to\infty} v_i^a(t) = V_{dc_{ref}}, \lim_{t\to\infty} u_i^{Ia}(t) = 0 \quad \forall i \in N \tag{13.10}$$

with $v_i^a = V_{dc_i}(t) + \int_{j \in M_i} u_i^{Va}(t)$, is feasible and stable. As a result, these attacks are termed as *stealth* attacks owing to identical convergence properties as in (13.5) even under the presence of attack elements. Hence, detection and mitigation of stealth attacks in a distributed network are important aspects to prevent the system from further instability or shutdown. A brief overview of the modeling of these attacks in DC microgrids is provided in Table 13.1.

Recently, event-triggered algorithms have been devised for DC microgrids to reduce the communication burden by estimating the desired states locally without use of communication [20, 21]. However, in the presence of external disturbances (categorized as *events*), the gradient projection algorithm used in [20] ensures that each control input remains within the bound by triggering the communication layer. However, stealth cyber attacks cannot be detected using the abovementioned event-triggered algorithms since it exhibits zero dynamics without affecting the system states. This has been theoretically verified in the next section.

Table 13.1 Stealth attacks in DC microgrids in [14, 15]

Affected counterparts	**Modeling**
Voltage [14]	$\mathbf{W}x^V_{attack}=0$
Current [15]	$\mathbf{W}x^I_{attack}=0$

13.2.1.2 Undetectability of stealth attacks to event-triggered algorithms [20]

Considering a theoretical state-space model under nominal operating conditions of the entire plant involving both converter and control dynamics in the presence of attacks, the estimated states can be obtained as

$$\begin{aligned} \dot{\hat{\mathbf{x}}} &= \mathbf{A}\hat{\mathbf{x}} + \mathbf{B}\mathbf{u}^a \\ \dot{\hat{\mathbf{x}}} &= \mathbf{A}\hat{\mathbf{x}} + \mathbf{B}[\hat{\mathbf{u}} + \mathbf{W}\mathbf{x}_{attack}] \end{aligned} \tag{13.11}$$

where $\hat{o}$ denotes the estimated projections of the concerned variables. More details regarding the projection of these variables can be referred from [20]. It should be noted that the initial conditions of (13.11) are acquired using the measured states and inputs. Using the difference between (13.11) and the actual states, an event-triggered error e_{ev_i} in i^{th} agent ascertains any disturbance from the steady-state operation by activating the communication channels, only when the binary variable ρ_i is triggered using:

$$\rho_i = \begin{cases} 1, \textbf{if } e_{ev_i}(t)(=\hat{x}_i(t) - x_i(t)) \geq \Upsilon \\ 0, \textbf{else} \end{cases} \tag{13.12}$$

where Υ is an activation threshold. To align the control scheme effectively, a full state feedback is employed to update the inputs with respect to the states using:

$$\hat{\mathbf{u}} = -\mathbf{K}\hat{\mathbf{x}} \tag{13.13}$$

where **K** is the feedback matrix. Substituting (13.13) in (13.11), we obtain the actual state-space model without external disturbances since the term $\mathbf{W}\mathbf{x}_{attack}$ exhibits zero dynamics for the modeled attack in (13.18), as per Remark II. Hence, it can be concluded that stealth attacks cannot be detected using event-triggered control mechanisms.

To provide a clear understanding, a case study is carried out in a DC microgrid with $N = 4$ agents in Figure 13.2 to demonstrate that stealth attacks cannot be detected using the event-triggered control mechanism for DC microgrids proposed in [20]. First, a balanced set of attack elements $\{-15, 0, 15, 0\}$ V are injected into the voltage control inputs in Figure 13.2 as per the attack model in [14]. After the attack is initiated, it can be seen that the average voltage estimates return back to the global voltage reference of 315 V. At t = 3 seconds, an increase in load is introduced to highlight identical dynamics and convergence, which make it difficult to distinguish between physical disturbances and stealth attack. Furthermore, a stealth attack of 4 A is conducted on the current sensor and outgoing communication links in unit II at t = 5 seconds. However, it can be seen in Figure 13.2 that the resulting control input made using the attacked sensors and other relevant quantities converges back to zero for stealth attacks. As a result, the event-triggering mechanism which gets activated in both the cases (at t = 1 and 3 seconds) does not provide a direct manifestation into the differentiation between stealth attacks and other disturbances. By maintaining this discretion, the attacker can launch a critical attack by increasing the magnitude of the injected attack elements, which activates the protective measures

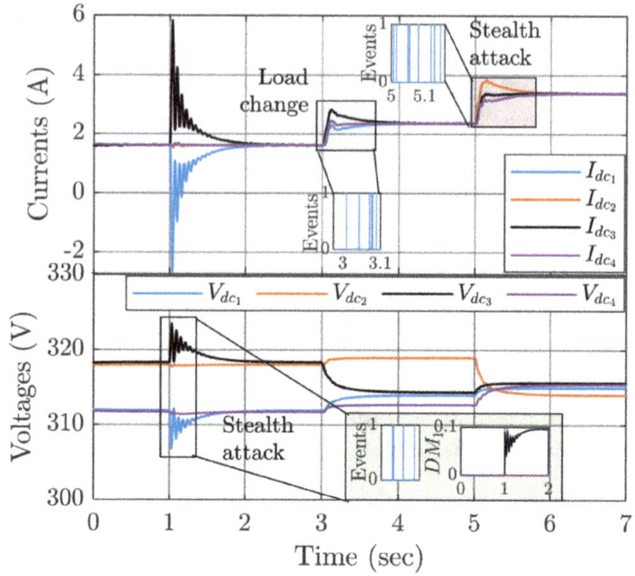

Figure 13.2 Stealth attacks in DC microgrid with N = *4 agents on average voltage control inputs [14] of units I and III at t* = *1 second and, current sensor and outgoing communication link [15] of unit II at t* = *1 second with attack II around t* = *5 seconds – conventional event-triggered philosophy [20] cannot discriminate between a load change and stealth attack since the objectives in (13.5) are met satisfactorily*

and consequently leading to shutdown. As a consequence, this case study necessitates the mitigation of stealth attacks using an authentic resilient mechanism, such that aforementioned risks can be prevented easily.

13.2.2 AC microgrids

As shown in Figure 13.3, the kth grid-forming converter (GfC) consists of a DC source (e.g., renewable energy or energy storage systems), an inverter bridge, an LCL filter, and a controller using local measurements. In the system shown in Figure 13.1 comprising of N agents, each communication digraph is represented via edges to constitute an adjacency matrix $\mathbf{A} = [a_{kj}] \in R^{N \times N}$, where the communication weights are given by $a_{kj} > 0$, if $(\psi_k, \psi_j) \in \mathbf{E}$, where $\mathbf{E}$ is an edge connecting two nodes with ψ_k and ψ_j being the local and neighboring nodes, respectively. Otherwise, $a_{kj} = 0$. $N_k = \{j | (\psi_k, \psi_j) \in \mathrm{E}\}$ denotes the set of all neighbors of kth agent. Furthermore, the in-degree matrix $\mathbf{Z}_{in} = \mathrm{diag}\{z_{in}\}$ is a diagonal matrix with its elements given by $z_{in} = \sum_{j \in N_k} a_{kj}$. Furthermore, the Laplacian matrix $\mathbf{L}$ is defined as $\mathbf{L} = \mathbf{Z}_{in} - \mathbf{A}$.

To improve their performance, neighboring GfCs' measurements, which are transmitted to the local GfC and vice versa, are used in a cooperative secondary controller to regulate their respective bus' average voltage $\bar{V}_k$ and frequency ω_k.

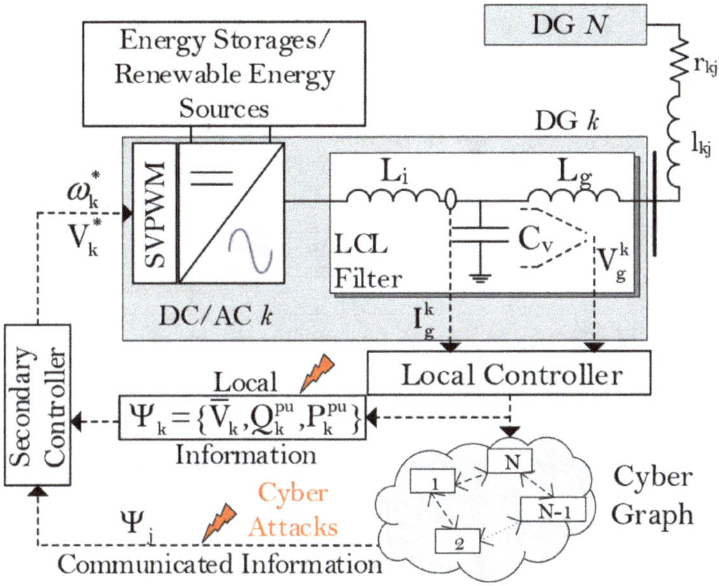

Figure 13.3 *Single-line diagram of a cyber-physical system consisting of* N *GfCs managed by a cooperative cyber topology – cyber attack (in red bolts) launched into local control input*

The control objectives of the cooperative controller can be mathematically represented as

$$\lim_{t\to\infty} \omega_k(t) = \omega^*, \quad \lim_{t\to\infty} \bar{V}_k(t) = V^*, \; \forall\, k \in N \tag{13.14}$$

where ω^* and V^* denote the global reference for frequency and voltage, respectively. Detailed control equations of cooperative secondary controller in AC microgrids can be referred from [31]. To achieve proportionate active power sharing along with frequency restoration, the primary layer droop control is modified into:

$$\omega_k(t) = \omega^* - m_k(P_k(t) - P_k^{ref}(t)) \tag{13.15}$$

where m_k, P_k, and P_k^{ref} denote the active power droop coefficient, measured active power, and secondary control active power reference in the kth agent, respectively. Basically, P_k^{ref} compensates for the error introduced by the droop coefficient in (13.15). This is done using:

$$\dot{P}_k^{ref}(t) = k_1(\omega^* - \omega_k(t)) + k_2 \sum_{j\in N_k} a_{kj}(y_j(t) - y_k(t)) \tag{13.16}$$

with k_1 and k_2 being positive variables and $y = mP$ Substituting (13.19) in (13.15) for all the agents and multiplying **L** in both sides, we obtain $\mathbf{L}\boldsymbol{\omega}(t) = 0$ for (13.14) to hold true.

However, the objectives in (13.14) can be misconstrued in the presence of cyber attacks on the frequency signal in the kth agent using:

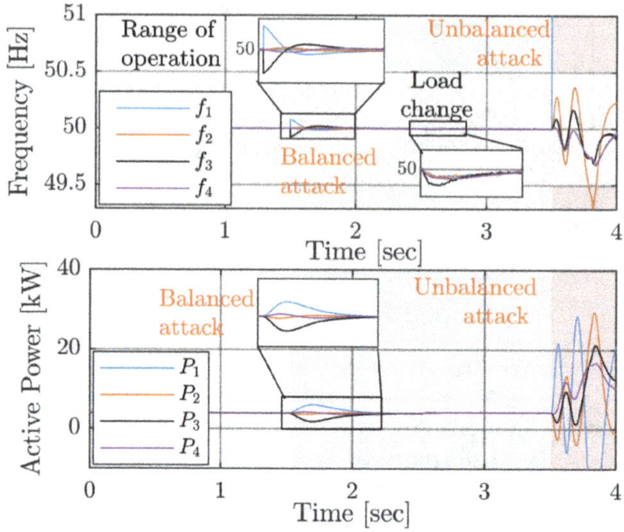

Figure 13.4 Case study for N = *4 GfCs – attacker conducts a balanced attack first to deceive the system operator and then conducts an unbalanced attack resulting in operation outside the allowable range*

$$\omega_k^f(t) = \omega_k(t) + \kappa\omega_k^a \tag{13.17}$$

where, $\kappa = 1$ denotes the presence of an attack element ω_k^a in the kth agent, or 0 otherwise. Furthermore, these attacks can be conducted in a *coordinated* manner to deceive the system operator using:

$$\dot{\boldsymbol{\omega}}(t) = -(\mathbf{L}\boldsymbol{\omega}(t) + \boldsymbol{\omega}^a) \tag{13.18}$$

where $\boldsymbol{\omega}$ and $\boldsymbol{\omega}^a$ denote column matrices of the measured frequency and attack signal for N GfCs, respectively.

Remark III: *Considering the attack model in (13.18), the attack can be termed as*

1. ***balanced attack, if*** $\dot{\boldsymbol{\omega}}(t) = 0$*. Such attacks always lead to a stable and feasible solution, thereby satisfying the objectives in* (13.14). *More details on the design of stealth attacks in AC microgrids can be referred from* [28].
2. ***unbalanced attack, if*** $\dot{\boldsymbol{\omega}}(t) \neq 0$*. They disregard the objectives in* (13.14).

Remark IV: Based on the definition of balanced attacks, it can be concluded that $\sum_{k=1}^{N}\omega_k^a = 0$ *for cooperative synchronization holds true. Conversely,* $\sum_{k=1}^{N}\omega_k^a \neq 0$ *for unbalanced attacks.*

A case study is carried out in Figure 13.4 on an AC microgrid with $N = 4$ GfCs to show the impact of balanced and unbalanced attacks on frequency control input. When an attack of $\boldsymbol{\omega}^a = 2\pi\{0.05, 0, -0.05, 0\}$ rad/s is introduced at t = 1.5 seconds, the frequency and active power of each agent converge back to the corresponding references, as defined in the control objectives. As per Remark III, all the necessary conditions are met, which certify it as a balanced attack. However, at t = 3.5 seconds, the attacker maintains this discretion and increases one attack element in $\boldsymbol{\omega}^a = 2\pi\{10, 0, -0.05, 0\}$ rad/s. As a result, it can be seen that ω_1 immediately goes outside the boundary of operation [49.5, 50.5] Hz (as highlighted in Figure 13.4) defined for the microgrid. As the frequencies reach close to the aforementioned threshold, it could unnecessarily lead to the activation of protective relays, which could cause shutdown of the microgrid.

13.3 Proposed event-driven resiliency against cyber attacks

13.3.1 DC microgrids

In this section, the detection philosophy along with the proposed countermeasures to operate normally during both steady-state and transient conditions in the presence of stealth attacks is discussed in detail.

13.3.1.1 Detection

The detection strategies for both the stealth attacks in Table 13.1 have been provided in Table 13.2. More details on its formulation can be referred from [14, 15]. Upon detection of attack elements, it is vital to remove these attacks, which may ultimately lead to system shutdown. It should be noted that the aforementioned detection criteria will perform satisfactorily, even under the presence of *uncoordinated* attacks. By definition, *uncoordinated* attacks can be defined as a set of attack elements, which do not follow synchronization theory in $\mathbf{W}\mathbf{x}_{attack} \neq 0$. As a consequence, it follows from (13.18) to conclude that $\mathbf{L}^T\mathbf{u}^a \neq 0$ for *uncoordinated* attacks.

Remark III: Using (13.12), it can be formalized that the set of detection criterion $DM^i = \{DM_1^i, DM_2^i\}$ in Table 13.2 can be defined as *events*, when they start operating outside the detection threshold $\Upsilon = \{\Upsilon_1, \Upsilon_2\}$, respectively.

Table 13.2 Detection criteria for stealth attacks in Table 13.1

Stealth Attack	Detection Criteria for i[th] Agent	Terminology
Voltage [14]	$h_i\left[\sum_{j\varepsilon M_i} a_{ij}(\Delta V_{1_j} - \Delta V_{1_i})\right]\left[\sum_{j\varepsilon M_i} a_{ij}(\Delta V_{1_j} + \Delta V_{1_i})\right] > \Upsilon_1$*	DM_1^i
Current [15]	$c_i\left[\sum_{j\varepsilon M_i} a_{ij}(I^j_{in_{ref}} - I^i_{in_{ref}})\right]\left[\sum_{j\varepsilon M_i} a_{ij}(I^j_{in_{ref}} + I^i_{in_{ref}})\right] > \Upsilon_2$†	DM_2^i

* *h*i is a positive quantity used for i^{th} agent.

† *c*i, $\mathbf{I}^i_{in_{ref}}$ denotes a positive quantity and the input current reference for *i*th agent.

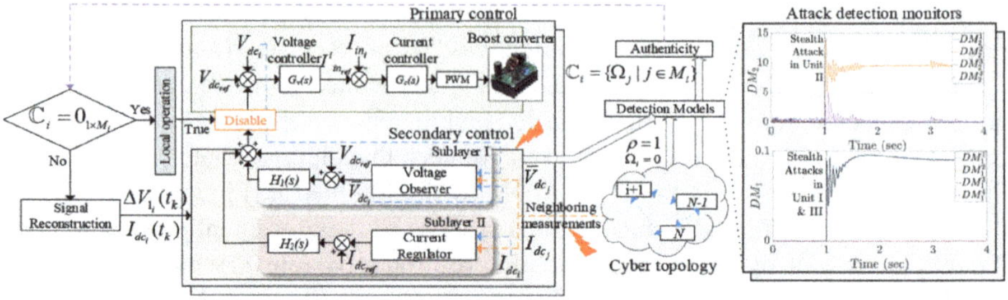

Figure 13.5 Proposed event-driven resilient control strategy to mitigate stealth attacks [14, 15] in DC microgrids with N agents

It is worth notifying that the detection thresholds are infinitesimal values, which are designed to disregard measurement noise to ensure accurate detection. As the detection criterion monitors (as shown in Figure 13.5) reveal the compromised section in each agent, an authentication signal Ω_i is generated for the particular counterpart (voltage/current) in i^{th} agent. It should be noted that the nature of authentication signal is binary, such that:

$$\Omega_i = \begin{cases} 0(\mathrm{F}), & \textbf{if } DM^i \leq \Upsilon \\ 1(\mathrm{T}), & \textbf{else} \end{cases} \tag{13.19}$$

Furthermore, as shown in Figure 13.5, a set of authentication signals from the neighboring agents $\mathcal{C}_i = \{\Omega_j | j \in M_i\}$ are also communicated to i^{th} agent. To simplify the representation of authentication for any signal, $\circ^{\mathrm{T}}$ and $\circ^{\mathrm{F}}$ will be used to symbolize true and false for local/communicated measurements, respectively, using (13.19).

13.3.2 Mitigation

As long as the event(s) hold true, the control variables used in designing DM^i are forced to follow the trajectories of noncompromised neighboring signals (with Ω_j labeled as true). As highlighted in Figure 13.5, if the set of authentication signals in i^{th} agent $\mathcal{C}_i$ is not a zero vector in the presence of attack elements, event-driven resilient signals are reconstructed to mitigate stealth attacks in Table 13.1 by using:

$$\Delta V_{1_i}(t_k) = \Xi_1(\Delta V_{1_j}^{\mathrm{T}}(t)) \tag{13.20}$$

$$I_{dc_i}(t_k) = \Xi_2(I_{dc_j}^{\mathrm{T}}(t)) \tag{13.21}$$

where, $\circ(t_k)$ (with k as the triggering instant) denotes the event-triggered samples of the respective signals generated when the triggering criterion in Table 13.3 is activated during stealth attacks. It is worth notifying that $\Xi(\circ)$ in (13.29)–(13.21) is a triggering function, which holds the input signal $\circ$ until the next instant of triggering. However, if $\mathcal{C}_i$ is a null vector, this implies that all the communicated measurements are compromised with attack elements and should be prevented from being used in i^{th} agent. Hence, this leads to localized operation of ith agent (as highlighted in Figure 13.5). Finally, this

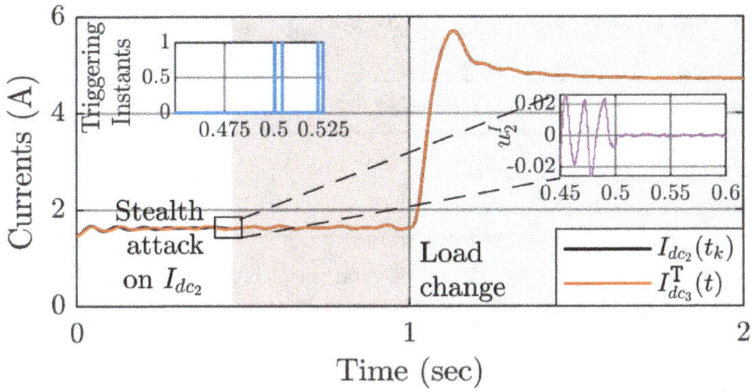

Figure 13.6 Performance of event-driven resilient mechanism for a stealth attack on $I_{dc_2}(t)$ *at t = 0.5 seconds (in the considered system in Figure 13.1 with* N = *4 agents) with the resilient signal reconstructed using* $I^{T}_{dc_3}(t)$

action is completed by substituting the event-driven resilient signals with the attacked signal based on the local authentication signal using:

$$\Delta V_{1_i}(t) = \Omega_1 \Delta V_{1_i}(t) + (1 - \Omega_i)\Delta V_{1_i}(t_k) \tag{13.22}$$

$$I_{dc_i}(t) = \Omega_i I_{dc_i}(t) + (1 - \Omega_i) I_{dc_i}(t_k) \tag{13.23}$$

Finally, the signals obtained in (13.24) and (13.25) are substituted into (13.4) and (13.1), respectively, to realize the mitigation of stealth attacks in DC microgrids. This philosophy exploits the basic theory of consensus law in (13.1) where all the participating elements are identical. Furthermore, this philosophy not only mitigates the attacks but also allows to operate normally under external disturbances such as load change, communication delay, etc. To prove its robustness in the presence of external disturbances, a case study is carried out for the considered system in Figure 13.1 with $N = 4$ agents following a ring-based cyber topology, where a stealth attack is injected into I_{dc_2} at t = 0.5 seconds. As soon as the attack is launched, Equation (13.25) is activated prior to detection of *events* in sublayer II of agent II in Figure 13.5. Upon signal reconstruction of event-driven a priori using $I^{T}_{dc_3}$, it can be seen in Figure 13.6 that the error convergence is held to zero owing to every triggering instants in Table 13.3. Furthermore, its performance aligns perfectly for external disturbances, such as load change at t = 1 second, thereby obeying (13.5).

13.3.3 AC microgrids

To detect the presence of attack elements in AC microgrids, we consider the vector representation of a balanced attack (defined in Remark I) upon substituting (13.15) in (13.18) to get:

$$\mathbf{L}\omega^* - \mathbf{LmP} + \mathbf{LmP}^{ref} + \omega^a = 0 \tag{13.24}$$

Table 13.3 Triggering criteria for stealth attacks in Table 13.1

Stealth Attack	Triggering Criteria for i^{th} Agent	Triggering Function
Voltage [14]	$\mathbf{L}\Delta\mathbf{V}_1^a \leq \Upsilon_1$*	Ξ_1
Current [15]	$\mathbf{L}\Delta\mathbf{V}_{in_{ref}}^a \leq \Upsilon_1$†	Ξ_2

*$\Delta\mathbf{V}_1^a$ denotes vector representation of ΔV_{1_i} with attack elements.
† $\mathbf{I}_{in_{ref}}^a$ denote vector representation of $\mathbf{I}_{in_{ref}}^i$ with attack elements.

where $\boldsymbol{\omega}_{N\text{x}1}^*$, $\mathbf{m}_{N\text{x}N}$, $\mathbf{P}_{N\text{x}1}$, and $\boldsymbol{\omega}_{N\text{x}1}^a$ denote the vector representation of ω^*, m_k, P_k, and ω_k^a, respectively. Since ω^* is constant for the whole microgrid, the first term in (13.24) can be eliminated with $\mathbf{L}$ being a symmetric matrix.

Furthermore, based on the timescale separation between the primary and secondary control layer, $\mathbf{LmP} = 0$, by virtue of the dynamics of primary layer with $\mathbf{L}\boldsymbol{\omega}$ already converging to zero. Hence, we get:

$$\mathbf{LmP}^{ref} + \boldsymbol{\omega}^a = 0 \tag{13.25}$$

Remark V: *Using Remark IV, every nonzero element(s) in* $\boldsymbol{\omega}^a$ *representing the attack signals will reflect an asymmetric relationship with* $\mathbf{LmP}^{ref}$ *for (13.25) to hold true. Hence, this asymmetry becomes the basis of detection for balanced and unbalanced attacks in cooperative AC microgrids.*

Accounting the asynchrony in $\mathbf{P}^{ref}$ under the presence of attacks, this chapter proposes a novel detection metric, termed as asynchrony index AI_k, to detect possible cyber attacks in the kth agent using:

$$AI_k = l_k[\sum_{j \in N_k} a_{kj}(P_j^{ref} - P_k^{ref})][\sum_{j \in N_k} a_{kj}(P_j^{ref} + P_k^{ref})] \tag{13.26}$$

Remark VI: *Using Remark V, any value beyond an adaptive detection threshold* γ *in (13.26) confirms the presence of an attack element in* kth *agent.*

To improve the transient performance under attacked scenarios, an adaptive detection threshold γ_k has been designed such that the detection mechanism can be given in

$$||AI_k|| > \underbrace{\beta \sum_{j \in N_k} ||a_{kj}(P_j - P_k)|| + \beta_0}_{\gamma_k} \tag{13.27}$$

where β and β_0 are positive values. It is worth notifying that the first term in γ_k and β_0 improves the detection performance during transients and steady-state condition, respectively. Since γ_k is designed using the active power measurement P_k, which is obtained from a low-pass filter with a cut-off frequency $\boldsymbol{\omega}_{ci}$ [31], the effect of measurement noise is reduced drastically. Following any disturbance in microgrids, the transients could affect the detection mechanism with a static threshold value. To address this issue, this chapter employs a state-dependent adaptive threshold to

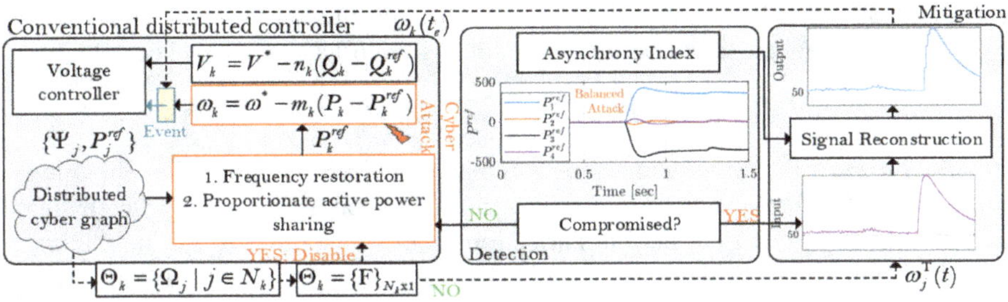

Figure 13.7 Proposed resilient synchronization strategy in k[th] *agent to mitigate balanced/unbalanced attacks in AC microgrid*

provide accurate detection performance during both static and dynamic conditions. Furthermore, the variance of noise in the measurements for a given system can be used as a good indicator to decide the minimum value of β_0 in advance.

Finally, to counteract against these attacks, this chapter uses an event-driven signal reconstruction-based mitigation strategy based on a follow-up signal from AI_k. Upon detection of the presence of attack element in k^{th} agent, an authentication label Ω_k is generated for frequency in k^{th} agent to alarm the presence of attack element (denoted as an *event* in this chapter). It should be noted that the nature of this authentication label is binary, such that:

$$\Omega_k = \begin{cases} 0(\mathrm{F}), & \textbf{if } \|AI_k\| > \gamma_k \\ 1(\mathrm{T}), & \textbf{else} \end{cases} \tag{13.28}$$

To simplify the representation of authentication of the signals in $x = \{\omega, P\}$, x^{T} and x^{F} will be used to symbolize true (uncompromised) and false (compromised) for the corresponding signals using (13.28), respectively.

As long as (13.28) holds true, the compromised frequency signal is forced to follow the trajectories of noncompromised neighboring signals (with Ω_j labeled as true). As highlighted in Figure 13.7, if the set of authentication signals Θ_k for k^{th} agent is not a zero matrix (matrix comprising of signals x, only labeled with F as per (13.28)) in the presence of attack elements, event-driven resilient signals can be reconstructed using:

$$\boldsymbol{\omega}_k(t_e) = \Xi[\omega_j^{\mathrm{T}}(t)] \tag{13.29}$$

where $\omega_k(t_e)$ (with e as the triggering instant) denotes the event-triggered samples of frequency in k^{th} agent. It is worth notifying that $\Xi[x_j(t)]$ in (13.29) is a triggering function, which holds the input signal until the next instant of triggering. In (13.29), the input signals are the frequency of neighboring measurements ω_j^{T} with authentication labeled as T. The event-driven signals are generated only when the detection mechanism in (13.27) holds true in the attacked agent. In particular, a trigger is generated when $\|AI_k\|$ hits the upper bounds of γ. A clear illustration of the event-driven resilient synchronization strategy is provided in Figure 13.7.

The resilient action is completed by substituting the event-driven resilient signals with the attacked signal based on the local authentication signal according to:

$$\omega_k(t) = \Omega_k \omega_k^f(t) + (1 - \Omega_k)\omega_k(t_e) \tag{13.30}$$

Finally, the signal obtained in (13.30) is replaced with (13.15) to realize the mitigation of *stealth* cyber attacks in AC microgrids. As soon as the reconstructed signal in (13.30) obeys (13.14), the authentication label transmitted to the neighbors is switched back to T.

13.4 Results

13.4.1 DC microgrids

The proposed event-driven resilient control strategy is tested on cyber-physical DC microgrid, as shown in Figure 13.1 with $N = 4$ agents for a global reference of 315 V. Each agent of equal power capacities comprising a DC source and DC/DC buck converter, operates to maintain an output voltage for a local reference $V^i_{dc_{ref}}$ at their respective buses. First, a sensitivity analysis to study the performance of the proposed strategy for different detection thresholds Υ is studied along with a study on variation of settling time of the attacked signals. Next, its performance validation for multiple stealth attacks under scenarios such as plugging out of converters, communication delay is carried out to verify the robustness of the event-driven signal reconstruction-based attack mitigation strategy. Moreover, it has been thoroughly verified in case studies that only attacks qualify as *events* to accommodate the resilience of DC microgrids. The simulated plant and control parameters are provided in Appendix.

A sensitivity analysis is carried out for the considered system to inspect the detection capability of the proposed strategy in Figure 13.8 for different values of Υ. A stealth voltage attack is performed at t = 0.3 seconds on agent III, as shown in Figure 13.8(a). As soon as the attack is launched, it can be seen that with increase in the value of Υ_1, the transient peak and the settling time to the optimal setpoint keep increasing. A similar performance can be observed for stealth current attack for different values of Υ_2 in Figure 13.8(b). Moreover, to provide resiliency against input and acquisition noise, Υ can be adjudged as small as possible, yet sufficiently larger than the measurement noise to avoid unnecessary triggering. Hence, the design of Υ is a deterministic task for each system, which highly affects factors such as accuracy and dynamic response. To simplify this task, the variance of noise in the measurements for a given system can be used as a good indicator to decide the minimum value of Υ in advance.

To encompass the relationship between triggering thresholds for different magnitude of attack elements and its physical impact, a case study is carried out in Figure 13.9 to variation of settling time $T_{settling}$ of the attacked control inputs. It can be seen that with increase in the magnitude of x_{attack} and Υ, $T_{settling}$ follows an increasing trend for both voltage and current attacks. It should be noted that the

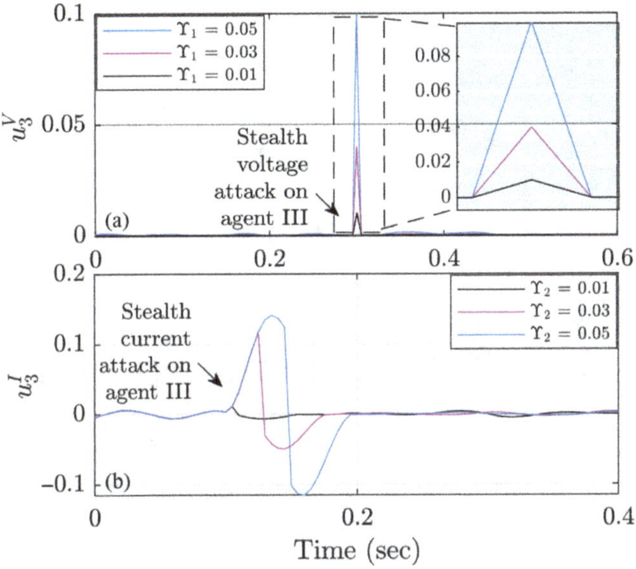

Figure 13.8 Sensitivity analysis of the proposed event-driven attack resilient mechanism in the considered system under the presence of stealth: (a) voltage and (b) current attack on agent III for different values of Υ_1 and Υ_2, respectively

consideration of x_{attack} in Figure 13.9 is limited to the respective upper and lower limits in (13.8) and (13.9). Moreover, better noise filtering capabilities of secondary sublayer II allow a lower value of Υ_2 as compared with Υ_1 to design the proposed event-driven resilient strategy.

Next, the performance of the event-driven resilient controller is compared with the case study for conventional event-triggering method (studied in Figure 13.5) in

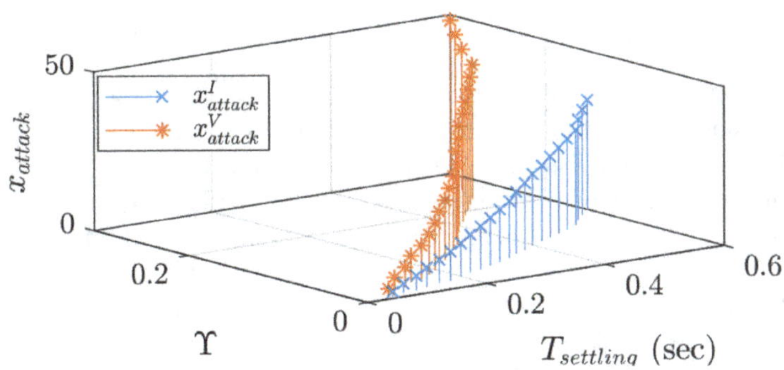

Figure 13.9 Variation of settling time $T_{settling}$ of the control input signals u^a in the considered system for different magnitude of attack elements x_{attack} and respective triggering thresholds Υ

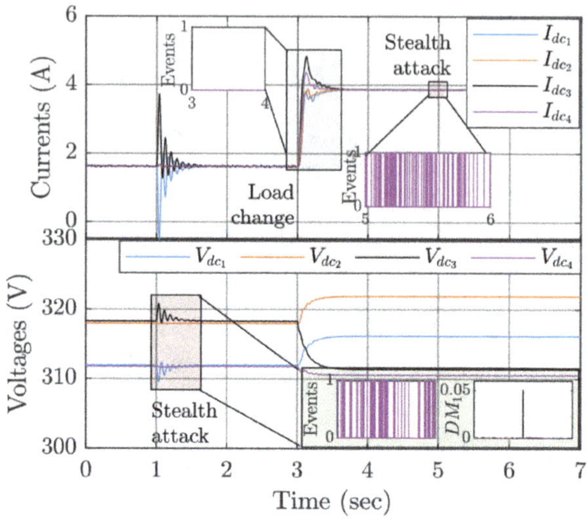

Figure 13.10 Performance of the proposed event-driven attack resilient controller in the presence of stealth voltage and current attacks in Figure 13.2 – it only detects attacks as events *and mitigates it using the triggering criteria immediately*

Figure 13.10. As it can be seen in Figure 13.10, the stealth voltage attack is successfully determined as *event* at t = 1 second, which triggers the signal reconstruction of $\Delta V_{1_1}(t_k)$ and $\Delta V_{1_3}(t_k)$, such that the triggered DM_1 remains within the detection threshold. However, for an increase in load at t = 3 seconds, the proposed controller does not detect any events; which justify its robustness to differentiate between physical disturbances and cyber attacks. Furthermore, at t = 5 seconds, a stealth current attack is launched on agent II, which led to increase in current from each agent, as studied previously in Figure 13.2. However, the proposed resilient controller ensures that the currents from each agents continue to operate at the same loading level as $I_{dc_2}(t_k)$ is reconstructed to follow consensus using the proposed mitigation strategy.

In the next case study, the performance of the proposed resilient controller is tested for instances when the authentication signal is switched from one agent to another. It can be seen in Figure 13.11 that a stealth current attack is conducted on agent III at t = 1 second, which triggers the mitigation philosophy as $\mathbb{C}_3$ is not a null vector. This implies that all the neighbors of agent III are transmitting true measurements. It is worth notifying that the selection of authentication signal from the set $\mathbb{C}_i$ is not governed by any priority labels. Using this hypothesis, agent II signals with authenticity labeled as $\Omega_2 = \mathrm{T}$ are activated immediately for signal reconstruction of $I_{dc_3}(t_k)$. Following up to monitor its performance to regard consensus during external disturbances, it can be seen in Figure 13.11 that the event-driven resilient signal $I_{dc_3}(t_k)$ still follows proportionate current sharing. However, when agent II is plugged out at t = 4 seconds, the outgoing communication links are disabled which

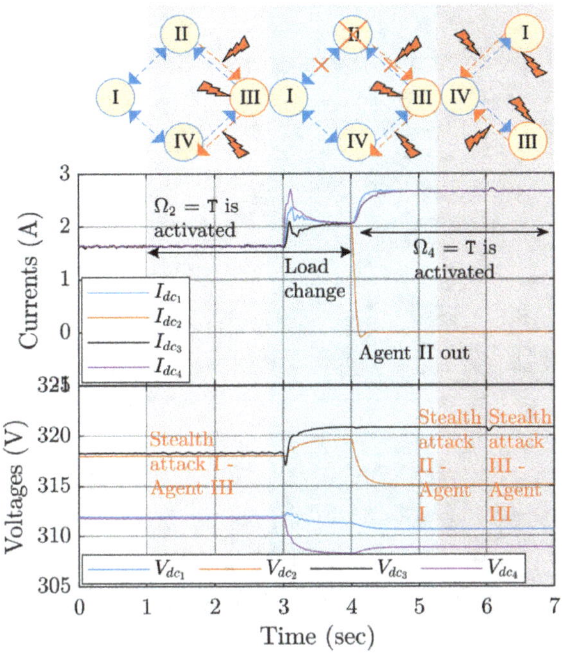

Figure 13.11 Performance of the proposed event-driven attack resilient controller in the presence of stealth voltage and current attacks in multiple agents with agent II plugged out at t = 4 seconds – resiliency is always achieved with authentication signal for agent III immediately switched from Ω_2 to Ω_4 when agent II is plugged out

restrict the transmission of signals to any of its neighbors. Consequently, agent III immediately switches from Ω_2 to Ω_4 for reconstruction of $I_{dc_3}(t_k)$ such that the remaining active agents share the load current equally. Moreover, when stealth current attacks II and III of magnitude 6 and 24 A are launched on agents I and III at t = 5 and 6 seconds, respectively, it can be seen that the sharing accuracy and consensus between agents are unaltered despite the magnitude of attack.

In the final case study, the performance of the proposed resilient controller is tested for multiple stealth current attacks under a maximum network communication delay of 125 ms in Figure 13.12. At first, when a stealth current attack is launched on agent III at t = 1 second; the attacked signal causes a momentary increase with the transient being eliminated as the authentication signal Ω_2 = T is reached after a delay of 125 ms to update the event-driven signal $I_{dc_3}(t_k)$ using (13.25). As this hypothesis is well-studied previously, the settling time intuitively increases to 0.1 seconds for a value of Υ_2 = 0.01. Furthermore, at t = 2.5 seconds, a stealth current attack is launched on agent II which creates a momentary increase and settles down as the resilient update of $I_{dc_2}(t_k)$ is received after a delay of 125 ms using Ω_1 = T. The

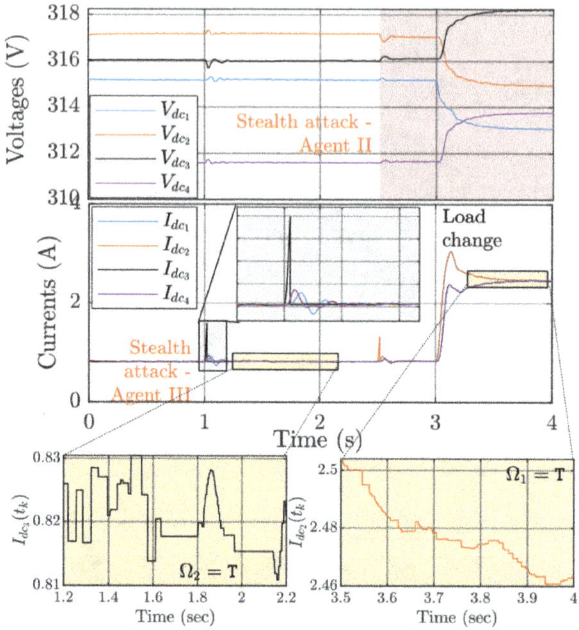

Figure 13.12 Performance of the proposed event-driven attack resilient controller in the presence of multiple stealth current attacks on agents II and III at t = 1 and 2.5 seconds under a maximum communication delay of 125 ms – the settling time increases due to delayed authentication updates from neighbors

robustness of the proposed controller can be ensured via a load change at t = 3 seconds, when the currents from each agent are proportionately shared. Furthermore, the discretized values from every triggering instant are zoomed in to show the trajectory which they follow during steady-state and dynamic conditions. Hence, the proposed event-driven resilient scheme is not limited to mitigating attacks for steady-state operation of converter(s) but is also flexible to operate for dynamic conditions such as load changes.

13.5 AC microgrids

The proposed detection strategy has been tested in a distributed AC microgrid (as shown in Figure 13.1) operating at a global voltage and frequency reference of 311 V and 50 Hz, respectively with $N = 4$ GfCs. Since each GfC is of equal capacity of 10 kVA, the droop coefficients m_k are equal and hence, active power will be shared equally. Using the local and neighboring measurements, the proposed resilient strategy shown in Figure 13.7 is modeled for every GfC to achieve resilient synchronization. All the parameters are provided in Appendix.

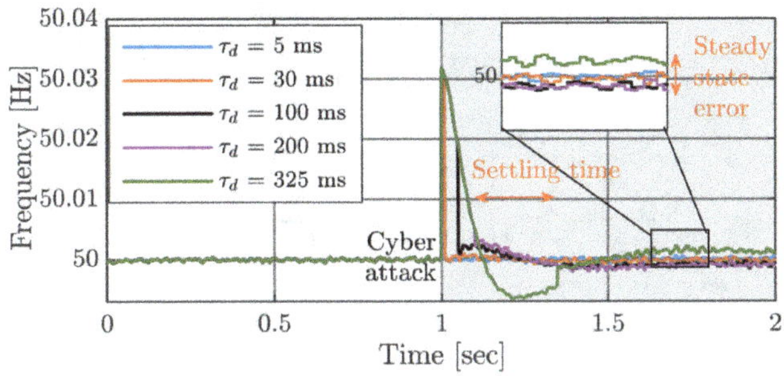

Figure 13.13 Performance of the proposed event-driven attack resilient controller to follow (13.14) for different values of communication delay τ_d *– settling time keeps increasing and even settle at another set point with increase in* τ_d

In Figure 13.13, the performance of the microgrid is tested for different values of communication delay τ_d. As soon as the attack is launched at t = 1 second, a *trustworthy* frequency signal from the neighboring measurements is communicated to the attacked unit. It can be seen in Figure 13.13 that with increase in the value of τ_d, the transient peak and the settling time to the reference set point keep increasing. However, for τ_d = 325 ms, it settles at another set point. This behavior can be explained owing to the microgrid's operation within the maximum communication delay τ_d^{max} allowable for the cooperative cyber graph [31]. As the delay in the network goes beyond τ_d^{max}, $\mathbf{L}\boldsymbol{\omega} = 0$ will not hold true anymore, thereby limiting the resilience capabilities of the proposed strategy.

To differentiate between other malfunctioning events such as line-to-ground (LG) fault, actuator fault, and cyber attacks, an evaluation theory for a GfC's response is presented in Figure 13.14 for a window of 100 ms. This has been analyzed using a frequency-average voltage characteristics at the GfC bus responding to the aforementioned events with the evaluation quadrant ranging from within

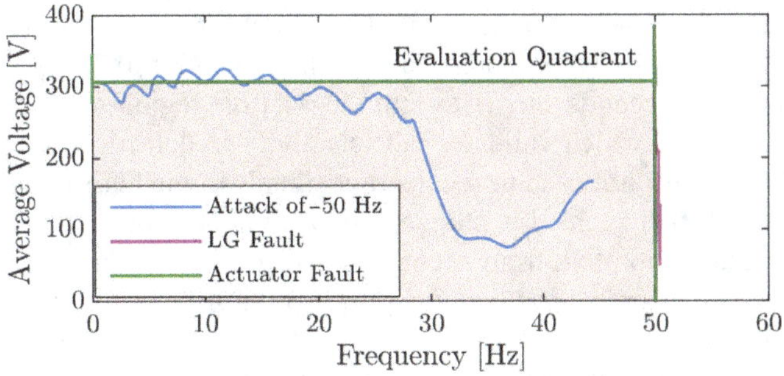

Figure 13.14 Response of GfCs to malfunctioning events – any trajectory inside the evaluation quadrant within 100 ms confirms the presence of cyber attack

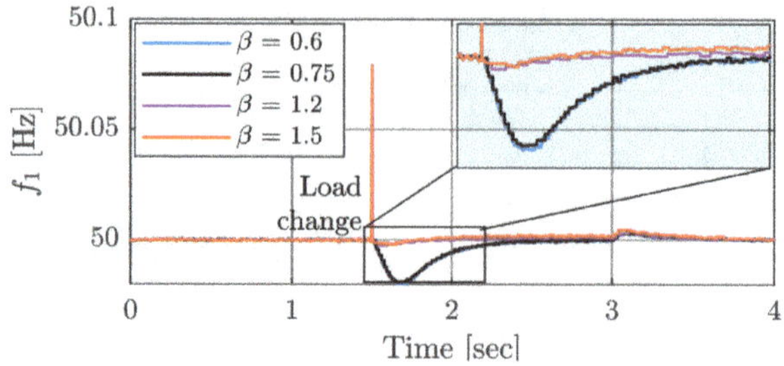

Figure 13.15 Accuracy of signal reconstruction for different values of β in the adaptive detection thresholdγ

XY limits: ($\{X_{min} = 0,\ X_{max} = \frac{\omega^*}{2\pi}\}$, $\{Y_{min} = 0,\ Y_{max} = V^*\}$). As it is evident from Figure 13.14, the inception of LG and actuator faults causes immediate drop in voltages and frequency to zero, respectively. As they form the boundaries of the defined quadrant, an unbalanced attack of –50 Hz causes the trajectory to fall gradually toward zero with movement inside the quadrant owing to the control layer dynamics of the GfC. Hence, a trajectory movement inside the evaluation boundary (within a certain time frame ≈100 ms) can be used locally as a substantial indicator to assist the proposed scheme in differentiating between faults and cyber attacks.

Furthermore, the accuracy of signal reconstruction is tested for different values of β in adaptive threshold γ. As explained earlier that the first term in γ is to provide accuracy in detection during transient conditions, it can be seen in Figure 13.15 that the accuracy can vary significantly when β is less or more than 1. When $\beta < 1$, the triggering instants become higher with increase in settling time. However, when $\beta > 1$, it can lead to peaky transients (up to 50.08 Hz) before the signal reconstruction is started, as the detection boundaries have increased leading to a decreased settling time.

Finally, the resilience capability of the proposed strategy is studied using only one *trustworthy* frequency input. In Figure 13.16, it can be seen that the proposed detection mechanism does not indicate any value above zero during a load change at t = 0.5 seconds. However, when a balanced attack of $\boldsymbol{\omega}^a = 2\pi\{0.75, 0, –0.75, 0\}$ rad/s is injected at t = 1.5 seconds, AI_k rises into the positive region suggesting the presence of cyber attacks, which validates the selectivity in detection of cyber attacks. As a result, ω_2^{T} and ω_4^{T} are used to reconstruct signals immediately for agents I and III, respectively. When agent II is plugged out at t = 3.5 seconds, the incoming and outgoing communication link from agent II is lost. As a result, agent I immediately traverses to ω_4^T to ensure resilient synchronization. Furthermore, at t = 5 seconds, when agent II is plugged in back to the microgrid, it can be seen that the frequency of all agents is restored back to the reference value with active power shared equally among each other. As agent II is uncompromised, ω_2^{T} is again made available for signal reconstruction. It can also be seen in Figure 13.16 that the average voltages of the active GfCs also restore back to the voltage reference upon plug-in of

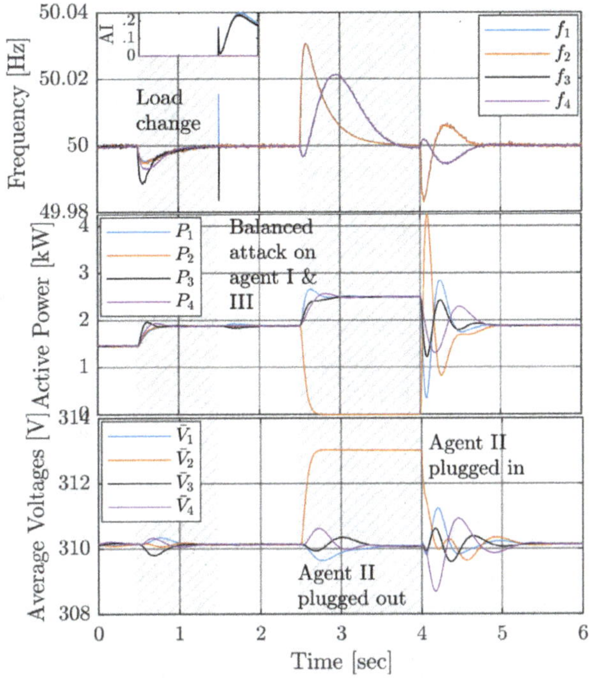

Figure 13.16 Performance of N = 4 GfCs with agent II plugged out and in at t = 2.5 and 4 seconds, respectively with agents I and III under a balanced attack at t = 1.5 seconds – resilient operation achieved only using ω_4^T

agent II. This justifies that the proposed scheme can provide resilient synchronization for up to $N-1$ attacked units.

13.6 Conclusions

A novel event-driven resilient control strategy is proposed to mitigate stealth cyber attacks in DC and AC microgrids. The proposed strategy is simple and offers the flexibility to build on the existing control framework without using any additional resources/information. Furthermore, it offers the maximum scale of resilience, since it is capable of establishing synchronization even with $N-1$ compromised converters. The robustness of this theory has been tested under multiple scenarios and malfunctioning events to restrict the triggering of *events* only under the presence of cyber attacks without interfering in the normal operation of microgrids. Since the basic philosophy of consensus theory complies with *identical* arrangements, this concept has been exploited to design the proposed controller. This strategy is also applicable for *uncoordinated* attacks and will be highly applicable for mission-critical application such as naval ships and electric aircrafts where security is a prime concern. As a future scope of work, this philosophy will be extended for heterogeneous system dynamics, wherein the system objectives will be different for grid-forming and grid-following applications of microgrids.

Appendix

DC microgrids

The considered system consists of four sources rated equally for 6 kW. It is to be noted that the line parameter R_{ij} is connected from i^{th} agent to j^{th} agent. Moreover, the controller gains are identical for each agent.

Plant: $R_{12} = 1.8\ \Omega$, $R_{14} = 1.3\ \Omega$, $R_{23} = 2.3\ \Omega$, $R_{43} = 2.1\ \Omega$.

Converter: $L_{se_i} = 3$ mH, $C_{dc_i} = 250\ \mu$F, $I_{dc_{min}} = 0$ A, $I_{dc_{max}} = 18$ A, $V_{dc_{min}} = 270$ V, $V_{dc_{max}} = 360$ V.

Controller: $V_{dc_{ref}} = 315$ V, $I_{dc_{ref}} = 0$, $K_P^{H_1} = 3$, $K_I^{H_1} = 0.01$, $K_P^{H_2} = 4.5$, $K_I^{H_2} = 0.32$, $G_{VP} = 2.8$, $G_{VI} = 12.8$, $G_{CP} = 0.56$, $G_{CI} = 21.8$, $V_{in} = 270$ V, $h = 1.2$, $c = 2.1$, $\Upsilon_1 = 0.02$, $\Upsilon_2 = 0.015$.

AC microgrids

Each GfC is equally rated with a capacity of 10 kVA. It should be noted that the controller gains are consistent for each GfC.

Plant: $N = 4$, $L_i = 1$ mH, $C_v = 10\ \mu$F, $L_g = 3$ mH, $r_1 = 0.8\ \Omega$, $r_{12} = 0.25\ \Omega$, $r_{23} = 0.75\ \Omega$, $r_{34} = 1.2\ \Omega$, $l_{12} = 2.4$ mH, $l_{23} = 1.8$ mH, $l_{34} = 1.5$ mH.
Controller: $\omega^* = 314.15$ rad/s, $V^* = 311$ V, $m = 0.0014$, $k_1 = 320$, $k_2 = 500$, $\beta = 0.5$, $\beta_0 = 0.01$, $l = 8.4$.

References

[1] Dragicevic T., Lu X., Vasquez J.C., Guerrero J.M. 'DC microgrids–Part I: a review of control strategies and stabilization techniques'. *IEEE Transactions on Power Electronics*. 2016;31(7):4876–91.

[2] Yazdanian M., Mehrizi-Sani A. 'Distributed control techniques in Microgrids'. *IEEE Transactions on Smart Grid.* 2014;5(6):2901–9.

[3] Sahoo S., Mishra S. 'A distributed finite-time secondary average voltage regulation and current sharing controller for DC microgrids'. *IEEE Transactions on Smart Grid.* 2019;10(1):282–92.

[4] Veitch C.K., Henry J.M., Richardson B.T., Hart D.H. Microgrid cyber security reference architecture. Sandia Nat. Lab. (Hierarch. SNLNM), Albuquerque, NM, Tech. Rep. SAND2013-5472; 2013.

[5] Dragicevic T., Lu X., Vasquez J.C., Guerrero J.M. 'DC microgrids–Part II: a review of power architectures, applications, and standardization issues'. *IEEE Transactions on Power Electronics*. 2016;31(5):3528–49.

[6] Sahoo S., Dragicevic T., Blaabjerg F. 'Cyber security in control of grid-tied power electronic converters—challenges and vulnerabilities'. *IEEE Journal of Emerging and Selected Topics in Power Electronics*. 2019;9(5):5326–40.

[7] Liu Y., Ning P., Reiter M.K. 'False data injection attacks against state estimation in electric power grids'. *ACM Transactions on Information and System Security*. 2011;14(1):1–33.

[8] Danzi P., Angjelichinoski M., Stefanovic C., Dragicevic T., Popovski P. 'Software-defined microgrid control for resilience against denial-of-service attacks'. *IEEE Transactions on Smart Grid*. 2019;10(5):5258–68.

[9] Pasqualetti F., Dorfler F., Bullo F. 'Attack detection and identification in cyber-physical systems'. *IEEE Transactions on Automatic Control*. 2013;58(11):2715–29.

[10] Danzi P., Stefanovic C., Meng L., Guerrero J.M., Popovski P. 'On the impact of wireless jamming on the distributed secondary microgrid control'. *IEEE GlobeCom Workshop*; Washington DC, 4–8 Dec; 2016.

[11] Liang G., Zhao J., Luo F., Weller S.R., Dong Z.Y. 'A review of false data injection attacks against modern power systems'. *IEEE Transactions on Smart Grid*. 2016;8(4):1630–8.

[12] Hahn A., Ashok A., Sridhar S., Govindarasu M. 'Cyber-physical security testbeds: architecture, application, and evaluation for smart grid'. *IEEE Transactions on Smart Grid*. 2013;4(2):847–55.

[13] Sun K., Esnaola I., Perlaza S.M., Poor H.V. 'Stealth attacks on the smart grid'. *arXiv preprint*. 2018.arXiv:1808.04184.

[14] Sahoo S., Mishra S., Peng J.C.-H., Dragicevic T. 'A stealth cyber-attack detection strategy for DC microgrids'. *IEEE Transactions on Power Electronics*. 2019;34(8):8162–74.

[15] Sahoo S., Peng J.C.-H., Devakumar A., Mishra S., Dragicevic T. 'On detection of false data in cooperative DC microgrids—a discordant element approach'. *IEEE Transactions on Industrial Electronics*. 2019;67(8):6562–71.

[16] Sahoo S., Peng J.C.-H., Mishra S., Dragicevic T. 'Distributed screening of hijacking attacks in DC microgrids'. *IEEE Transactions on Power Electronics*. 2019;35(7):7574–82.

[17] Sundaram S., Hadjicostis C.N. 'Distributed function calculation via linear iterative strategies in the presence of malicious agents'. *IEEE Transactions on Automatic Control*. 2011;56(7):1495–508.

[18] Beg O.A., Nguyen L.V., Johnson T.T., Davoudi A. 'Signal temporal logic-based attack detection in DC microgrids'. *IEEE Transactions on Smart Grid*. 2019;10(4):3585–95

[19] Sundaram S., Gharesifard B 'Consensus-based distributed optimization with malicious nodes Monticello, IL, IEEE, 2015. pp. 244–49.

[20] Sahoo S., Mishra S. 'An adaptive event-triggered communication-based distributed secondary control for DC microgrids'. *IEEE Transactions on Smart Grid*. 2018;9(6):6674–83.

[21] Moreira L.G., Groff L.B., Gomes da Silva J.M.G. 'Event-triggered state-feedback control for continuous-time plants subject to input saturation'. *Journal of Control, Automation and Electrical Systems*. 2016;27(5):473–84.

[22] Sahoo S., Peng J.C.-H. 'A localized event driven resilient mechanism for cooperative microgrid against data integrity attacks'. *IEEE Transactions on Cybernetics*. 2020;51(7):3687–98.

[23] Sahoo S., Dragicevic T., Blaabjerg F. 'Resilient operation of heterogeneous sources in cooperative DC microgrids'. *IEEE Transactions on Power Electronics*. 2020;35(12):12601–5.

[24] Nasirian V., Moayedi S., Davoudi A., Lewis F.L. 'Distributed cooperative control of DC microgrids'. *IEEE Transactions on Power Electronics*. 2015;30(4):2288–303.

[25] Abhinav S., Modares H., Lewis F.L., Ferrese F., Davoudi A. 'Synchrony in networked microgrids under attacks'. *IEEE Transactions on Smart Grid*. 2017;6(9):6731–41.

[26] Zhu M., Martínez S. 'Discrete-time dynamic average consensus'. *Automatica*. 2010;46(2):322–9.

[27] Pasqualetti F., Dorfler F., Bullo F. 'Control-theoretic methods for cyberphysical security: geometric principles for optimal cross-layer resilient control systems'. *IEEE Control Systems*. 2015;35(1):110–27.

[28] Lu L.-Y., Liu H.J., Zhu H., Chu C.-C. 'Intrusion detection in distributed frequency control of isolated microgrids'. *IEEE Transactions on Smart Grid*. 2019;10(6):6502–15.

[29] Zuo S., Beg O.A., Lewis F.L., Davoudi A. 'Resilient networked AC microgrids under unbounded cyber attacks'. *IEEE Transactions on Smart Grid*. 2020;11(5):3785.94.

[30] Zhou Q., Shahidehpour M., Alabdulwahab A., Abusorrah A. 'A cyber-attack resilient distributed control strategy in Islanded microgrids'. *IEEE Transactions on Smart Grid*. 2020;11(5):3690–701.

[31] Rana R., Sahoo S., Peng J.C.H., Mishra S. 'Performance validation of cooperative controllers in autonomous AC microgrids under communication delay'. Presented at 2019 IEEE PES General Meeting; Atlanta, GA.

[32] Sahoo S., Dragičević T., Blaabjerg F. 'An event-driven resilient control strategy for DC microgrids'. *IEEE Transactions on Power Electronics*. 2020;35(12):13714–24.

[33] Sahoo S., Dragičević T., Blaabjerg F. 'Resilient operation of heterogeneous sources in cooperative DC microgrids'. *IEEE Transactions on Power Electronics*. 2020;35(12):12601–5.

[34] Sahoo S., Dragičević T., Blaabjerg F. 'Multilayer resilience paradigm against cyber attacks in DC microgrids'. *IEEE Transactions on Power Electronics*. 2021;36(3):2522–32.

[35] Sahoo S., Blaabjerg F. 'A Model-free predictive controller for networked microgrids with random communication delays'. *IEEE Applied Power Electronics Conference and Exposition (APEC)*; Phoenix, AZ, 14–17 Jun; 2021. pp. 2667–72.

[36] Sahoo S., Peng J.C.-H. 'A localized event-driven resilient mechanism for cooperative microgrid against data integrity attacks'. *IEEE Transactions on Cybernetics*. 2021;51(7):3687–98.

[37] Sahoo S., Yang Y., Blaabjerg F. 'Resilient synchronization strategy for AC microgrids under cyber attacks'. *IEEE Transactions on Power Electronics*. 2021;36(1):73–7.

Chapter 14

Cybersecurity in future energy systems: outlooks and recommendations

Subham Sahoo[1], H. Alan Mantooth[2], and Frede Blaabjerg[3]

This chapter provides future outlooks and recommendations/perspectives on the digitization of future energy systems. Cyberattacks are often viewed as a low-probability, high-consequence event. Complacency in the face of changing information and the information on the scope of the cybersecurity risk is growing ever more concerning. When confronted with a complex problem like cybersecurity, finding a solution can be done by visualizing the problem in five elements: vulnerability, threat, consequence, probability, and response. This chapter concludes with our remarks on the obstacles and suitable practices/guidelines, which are most suited for different sectors of grid operation.

14.1 Introduction

Smart grid modernization ensues as upgrades to electric power infrastructure are added. Substations controlling transmission voltage are being automated with advanced switching capabilities to enhance current flows and control of the grid. The speed inherent in the smart grid's enabling digital technologies may also increase the chances of a successful cyberattack, potentially exceeding the ability of the defensive system and defenders to comprehend the threat and respond appropriately. Machine-to-machine interfaces enabled by artificial intelligence are being investigated to improve cybersecurity [1].

Many countries have already started to put in place policies, rules, and practices that address these major trends affecting their power sectors, offering important lessons [2, 3]. While each trend or threat to the system might require a specific, tailored security response, several overarching action areas can serve as the basis for achieving more appropriate electricity security frameworks for the future. These are as follows: institutionalizing responsibilities and incentives; identifying risks; managing

[1]Department of Energy, Aalborg University, 9220 Aalborg, Denmark
[2]Department of Electrical Engineering, University of Arkansas, Fayetteville, AR, USA
[3]Department of Energy, Aalborg University, 9220 Aalborg, Denmark

and mitigating risks; monitoring progress; and responding to and recovering from disruptions.

Power systems need to improve preparedness against risks across the electricity supply chain. In this regard, system assessments in areas such as adequacy and cyber resilience, long-term planning exercises, the setting of standards, and sharing of best practices all play important roles. To mitigate risk, policymakers also need to consider establishing market frameworks that provide appropriate investment signals to holders of assets that provide system security and flexibility. They also need to build capacity in new areas like cybersecurity.

14.1.1 Energy transition

1. Policymakers assess where increased diversity of the power mix could ensure resilience against social, geopolitical, market, technical, and environmental risks.
2. Regulators and system operators: consider all flexibility sources as options to satisfy adequacy in long-term planning.
3. Regulators and system operators: set rules that reward energy sources for their actual contribution to secure operation, instead of an expected or average contribution.
4. System operators: develop grid codes to future-proof connection requirements.
5. Regulators create investment frameworks to take advantage of smart grid infrastructure, enabling a higher degree of visibility and controllability of demand response, storage, and variable renewable energy (VRE) sources.
6. System operators: review and adapt historic load-shedding plans in the context of embedded generation, digitalization of the entire value chain, and greater economically viable demand response.

14.1.2 Cyber resilience

1. Policymakers and industry: provide accessible tools and guidance on cyber resilience best practices.
2. Utilities implement proper risk management strategies to identify the capabilities and risks of their systems from both IT and OT perspectives. Establishing a clear risk management strategy can help prioritize areas of work and investment decisions to maximize benefits.
3. Policymakers, standards bodies, industry, and researchers: develop facilities to test and validate the effective implementation of cybersecurity measures and controls.
4. Policymakers and standards bodies: consider certification of products and services by carefully analyzing criticality, enforcement options and market impact.
5. Policymakers and industry: develop capacity building for cybersecurity to ensure skills and resources evolve appropriately. This involves achieving buy-in and a basic understanding across the entire organization. Mandatory training and certification of critical staff should be considered.

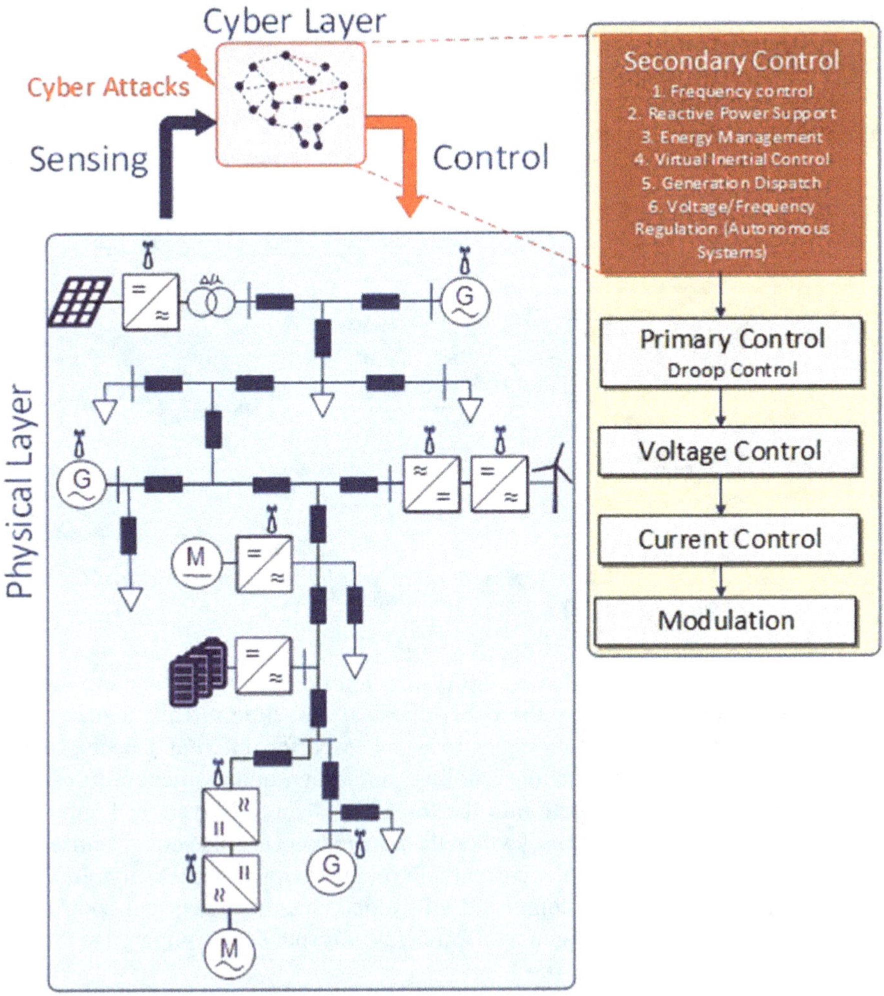

Figure 14.1 *Cyber and computation layer acting as the brain of power electronics-based power systems—false commands as cyberattacks can cause stability problems or even shutdown*

As shown in Figure 14.1, the cyber and the computation layer become the brain for the physical power electronics layer. As the cyber layer is responsible for carrying out secondary control functions in the grid, it could affect a lot of ancillary services such as frequency control, reactive power support, virtual inertial control, economic dispatch, voltage and frequency regulation, etc. Any falsified commands in form of control inputs will finally affect the physical layer operation and cause problems in many ways. Furthermore, the physical system can have many anomalies, which can also be replicated by such cyberattacks. Hence, a framework to

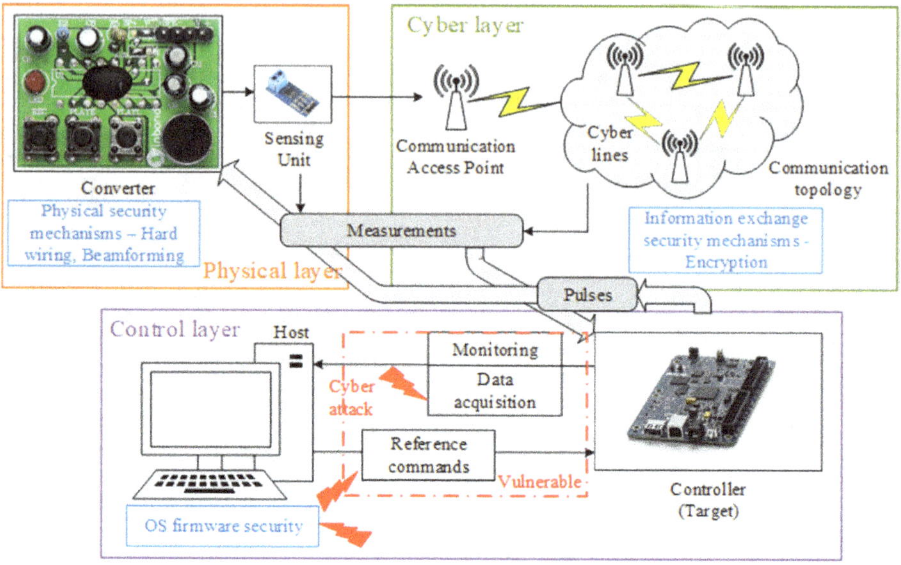

Figure 14.2 Vulnerability assessment of cyber-physical power electronic systems

determine and distinguish between the system malfunctioning events and cyberattacks is mandated to delineate the cyber-physical modeling threats. Moreover, its impact on the system operation needs to be assessed for different aspects, such as power systems control, protection, stability, reliability, and resiliency. In essence, cyberattacks can also propagate into the lower-level control layers in Figure 14.1. However, that is only feasible only when the upper layers are also compromised. For example, any compromise in the current control layer would necessarily require the upper layer controllers to be compromised. Nonetheless, if the current controller is compromised, such attacks can incur a much faster response as compared to the ones in secondary controllers.

14.2 Vulnerabilities, outlooks, and recommendations

In this section, we discuss the vulnerabilities in cyber-physical power electronic systems. As it can be seen in Figure 14.2, there are several vulnerabilities in the cyber, physical, and computational layers. Although several security hardening mechanisms are already present, new vulnerabilities emanate based on newly configured updates. Interestingly, this highly desirable capability may require new cybersecurity methods, since the current methodology, based heavily on encryption and authentication technology, will not be suitable unless the encryption and authentication algorithms are extremely fast [4]. Some older technologies can impose large latencies to get the needed unencrypted data to the proper points to reconfigure the system or adjust the control. These large latencies can destabilize the control loops.

It is also desirable to employ very low complexity sensor nodes in these Internet of things (IoT) systems, further limiting the nodes' ability to efficiently encrypt data. Serious security concerns for IoT systems have already been demonstrated, and even more issues are expected in the future, e.g., self-driving cars could become dangerous weapons if adequate solutions are not developed.

Many researchers are focused on finding new cybersecurity technologies for the IoT that is not focused on current encryption technology. One possibility (using a CPS-based approach) is to use the sensors available in IoT applications, along with any possible models of the physical system being monitored, to detect cyberattacks on these IoT sensor systems. There are other novel IoT security approaches being investigated. While most cybersecurity algorithms are applied at the so-called network layer in the communications layering scheme used in all communication systems today, another promising method for IoT cybersecurity operates at the lowest level, called the *physical layer*. Physical layer security, usually studied by electrical engineers, enables many security methods not possible at the network layer, such as beamforming. Yet another approach is to develop low-complexity encryption or authentication methods [5].

Another vulnerable layer that receives measurements from both physical and cyber layers is the computational layer, as shown in Figure 14.2. In view of the contributors, this is one of the critical layers, which has several degrees of communication with the outer world and can be attacked by an adversary upon exploiting endless possibilities. However, the security mechanism protecting this layer is the operating system firmware security. Hence, security hardening algorithms especially in the control layer, which determines any false triggers and timely actions, are necessary. Hence, this book addresses these concerns by providing an in-depth summary of scientific contributions at each layer for technological progress in cybersecurity of future energy systems.

Finally, we discuss the usage of machine learning/artificial intelligence (AI/ML) for cybersecurity of future energy systems. As shown in Figure 14.3, although approach 1, ascribed to the design of data-driven algorithms for anomaly identification, cyber-attack detection, and mitigation, can be more accurate; however, their accuracy can depend entirely on the quality of data. It can be seen that apart from the traditional sensing equipments, the cloud can add to one of the vulnerabilities, which can ultimately be accessed by the attacker. Under such circumstances, the attacker can inject false data into historic datasets and manipulate its statistical properties. Finally, if these data are arranged later for condition monitoring of the system or any historical analysis, it will always be incorrect as we operate on bad data analysis. Hence, if these data are used for training any data-driven regression or classification tool, it will always end up with *false alarms* in the cybersecurity diagnosis mechanism. However, approach 2 ascribed to physics-informed diagnosis methods is more reliable as the sensing equipment can be the only vulnerable component in this case. Hence, although the results cannot precede the accuracy of approach 1 (with unbiased data), its output can always be explained with a high degree of confidence. As a result, approach 2 becomes fairly reliable as compared to approach 1. This requires further work on improving the explainability [6] and accuracy in approaches 1 and 2, respectively.

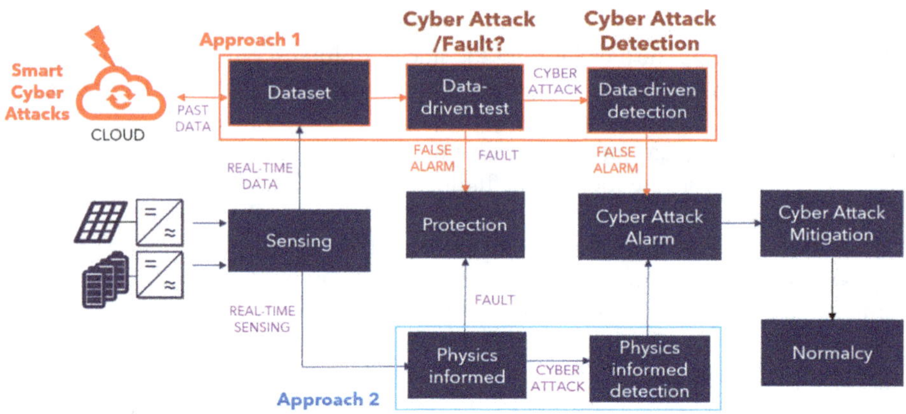

Figure 14.3 Using AI/ML for cybersecurity diagnosis in power electronics—boost or boom?

Advances in cyber-physical power electronic systems should enable capability, adaptability, scalability, resilience, safety, and security far beyond what is available in today's embedded systems [7]. Further innovation requirements range from the IP exchange layer to the non-IP (field control system) and front-end network layer. Essentially, blockchain can also play a big role in devising a distributed chronological order of data records, which provides formal guarantees from tampering. New technologies for developing smart contracts, software/firmware, and hardware security can also be established especially for grid-tied PV systems and EV chargers. Furthermore, more consequence-driven cybersecurity solutions need to be developed that determine vulnerabilities accordingly and ensure grid resilience.

14.3 Book summary

This book covers a wide range of cybersecurity topics on future energy systems. A brief schematic of the technical flow among different chapters can be seen in Figure 14.4. Up first, the impact of cyberattacks and its project on steady-state and dynamic stability of power systems is discussed in Chapter 1. Then, the modern power systems with advanced monitoring functions and advanced power conversion technologies are introduced in Chapter 2, mandating the need for communication in every application. Chapter 3 introduces the risks in microgrids and active distribution networks used in mission-critical applications with a specific assessment of the risky sectors. In view of this development, Chapters 4 and 5 provide risk assessment strategies and evaluation of the security arrangements in the current system. Chapter 6 then condenses these threats in microgrid systems considering the practical hypothesis and analogies from Chapters 4 and 5. Chapters 7 and 8 then address and evaluate the impact of the existing protection and flexibility infrastructure in microgrids, respectively. Based on the attack modeling and emulation techniques

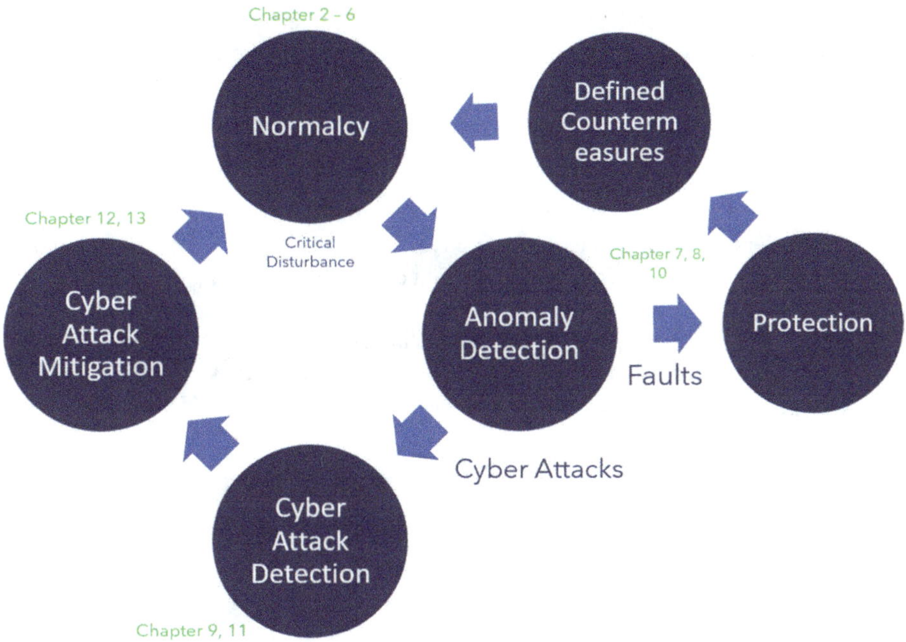

Figure 14.4 *Chapters' organization based on different cybersecurity elements in this book*

in Chapter 9, new detectability and observability tests are studied in Chapter 10. Finally, the rest of the chapters discuss different attack removal techniques considering different vulnerabilities in the cyber-physical layer. In addition, Chapter 12 also provides a brief overview of cybersecuity testbeds for co-simulation analysis of microgrids and EV charging infrastructure. In this way, this book provides an overall comprehension of the framework required for understanding cybersecurity and its basic concepts specifically for microgrid systems. In addition, this book ensures an educational design concept to serve as a mini-handbook for educating readers about the cybersecurity innovations in microgrids and power electronic systems in general.

References

[1] Sahoo S., Dragicevic T., Blaabjerg F. 'Cyber security in control of grid-tied power electronic converters–challenges and vulnerabilities'. *IEEE Journal Emerging and Selected Topics Power in Electronics*. 2019.
[2] Sabillon R., Cavaller V., Cano J. 'National cyber security strategies: global trends in cyberspace'. *International Journal of Computer Science and Software Engineering*. 2016;5(5):67.
[3] Boeke S. 'National cyber crisis management: different European approaches'. *Governance*. 2018;31(3):449–64.

[4] Balda J.C., Mantooth A., Blum R., Tenti P. 'Cybersecurity and power electronics: addressing the security vulnerabilities of the Internet of things'. *IEEE Power Electronics Magazine*. 2017;4(4):37–43.

[5] Yu P., Perazzone J.B., Sadler B.M., Blum R.S. 'Authenticated side channel via physical layer fingerprinting'. *Proceedings IEEE Conference Communcation and Network Security*, Oct. 17–19; 2016.

[6] Sahoo S., Wang H., Blaabjerg F 'On the explainability of black box data-driven controllers for power electronic converters' *2021 IEEE Energy Conversion Congress and Exposition (ECCE)*; Vancouver, BC, Canada, IEEE, 2021. pp. 1–5.

[7] Mazumder S.K., Kulkarni A., Sahoo S., *et al.* 'A review of current research trends in power-electronic innovations in cyber–physical systems'. *IEEE Journal of Emerging and Selected Topics in Power Electronics*. 2021;9(5):5146–63.

Index

www.ingramcontent.com/pod-product-compliance
Lightning Source LLC
LaVergne TN
LVHW011847200926
848484LV00002B/408

9781839533310